The LICIT LIFE *of* CAPITALISM

HANNAH APPEL

The LICIT LIFE *of* CAPITALISM

US OIL IN EQUATORIAL GUINEA

DUKE UNIVERSITY PRESS Durham and London 2019

Designed by Aimee C. Harrison
Typeset in Minion Pro and Helvetica Std.
by Copperline Book Services

Library of Congress Cataloging-in-Publication Data Names:
Appel, Hannah, [date] author.
Title: The licit life of capitalism : US Oil in Equatorial
Guinea / Hannah Appel.
Description: Durham : Duke University Press, 2019. |
Includes bibliographical references and index.
Identifiers: LCCN 2019013458 (print)
LCCN 2019016274 (ebook)
ISBN 9781478004578 (ebook)
ISBN 9781478003656 (hardcover : alk. paper)
ISBN 9781478003915 (pbk. : alk. paper)
Subjects: LCSH: Oil industries—Economicaspects —
Equatorial Guinea. | Petroleum industry and trade—
Equatorial Guinea. | United States—Foreign economic
relations—Equatorial Guinea. | Equatorial Guinea—
Foreign economic relations—United States. |
Capitalism—EquatorialGuinea.
Classification: LCC HD9578.AE6 (ebook) |
LCC HD9578.AE6 A674 2019 (print) |
DDC 338.8/8722338096718—dc23
LC record available at https://lccn.loc.gov/2019013458

This title is freely available in an open access edition
thanks to the TOME initiative and the generous support
of Arcadia, a charitable fund of Lisbet Rausing and Peter
Baldwin, and of the UCLA library.

Cover art: Photo by Ed Kashi.

No theory of history that
conceptualized capitalism as
a progressive historical force,
qualitatively increasing the
mastery of human beings over
the material bases of their
existence, was adequate to
the task of making the experiences of the modern world
comprehensible.
—Cedric J. Robinson,
*Black Marxism: The Making
of the Black Radical Tradition*

CONTENTS

ACKNOWLEDGMENTS

Kié! Di man, what a journey. To begin in the middle of things, this book started as a dissertation at Stanford University. Jim Ferguson—advisor, professor, thinker—turned me into an anthropologist, and his book *Global Shadows* sent me to Equatorial Guinea. Sylvia Yanagisako is a force of nature. Over the years, she has offered sustained and rigorous feedback on my work and drawn me into her intellectual orbit around the anthropology of capitalism. Paulla Ebron's was the voice I heard on the phone in 2004, welcoming me to Stanford. Since then, she has been a beacon in the ways of imaginative scholarship and meaningful pedagogical practice. I was lucky to overlap with Amita Baviskar during my first year at Stanford, and her Introduction to Political Economy course remains the most formative class I have ever taken. Nikhil Anand, Elif Babul, Robert Samet, and Rania Sweis: cohort4eva! Expanded cohort love and thanks to Ramah McKay, whose wisdom, introspection, and friendship saw me through the hardest of times; and to Austin Zeiderman, Maura Finkelstsein, Lalaie Ameeriar, and Tomas Matza—in the cohort if not of it!

A postdoctoral position with Columbia University's Committee on Global Thought was my first step into the wider world of academia. That the postdoc coincided with Occupy Wall Street forever changed me as a scholar and activist, and I am so thankful for that opportunity and for those who defended it: Partha Chatterjee, Mamadou Diouf, Cassie Fennell, David Graeber, Timothy Mitchell, Tavia Nyong'o, Michael Ralph, Saskia Sassen, and Joseph Stiglitz, in particular. Moving from there to UC Berkeley's Ciriacy Wantrup fellowship, I had the good fortune to have Michael Watts as an astonishing mentor. There are few people I'd rather hang out and talk politics with than the inimitable MW. Over the years of this manuscript's life, Andrew Barry, David Bond, Jessica Cattelino, Karen Ho, Enrique Martino, Bill Maurer, Kristin Peterson, Michael Ralph, Suzana Sawyer, Tusantu Tongusalu, Anna Tsing, and Jerry Zee have also given me tremendous support and feedback.

I give thanks also to the Caribbean Cultural Studies program at the University of the West Indies, Mona Campus. If it weren't for my intellectual coming-of-age in that Kingstonian space, I don't know where I'd be. Big up to Joe Pereira, Carolyn Cooper, Hubert Devonish, and Michael Witter, and to Barry Chevannes who joined the ancestors too soon. I give thanks especially to Sonjah Stanley-Niaah, who inaugurated me into the LMU (Lawless, Militant, and Ungovernable) crew, and to Jahlani Niaah. Ashé. And thanks also to Stephanie Black, whose film *Life + Debt* sparked my interest in the intersection of political economy and creative endeavor.

And then, incredibly, I got a job that changed my life. Thank you to Akhil Gupta for his work in bringing me to UCLA and for welcoming me so warmly along with Purnima Mankekar. As my official and unofficial mentor, Jessica Cattelino has guided me through everything from bureaucracy to posthumanism to where my kids should go to school. Jessica, I can't thank you enough. Shout out to the other two of the Fab4—Norma Mendoza Denton and Jemima Pierre. Profound thanks to Shannon Speed, Jemima Pierre, and Jessica Cattelino for my education in antiracist administrative labor. In so many ways, UCLA has been a foundational reeducation, my exposure to the Black Radical Tradition chief among them. The scholarship, mentorship, and friendship of Jemima Pierre, Peter Hudson, Cheryl Harris, and Robin D. G. Kelley in particular have revolutionized my thinking. Thanks also to Marcus Hunter and Shana Redmond for their generosity and kindness, and to Ananya Roy, whom I had admired from afar for so long and who (up close) has turned into a dear friend and mentor. I also thank friends and scholars at UCLA and beyond—Tendayi Achiume, Can Aciksoz, Aomar Boum, Kamari Clarke, Erin Debenport, Laurie Hart, Kelly Lytle Hernandez, Tami Navarro, Sherry Ortner, Amy Ritterbusch, Jason Throop, Tiphanie Yanique, Alden Young, and Noah Zatz for being all kinds of awesome. And to my conspirators at the Debt Collective, with whom I have worked since 2012: together we have learned more about capitalism than I ever cared to know, and from you I draw strength and inspiration.

This project has benefited from several grants and fellowships, including support from the National Science Foundation, the Social Science Research Council, the American Council of Learned Societies, the Wenner Gren Hunt Fellowship, Fulbright Hays, and the Hellman programs. It was the always-enthusiastic Elizabeth Ault at Duke University Press who turned this from a (never-ending) project into a book manuscript, and I thank her for her faith in my ideas. Kate Herman was also a tremendous, gracious help as I fumbled through the more technical parts of manuscript preparation.

Finally, to Guinea, *querida Guinea*. Thank you to Tusantu Tongusalu and Angela Stuesse for encouraging my early interest in this project and facilitating my first connections in Equatorial Guinea. This book cowers in the shadows of your unswerving activism. I hope you find in it something of the promise I know you hoped for. I wish Equatorial Guinea was a place that allowed me to name names, but instead I invoke friends and mentors with the elliptical references and inside jokes that all fieldwork produces—to the EITI crew, to all the rig workers and Equatoguinean industry personnel who shared their days with me, and to all the industry migrants who let me into their peripatetic lives in the oil diaspora. To *los de Hacienda y Presupuestos*, and the folks at DAI and L, in particular, who were actually happy when I announced myself as an anthropologist. To the woman I refer to as Isabel—my fieldwork boss and sister—without your trust in me, my research would have been a fragment of what it became. To the man I refer to as Eduardo, K2R, Sonrisa, Bocadillo, and the other fearless ones, the arc of history bends your way. Keep pulling. To Elo, who listened and supported. To Benita Sampedro, Peter Rosenblum, and Alicia Campos, your work and convictions echo through my own. And finally, to *el bonobo* y *el príncipe*: to the former, I give all my thanks as a *gemelo del alma, del cerebro, y del arte sino de la política*; and to the latter, nothing that I can give is thanks enough.

I was told repeatedly in the field that there is no word for "thank you" in Fang. Popular etymology had it that among family and clan members, the generosity of giving and receiving didn't need to be acknowledged because it suffused daily life. I am blessed to come from a similar clan, and there are no words for how much they have given me. Mom, Oona, and Maureen, I love you with the plainness with which you love me. And James, this project pushed the boundaries of our love, and that we expanded in response is our greatest triumph. To our boundless love, and to Thelonious and Ocean, whom we have since welcomed into it. And to justice—boundless love in public.

INTRODUCTION

Dawn in Bata, Equatorial Guinea's second city. At 6:00 a.m., I stood outside the headquarters of a large US-based oil company with a small group of others—a Spanish woman, a man from Louisiana, and two Equatoguinean men—waiting to "go offshore" by helicopter. We stood quietly and not quite together, separated by the early hour and by not knowing if we were all there for the same purpose. Eventually, an Equatoguinean driver pulled up in a company bus. As we boarded, he requested our identification passes to electronically register each of our exits from the compound, and then drove us to the company's private wing of the airport. After an airport worker searched our bags, we sat in a small room to watch a safety video on the importance of in-flight protective equipment and what to do if our helicopter were to catch fire in midair. At liftoff, the helicopter rose effortlessly as the city of Bata spread out beneath us. Further from shore, looking back, the Ntem River marked the edge of the continent. After a while, sights and sounds faded into the calm of the open ocean seen from above and the gently vibrating lull of the helicopter through noise-canceling headphones.

Eventually, a bright flame appeared in the distance, attached to an indistinct industrial atoll—a rig. Just as the rig came into view, the helicopter banked left to land briefly on what looked like an aircraft carrier, leaving the Spanish woman on what was, in fact, a Floating Production, Storage, and Offloading (FPSO) vessel. With the production rig visible some hundreds of yards away across the water, the FPSO was animated by its own large flare, burning the crude's gaseous by-products. Both the rig and this vast, self-propelling, ship-like structure floated above a field producing 100,000 barrels of crude oil per day. Every ten days, a tanker pulled alongside the FPSO and left with one million barrels of oil. From subsea hydrocarbon deposits, to the rig, to the FPSO, to the tanker, and finally to market, Equatorial Guinea's oil production chain was clearest to me by helicopter, far off the country's shores.

Capitalism is not a context; it is a project.[1]

This book offers an ethnographic account of the daily life of capitalism. It is both an account of a specific capitalist project—US oil companies working off the shores of Equatorial Guinea—and an exploration of more general forms and processes (the offshore, contracts, infrastructures, something called "the" economy) that facilitate diverse capitalist projects around the world. Each of these forms and processes, which organize the book, chapter by chapter, is both a condition of possibility for contemporary capitalism and an ongoing entanglement with the raced and gendered histories of colonialism, empire, and white supremacy out of which capitalism and *liberalism* emerged. Indeed, the book explores the relationship between the liberal modernity claimed by US oil companies—contractual obligation, market rationality, transparency—and the racialized global inequality that radically delimits the ways in which Equatorial Guinea and other postcolonial African countries might engage with multinational oil companies. Just as racism, patriarchy, and dispossession are not exceptions to liberalism, but constitutive of it (James 1963; Hartman 1997; Makdisi 1998; Chakrabarty 2000; Mills 2003, 2017; Stoler 1995, 2010; Mehta 1997, 1999; Byrd 2011; Lowe 2015), so too, this book argues, must we shift our critical understanding of capitalism from one in which "markets" merely deepen or respond to postcolonial inequality, to one in which markets are *made by* that inequality.[2] In Equatorial Guinea and around the world, accreted histories of racialized disparity "proxy" (Ho 2016) for rational, neutral market behavior—"the rules of the economy." Global markets, the oil market chief among them, do not merely take advantage of these circumstances; they are constituted by them.

This view from the helicopter window—through which Equatorial Guinea seems to recede; in which hydrocarbons seem to move effortlessly from one infrastructural node of the commodity chain to another; and where a space referred to as "offshore" seems to be a literal watery stage for placeless economic interaction—requires a tremendous amount of work. From manual, managerial, domestic, and political labor; to material infrastructures and technologies; to the legal, ethical, and affective framing processes required to lubricate the passage of oil and gas to market, the apparent smoothness of the offshore is made and remade in the quotidian project that is hydrocarbon capitalism in Equatorial Guinea. The view was redolent with qualities often thought to be intrinsic to capitalism: standardization, replicability, technical mastery, and the disembedding of economic interaction from social context. In contrast, the view from fourteen months of fieldwork in and around Equatorial Guinea's oil industry demonstrated nothing more than the work required to produce tenuous and contested approximations of those ostensibly intrinsic qualities. This book describes these work-intensive processes as I found them in Equatorial Guinea.

Yet the view from the helicopter window is not only misleading; it is also productive.

If anthropology (at least in the poststructural moment, if not before) has concerned itself with rescuing local specificity and complexity from the abstracting distance of views like this one, this book is equally invested in understanding—ethnographically, theoretically, and politically—what these kinds of views do in the world. These views are not merely "wrong" in any narrow sense. On the contrary, they are performative in that they generate durable material and semiotic effects in the world.[3] Insofar as anthropology and critical theory approach these abstracting views as fodder for deconstruction—to show contingency, complexity, heterogeneity, or locality "within" or "beneath" them—we fail to account for their performative work in the world. We seem to suggest that "mere" appearances are easily undone by ethnographic intimacy. On the contrary, something widely recognized as global capitalism persists despite that kind of deconstructive work. How? Ethnography can help us follow the work required to create the "as ifs" on which capitalism has so long relied: abstraction, decontextualization, and standardization. In this book, I take these as ifs *themselves* as ethnographic objects, aspirational processes, and political projects that we can follow in the field. Rather than recovering the complexity and friction effaced by the view from the helicopter window, then, this ethnography accounts for how things come to *seem* smooth, how the US oil and gas industry works to *seem* separate, distanced, and outside of local life in Equatorial Guinea. As I will go on to chronicle, many of the people with whom I worked—itinerant oil company management in particular—were preoccupied each day with this work of abstraction and distancing: how to ensure that the production and export of oil from Equatorial Guinea might seem detached from local lives, histories, and landscapes.[4]

To use ethnography in this way—to follow the work of standardization, decontextualization, and distancing—allows us to attend to capitalism as a project; to show how it is at once uneven, heterogeneous, and contested and, at the same time, proliferative, powerful, and systemic. Holding these analytic poles in tension, as equally empirically true in the world, asks us to account for their simultaneity. How is it that both can be true? As with any project, capitalism's apparent coherence and momentum take work. This book offers an account of some of our world's most powerful corporations— US oil firms—and those who work with, alongside, and against them as they undertake this work in Equatorial Guinea. To be clear then, this book is not, in any simple way, an account of local inflections or instantiations of capitalism. Rather, it asks after the force and fulsomeness with which capitalism, in fact, seems to do all the things it is supposed to do: standardize, abstract, distance, and decontextualize. How can we account for these phenomena ethnographically, showing—despite the frictions and seams—how this work gets done?

Because this book's analytic trajectory follows the industry's work toward apparent distance and standardization, it is not *about* Equatorial Guinea in the conventional ethnographic sense. This is why I begin with a departure story of sorts—the helicopter leaving Bata for the offshore—rather than with the expected arrival story; this is the directionality of sociopolitical life and work I explore in the book. The ways in which this book is and is not about Equatorial Guinea are also choices about a certain kind of ethnographic refusal (Ortner 1995; Simpson 2014) on the one hand, and an ethnographic insistence on the other. Like Simpson (2014, 105) with the Iroquois (although very differently positioned as a white North American anthropologist), I refuse the "previous practices of discursive containment and pathology" that have plagued white textualizations of Equatorial Guinea. I refuse them not only because of their internal flaws, but also because these accounts "have teeth, and teeth that bite through time" (Simpson 2014, 100). My oil company interlocutors used white textualizations of Equatorial Guinea, and Africa more broadly, to justify the violence of their industry's daily practices—from contracts that contravene Equatoguinean sovereignty to economic theory that locates the reliably grotesque local outcomes of oil production solely within the "pathological African state." The industry used anthropology, history, economics, and political science to efface the agency of transnational corporate capitalism and to distance itself still further from that by which it was surrounded and to which it gave shape. The teeth of knowledge production, in the mouths of some of our world's most powerful corporations, indeed bite through time. Thus, this book does not offer a general ethnographic description of Equatorial Guinea (as if such a thing were possible), but a specific political history of the conditions of possibility that made a certain form of hydrocarbon capitalism possible.

This form of ethnographic refusal also contains an ethnographic insistence. If knowing, and if anthropological knowing in particular, has been a mode of power (Asad 1979; Said 1978, 1989; Foucault 1980), then this book advocates knowing more about that over which we need more power. It is *capitalism*—its ideologies and institutions, people and dreams, ecologies and erasures—that is my ethnos. Through that commitment, I stumbled upon capitalism's intimacy with liberalism, and that too became an ethnographic object. More precisely, I found liberalism *in the field*, or what Sartori (2014) calls vernacular liberalism: "the movement of liberal concepts beyond the rarified domains of self-conscious political theory ... into wider worlds" (7). Specifically, I follow the ways in which oil company management and, to a lesser extent, Equatoguinean state actors use law, contracts, economic theory, and market rationality not only as powerful tools in and of themselves, but also as a felicitous moral architecture through which to sanction capitalist practices. Liberalism here "is not a thing. It is a moving target developed in the European empire and used to secure power in the contemporary world. It is located nowhere but in its continual citation as the motivating logic and aspiration of dispersed and competing social and cultural experiments" (Povinelli 2006, 13). Both liberalism and capitalism are always made through and with the things that anthropology has long been so good at capturing—specific people and histories, places and politics, landscapes and livelihoods. This is no less true in Equatorial Guinea, despite the fact that it is precisely these entanglements that the industry works so hard to sever.

Thus, this book *is* about Equatorial Guinea insofar as it is the historical specificity of that country leading up to US corporations' discovery of oil and gas which made the industry's work toward disentanglement so apparent. Equally relevant to the story is the historical specificity of the US-based transnational oil and gas industry in the mid-1990s, the moment it discovered subsea hydrocarbons in Equatorial Guinea. *Both* histories—similarly steeped in secrecy, suppression, and violence—come to shape the project of petro-capitalism in the country. In the mid-1990s, Equatorial Guinea was governed by an authoritarian regime on its last legs, ready to acquiesce to nearly any industry condition in exchange for complicity and support. At the same moment, the industry was reeling from the rise of the global environmental movement, increasingly public breakthroughs in climate science, and the swelling power of transnational nongovernmental organizations (NGOs) (Kirsch 2014). In addition, Shell's ongoing catastrophe in neighboring Nigeria—involving everything from the killing of Ogoni activists to the visible dispossession and despoliation of the Niger Delta (Adunbi 2015; Saro-Wiwa 1992; Watts 2004)—had made that case a model failure in the industry by the time investment in Equatorial Guinea began, not to be repeated at all costs. In this moment, respective histories of secrecy, the active suppression of information, and global pariah status in *both* Equatorial Guinea and the US-based oil and gas industry came together in resonant frequency, amplifying the silence and intimacy that has come to characterize their complicity.

Today, Equatorial Guinea is widely considered to have one of the most corrupt dictatorships in the world. The global oil and gas industry is similarly disreputable. How, then, at this intersection, are hydrocarbons so reliably transformed from subsea deposits into everything from gas to lipstick to futures prices? How is capitalism, in its own image, reliably reproduced at the intersection of an industry and a dictatorship (now the longest-standing in the world) that are equally notorious, illiberal, and constituted by histories of violence, destruction, suppression, and agnotology? In Equatorial Guinea and beyond, the oil and gas industry consistently escapes consequential responsibility for local outcomes, despite profound political, environmental, economic, and social entanglements in each and every supply site. *How*? This is the puzzle that this book seeks to address by focusing ethnographically on what I call the "licit life of capitalism"—contracts and subcontracts, infrastructures, economic theory, corporate enclaves, "transparency"—and the forms of racialized and gendered liberalism on which it relies for its moral architecture. These practices have become legally sanctioned, widely replicated, and even ordinary, at the same time as they are messy, contested, and, to many, indefensible.

Before setting out to understand the licit life of capitalism, this book's ethnographic project, we must first understand that which *the licit* is set up to manage, to distance itself from, and to frame out of the picture. To illustrate this, I start with a scene from the field that conveys the intimacy of absolute rule and transnational oil firms, before moving back briefly into Equatorial Guinea's colonial and postcolonial history to give a sense of the sociopolitical world which US oil companies entered—and then altered—starting in the late 1990s.

ON EQUATORIAL GUINEA

You get the land but you don't provide a lot of jobs, you may be destroying the environment, and most of the profit goes to international capital. The companies don't have a strong case to sell to local communities, so they come to not only accept highly centralized government but to crave it. A strongman president can make all the necessary decisions. It's a lot easier to win support from the top than to build it from the bottom. As long as we want cheap gas, democracy can't exist.
—Ed Chow, longtime Chevron executive, quoted in
Ken Silverstein, *The Secret World of Oil*

Elena, an Equatoguinean friend, called one afternoon to invite me to an outdoor dinner at a Spanish-style tapas place. Our dinner companions were three other people I didn't know well—two visiting American lobbyists employed by the Equatoguinean government, whom I had met briefly on one of their earlier visits, and an Equatoguinean woman I'd never met who was introduced to me at the beginning of dinner as "an entrepreneur." The five of us ambled through normal (for Equatorial Guinea) dinner conversation. The woman had a new iPhone, and we talked about the recent statistic that Equatorial Guinea had the highest per capita percentage of iPhone users in the world. We also discussed the construction boom and how bad the harmattan was expected to be this year. Soon the conversation turned toward my research, and the two American men and the Guinean woman[5] began asking me a series of questions about my project: "How is it going? Who are you interviewing? What are you finding out? How do you get your information?" I answered with my usual mix of candor and vagueness. "It's going well. I interview locals and expats who work in the oil industry. I'm finding out that things are more complicated than they seem." As the question-and-answer session continued, Elena began to press her foot on mine under the table. I wondered if she had mistaken my foot for the table base, and I gently moved my foot from under hers as I continued talking. Gradually talk turned to politics, and I was careful, as always, to be my best noncommittal self as I listened to what the others had to say (also noncommittal, vapid statements) and responded with vagaries of my own. "How well adjusted I am to living in a paranoid dictatorship," I thought to myself as I again moved my foot from under Elena's.

We finished our beers and said our goodbyes, and I got in Elena's car to head home. I was in trouble. The "entrepreneur," it turns out, while she did have her own store, also worked for national intelligence, and, of course, Elena knew this because everyone knows everyone in Malabo. She was not able to tell me this before we arrived, however, not knowing who would be attending the dinner. As Elena yelled frantically at me in the car about how naïve I was to talk about my research, I tried to stutter in protest that I intentionally said nothing political or dangerous, and that in terms of politics, I had also been vague and effectively said nothing. She said that it didn't matter. They can take any little piece of information and twist it the wrong way. "A banana," as another informant put it, "is a stone." And worse, it was not only me that I was endangering, but also *her*. "They killed a French guy and framed his Guinean friend for the murder," she said. "I would be blamed for your death!"

Rattled by Elena's fear and anger at my ignorance (her foot was an intentional, repeated effort to shut me up), and wondering about the actuality of it all—death by research, friends framed for my death—the next day I approached Isabel, my closest friend in the field who was also rising through the ranks of the government. I was wide-eyed, agitated, and incredulous as I told her my story about Elena's anger; about being told that the intelligence operative was an "entrepreneur"; and about how I had answered questions about my research vacuously. Isabel listened calmly, nodding slowly, saying nothing. When I finished, looking at her expectantly, she returned my gaze with a quiet, knowing smile and a silence that seemed to last forever. And then she said, "Welcome." Having returned to Equatorial Guinea only six years earlier from a life abroad, Isabel said that the same thing happened to her upon her return. She told me it was valuable experience for my research "to experience the fear we all live in." "If you're not involved with locals," she said, "you'll never experience it. You have to figure out how to write about this."

As Elena's foot, Isabel's "Welcome," and Ed Chow's words that begin this section all suggest, there is a mutually beneficial relationship between absolute rule and transnational oil firms. It is a relationship characterized by impunity and secrecy on both sides, and by a form of collusion Anna Tsing (2005) has described as franchise cronyism, "in which foreign funds support the authoritarian rule that keeps the funds safe.... In exchange for supplying the money to support national leaders who can make the state secure, investors are offered the certainties of the contract, which ensures titles to mineral deposits, fixes taxation rates, and permits export of profit" (69). The licit life of capitalism—the industry's striving for capitalism in its own image—is uniquely evident in Equatorial Guinea precisely because of the specific political histories of the place, histories that led dramatically to the fear in which Equatoguineans had long lived by the time US oil companies came to town.

A BRIEF HISTORY

While Spain had technically gained imperial rights to "Spanish Guinea" in 1777, it was not until the late nineteenth century that Spanish sovereignty was fully recognized on Bioko Island (then called Fernando Pó), and it was not until the beginning of the twentieth century that Spanish missionaries had even *seen* the interior of Río Muni, let alone established administrative rule or systems of economic extraction of any kind (Ndongo-Bidyogo 1977; Nerín 2010; Martino 2012). While Spanish administrative presence was minimal in the early years of colonialism, foreigners of various nationalities out to make money were not, at least on the main island. Bioko remained a crucial, dynamic economic site, characterized as a "watering hole of explorers, traders and missionaries" at the end of the nineteenth century (Fegley 1989, 13). Río Muni, on the other hand, receded still further from its earlier small role in the slave trade (Aranzadi, forthcoming). Where Bioko was an economically and politically strategic holding, with increasing Spanish presence if not rule, Río Muni, at least for a time, "was viewed as a magnet for the border population [from Gabon and Cameroon] because it was a place where censuses, native taxation, levies, and native justice [were] unknown. According to one French Official, it was possible for Africans to live 'in complete freedom' in Rio Muni" (Sundiata 1990, 34).[6] Cameroonian author Ferdinand Oyono (1966) says as much in the opening of his novel *Houseboy*:

> It was evening. The sun had gone down behind the peaks. The deep shadow of the forest was closing in around Akomo. Flocks of toucans cut the air with great wingbeats and their plaintive calls died away slowly. The last night of

my holiday in Spanish Guinea came stealthily down. Soon I would be leaving this country used by us "Frenchmen" from Gabon and Cameroon as a place to slip away for a break whenever things became a little strained between ourselves and our white compatriots. (3)

When the Spanish Civil War broke out in 1936, most Spanish settlers in Equatorial Guinea were passively associated with the anti-Franco Popular Front, but their resistance was easily overwhelmed by troops sent from Spain. By the end of the year Spanish Guinea was securely in Franco's hands, "contributing money, raw materials and food to the long and bitter campaign against the republic" (Roberts 1986, 543). Franco's victory in 1939 marked a dual shift in the daily life of colonial rule in Equatorial Guinea, creating more metropolitan interest and investment on the one hand, and more oppressive, violent, and sharply racist rule on the other. Remembering his schooling in the 1940s and 1950s, Equatoguinean journalist and Fanon scholar Donato Ndongo-Bidyogo writes: "'Are we Spanish?'—the teacher would ask the class—'we are Spanish by the grace of God!!' . . . Entering school in the morning you had to stand in formation, do five or ten minutes of military gymnastics, sing 'Cara al Sol' [the anthem of Franco's Falangist party] while saluting. 'I am a Falangist, I will be a Falangist until I die or overcome. Long live Spain!'" (1977, 66). While on school grounds, students were required to speak Spanish exclusively, regardless of their age or how long they had been studying the colonial language. "Those that disobeyed or could not communicate sufficiently were lashed, or made to kneel for hours on gravel. This was not cultural assimilation. This was cultural assimilation at gun point" (66). Colonialism under Franco was radically and unpredictably violent for black Equatoguinean adults as well, whose movements around the island and mainland, or in between, were de facto forbidden but de jure governed by a pass system. Equatoguineans could be beaten, jailed, and killed at any time without recourse. The arbitrary violence that characterized Franco's colonial fascism—authoritarian dictatorship, military rule, forced labor, radical limitations on movement, and rampant executions—would later come to characterize postcolonial rule in Equatorial Guinea as well.

By the late 1950s and early 1960s, anticolonial sentiment and organizing was growing in Equatorial Guinea and across the continent. After unsuccessfully trying to co-opt the majority of nationalist Equatoguineans, the colonial administration proposed a vote for *autonomy* (not independence), which Equatoguineans passed, thus establishing a General Assembly of colonial administrators who, in turn, named a ten-member Consejo de Gobierno, or Government Council of Equatoguineans. This council included, among others, Francisco Macías Nguema, the former mayor of Mongomo (an inland continental district on the border with Gabon), and other authority figures from within what Mamdani (1996) has characterized as the native bureaucracies of indirect rule: those Equatoguineans who had been enlisted by the colonial government as "traditional," often meaning rural, legal, and fiscal authorities. Ndongo-Bidyogo (1977) argues that the Spanish were grooming the Equatoguineans named to the Government Council for their emergent role as the national bourgeoisie, to serve as mediators with Spain both politically—where their complicity guaranteed autonomy but not independence—and economically—wherein council members guaranteed Spain continued access to local riches and resources. In exchange, council members "were given exorbitant salaries, a Mercedes Benz, a chalet replete with servants paid by Spain, and control over the national budget" (105).

Here too, the conflation of public office with private gain that began under colonialism (Martino 2018b) set an important foundation for the expectations and norms of postcolonial Equatoguinean regimes to come. The Governing Council adopted a predictably pro-Spanish line, but Macías Nguema, in particular, began to separate himself ideologically, refusing to accept the Spanish agenda and beginning to talk about opposition to neocolonialism. In February 1968, only four years after the autonomous regime began, Equatoguinean politicians demanded independence at a constitutional conference in Spain, and Franco's regime passed a decree suspending the renewal of autonomous status. Equatorial Guinea held its first election as an independent nation-state on October 12, 1968, and Francisco Macías Nguema was elected president.

Trouble started almost immediately. In the month following independence, Spain promised financial help that never came. Records from the cocoa, coffee, and timber exports of 1968 showed that there should have been roughly $43 million in the bank (in 1968 dollars; roughly $300 million in 2017), but the national accounts were empty. The Spanish had stolen the money. Macías's relationship with Spain deteriorated rapidly. The Spanish settlers who remained began to openly provoke the newly independent government in an effort, Ndongo-Bidyogo (1977) argues, to precipitate a confrontation and "justify the intervention of the fully armed 270 members of the Spanish army still in the country" (154). As Equatoguineans began to speak up about corporal punishment, as well as racist language on plantations and in Spanish-owned businesses, the Spanish ambassador responded by threatening to withdraw the Spanish doctors, engineers, teachers, administrators, and media operators who continued to run the country's basic social infrastructure.

Not three months after assuming the presidency, Macías (quite reasonably) began to suffer from what Ndongo-Bidyogo (1977) called the "paranoia and psychosis" of assassination attempts and coups. And indeed, in February 1969, Spanish soldiers occupied the airport and media production centers, distributing arms to all remaining whites, who then patrolled the streets. A mere four months after independence, then, the Spanish organized a coup attempt, provoking Macías to declare a state of emergency, still referred to today as *la emergencia*. Macías asked all Spanish settlers who remained in the country to leave, including the missionaries who ran schools and orphanages, which left Equatorial Guinea largely without technical experts. Algerians came on technical missions as doctors and nurses, but they didn't speak Spanish and struggled to serve a population in a moment of chaotic transition. Schools were closed; children roamed the streets in large numbers; food imports were disrupted; and many people left their towns for the cities of Malabo and Bata, hoping to find more institutional stability. Thousands of Nigerians, who had long provided much of the manual labor on cocoa plantations, also began to leave, spurred by Macías's decree forbidding wage remittances to support the Biafran secession. In short, both the technocracy and the manual labor that had sustained Equatorial Guinea during the colonial era disappeared essentially overnight.

In the wake of the Spanish coup attempt, March 1969 marked the official beginning of a period Guineans to this day call *la triste memoria*, the sad memory. The few doctors who remained from Algeria, Egypt, and Nigeria began to leave in response to orders from Macías not to cure ill people considered "counterrevolutionaries." Jails began to fill with "persons of suspicion"—people Macías perceived to be political opponents—most of whom died in prison. By Christmas 1969, Macías had jailed, tortured, and killed all politicians he perceived to be against him. All incoming mail was searched and censored, on penalty of death to the intended recipient should the censors dislike what they read. Spain responded with a press war, calling openly for another coup, to which Macías responded by launching a campaign against all Equatoguineans who were in Spanish universities or who had ever studied there. "Intellectual" became a word punishable by jail or worse. In response, Franco's Ministry of External Affairs switched course, using the Law of Official Secrets to declare all information about Equatorial Guinea and its relationship with Spain *materia reservada*—strictly confidential and not to be covered by any media. At this point, "Equatorial Guinea virtually dropped out of the news. Macías closed down most of the press, instated severe censorship and banned all foreign journalists. Visas became very difficult to obtain. After 1970 there was not one reliable economic figure, government statistic or census report to be found in the country. . . . The Franco regime further aided and abetted Macías by maintaining strict silence from the beginning" (Fegley 1989, 72).

In the years that followed, Macías began to jail his own ministers; publicly execute people who had served in the pre-independence government; and persecute, detain, and execute clergy. In 1970, he outlawed political parties and created PUNT—el Partido Unico Nacional de Trabajadores, or the Unified National Workers Party. The youth wing of PUNT—Juventud en Marcha con Macías (Youth Marching with Macías)—was given free rein to accuse and attack others with impunity. In policies reminiscent of Falangist colonial practice and in active dialogue with Maoist practice of the day, military drills became compulsory for the entire population, including children as young as five and pregnant women, and Equatoguineans were forbidden from traveling within or leaving their country freely. Having declared himself president for life in 1972, Macías once again enclosed Guineans in their homeland, "by laws so similar to those from the colonial period that one could hardly note a difference" (Ndongo-Bidyogo 1977, 215). With economic production nearly at a standstill in the wake of the first Nigerian exodus, Macías decreed compulsory labor from Equatoguinean citizens (unpaid labor from all men).[7] Those able to escape streamed across the borders. In response, Macías redoubled the compulsory labor act and jailed or killed people caught escaping.

Nominally socialist, Macías cultivated relations with Cuba, China, and the USSR throughout the 1970s, although his regime alienated each in turn. By 1975, doctors and medical care were officially outlawed. (Most medical professionals in the interceding years had been Cuban.) Amidst cholera outbreaks, a resurgence of leprosy, and a population either fleeing or dying, Macías stopped the circulation of all boats to prevent further escape, thereby indiscriminately prohibiting all crafts used for fishing or those that fitfully brought medicines and food to the smaller islands of Annobón, Corisco, and the Elobeys. When, after a year's time, a group finally sailed to Annobón, half of the inhabitants were dead; the other half were transferred to Bioko as enslaved labor. "Spain, which could've intervened in this growing isolation not only as the former colonial power but more importantly as home to the best educated Equatoguineans, retained an ironclad silence—a long chain of international complicity. Even those [Equatoguineans] who wanted to organize an opposition in Spain were radically stifled by the repression and secrecy of the Franco regime" (Ndongo-Bidyogo 1977, 270). In January 1976, Macías refused a mass repatriation request from the remaining Nigerian laborers. Nigeria responded by sending war ships into the waters surrounding Malabo, and Macías fled from the capital on Bioko Island to the interior of the continental region. *From this moment on, President Macías never returned to Bioko Island, or to the capital city.* Having burned, razed, and evacuated the towns of his perceived enemies, Macías was "pursuing the phantasms in his ill and tormented mind. . . . 'When the opposition comes,' he declared, 'they will find nothing but ashes'" (Ndongo-Bidyogo 1977, 273).

By 1978, at least 20,000 people had been killed in a country with a population of roughly 300,000. Another one-sixth of the population was forcibly recruited as slave labor on cacao and coffee plantations and in timber yards. One out of every three Equatoguineans had become a refugee (Fegley 1989). An estimated 60,000 fled to Gabon; 30,000 to Cameroon; and several thousand to Nigeria. By 1978, roughly 6,000 Equatoguineans lived in Spain.

During my time in Equatorial Guinea between 2006 and 2008, in talking about la triste memoria, friends and informants agreed that Macías suffered from serious mental illness (many mentioned schizophrenia) that worsened progressively and monstrously during his decade as president (Sundiata 1983). Very few people were willing to condemn him individually, noting that by the end of la emergencia, he was completely incapacitated mentally and refused to leave his continental compound in Nsork. For years toward the end of his rule, the capital city, and indeed the country, were effectively no longer under his command; others were carrying out the terror. By 1975, all sophisticated weaponry, vehicles, aircraft, and boats were held under the control of three or four commanders. Leading these was Lieutenant Colonel Obiang Nguema Mbasogo, who was "virtually the ruler of [Bioko] while Macías isolated himself" (Fegley 1989, 162). A graduate of the Spanish military academy at Zaragoza, Obiang was among Macías's closest adherents. As Deputy Minister of Defense, he was in charge of the penitentiary system on Bioko Island, including both local precincts and the infamous Blackbich (Black Beach) Prison where so many had died. In these capacities, Obiang "spoke and acted with the authority of the president and personally saw that his punishments were carried out" (Fegley 1991, 162; Liniger-Goumaz 1989). And it was Obiang who, in 1979, overthrew President Macías in a coup. Thanks to US oil firms, Obiang remains president to this day.

After overthrowing Macías, Obiang announced that he would rule the country with the Consejo Militar Supremo, or Supreme Military Council

(SMC), which would also put Macías on trial in a courtroom hastily laid out in Malabo's Marfil Cinema. (Law and its infrastructures had dissolved under Macías, a fact that becomes unduly important once US oil companies arrive.) Charges against Macías initially included genocide, mass murder, and the embezzlement of public funds; however, because those running the Supreme Military Council, including Obiang himself, were directly implicated in those atrocities, the charges were quickly limited to 101 proven murders. When the trial eventually proceeded, the accusations were limited further still to the period between 1969 and 1974, "after which time most members of the SMC were involved in the terror" (Fegley 1989, 167). Found guilty of 101 murders, Macías was executed by a firing squad and Obiang took power.

Despite the fearsome continuities between the two regimes, Obiang's coup brought immediate and meaningful changes to Equatoguineans' daily lives. After roughly nine years of school closures, my friends and interlocutors remembered all of a sudden being able to go to school. Churches too, which had been forbidden and closed, were reopened. People recalled to me dressing up and going to church again, with long lines for new and retroactive baptisms. Economically, there were immediate changes as well. One could go to the market and buy chicken and pork, rice and oil, bread and candy. An Equatoguinean friend who had been a child at the time remembered this moment as a switch from plastic sandals to the availability of sneakers. Indeed, foreign aid poured in with the advent of Obiang's rule, including millions of dollars from the Spanish, as well as large multilateral loans. Obiang released thousands of prisoners and received the first resident ambassador from the US in 1981. Declaring his regime's nonaligned openness to aid from the East and West, Obiang began the process of joining the French-aligned Central African Economic and Monetary Community (CEMAC) in 1982, and the nation's currency changed officially from the ekuele to the CFA franc in 1985.

But formidable continuities lingered. The country was under military rule without foreseeable end; the press and political parties remained illegal. Open political dissidents, including Eugenio Abeso Mondu and Pedro Motu, were killed in Obiang's early years. After a decade of Obiang's rule, Equatorial Guinea was drowning in multilateral debt and bereft of any political freedoms. A handful of people who were not politicians, but university professors, doctors, and engineers—Placido Mico, Pablo Mba, Fernando Abaga, and Jose Luis Mvumba—began to mobilize against political killings and military rule with the clandestine distribution of pamphlets containing information about the current regime. Their movement built toward the presidential elections of 1992 when Severo Moto ran in opposition to Obiang, after which Moto was arrested, incarcerated, and released. The terrain for oppositional mobilization was rocky at best; however, between 1992 and 1995, aided by loan conditionalities that required the superficial legalization of political parties, Convergencia para la Democracia Social (Convergence for Social Democracy, or CPDS) gained power. Then in 1995, the Plataforma de Posicion Conjunta (an opposition coalition) won a majority in the municipal and parliamentary elections. In other words, nearly two decades after Obiang took power, the conjuncture of deep debts and an opposition coalition looked like it might finally unseat the dictatorship. And indeed, 1995 was a watershed year, although not in the way this election victory indicates. The US government had closed its embassy in the country that year, in part to protest human rights abuses. Nearly simultaneously, the company then-called Exxon discovered that the Zafiro oil field had production capacities three times greater than the company's entire worldwide output of oil and gas at the time. "The following year, in advance of the presidential election, ExxonMobil's petro-dollars bankrolled the involvement of a US lobbyist who helped legitimize a rigged contest in which Obiang claimed 97.8 percent of the vote from the same constituency that only months earlier had opted overwhelmingly for the opposition coalition" (Alicante 2017). Indeed, the Exxon-funded group, the Institute for Democratic Strategies, played a pivotal role in the manufacture of Equatorial Guinea's 1996 presidential election (Shaxson 2008). "And that" an opposition member of parliament put it to me succinctly, "was when petroleum started. Petroleum was like a life jacket for the regime, an oxygen balloon to help it float." An oxygen balloon for dictatorship and a lead weight for democracy.

———————————

Since the discovery of commercially viable hydrocarbon deposits in Equatorial Guinea in the mid-1990s, the country has received nearly $100 billion in capital deployment from US oil and gas companies alone. Among Africa's most important oil producers, the long-impoverished microstate is now at the center of the petroleum industry's "new Persian Gulf." At its peak in 2009, Equatorial Guinea exported ninety thousand barrels of oil per day to the US alone (US Energy Information Administration [USEIA] 2016) and is today the richest country per capita on the African continent. Production sharing contracts worth billions of dollars annually to companies and the state alike require protracted negotiation and complicity between US oil companies and Obiang's authoritarian regime, which, at forty years strong (as of 2019), makes Obiang the longest continuously serving leader in the world today. For this repressive regime, once crippled by external debt burdens and threatened by an opposition coalition, US oil and gas contracts have been an unparalleled state-making project. In exchange for a funded regime, the Equatoguinean government must negotiate with oil companies to change local environmental, labor, or taxation laws that might affect those companies' profit margins. Ostensibly progressive laws requiring 35 percent local ownership of all foreign assets are abided with highly placed Equatoguineans serving as well-paid "associates" (*socios*) for foreign companies. The Equatoguinean state and US oil companies unevenly share governance and sovereignty in a complicated and profoundly unequal relationship of corporate–sovereign interdependency (Cattelino 2011; Mitchell 1991).

If the political landscape has been transformed by the oil and gas industry, so too has Equatorial Guinea's physical landscape, which has transmogrified at a hallucinogenic pace. Offshore gas flares blaze against the nights' dark skies in an uninterrupted string that seems to stretch from Nigerian waters all the way down. La Planta screams into view as planes land in Malabo's airport. Dazzlingly bright, the natural gas and methanol plant is a tangled, illuminated kingdom of small and large pipes, with some pipes big enough to fit a car inside, connecting metal vats and silos and containers and wires and more pipes and conveyor devices and cranes, all weaving in and out of one another. It seems the plane will scrape its metal belly on the highest reaches of the plant. The small capital city in the distance is dim and receding, or at least it was when I first started research in 2006. Yet there, now, contractual clauses built entire cities as if overnight (Appel 2012d). Malabo II sprouted beside colonial Malabo and, dotted with Chinese and Egyptian construction workers, asphalt extended filament-like in all directions (Mba 2011). Stadiums, palaces, skyscrapers, conference centers, hotels, and vast housing and apartment complexes rose from red dirt exposed beneath equatorial green only days before. In 2013, Equatorial Guinea saw more investment as a percentage of gross domestic product (GDP) than any other country in the world (Harrison 2013; Appel 2018a).

This extraordinary intensity of infrastructure investment has entirely remade the small country's property regime, as the president publicly expropriates his own substantial holdings "in the name of development," while oil and gas companies rent what is still widely considered "his" land. *Los de a pie* (the masses; literally, those on foot or those who walk) are expected to equate their dispossession with the president's hollow act. Gated residential and corporate enclaves for migrant industry personnel spring up in these spaces, serviced by their own sewage, septic, electricity, telecommunications, and food procurement systems (Ferguson 2006, chapter 2). The infrastructures of both hydrocarbon production and development—from rigs to hydroelectric dams, gated corporate enclaves to freshly paved roads and entirely new cities—become key sites in which companies and the state negotiate the ethical and political entanglements of hydrocarbon capitalism (Appel 2012d).

In 2014, petroleum revenue (from crude oil and gas condensate production) accounted for roughly 90 percent of Equatoguinean government income and over 90 percent of total exports (IMF 2015). As Equatorial Guinea's reserves decline toward exhaustion and the global price of oil continues to fall, both of these figures are down from 2008's numbers of 98 percent of government revenue and 99.3 percent of the nation's exports (IMF 2010; República de Guinea Ecuatorial 2010). While local employment in both the service sector and construction expanded marginally in Equatorial Guinea's boom times, the oil and gas industry remains the only large employer other than public administration, and work therein schedules Guineans' daily lives, putting them in security guard or maid uniforms, or sending them to offshore platforms for weeks at a time. The industry has enabled some Equatoguineans to return from earning degrees abroad and work as government liaisons or accountants, while it has enabled others to engage in sex work and window washing. Corporate social responsibility (CSR) programs subcontracted to international development firms fan across cities and towns, offering education reform, malaria control, the provision of hospital equipment, and neighborhood drinking wells.

If US oil companies immeasurably stabilized Obiang's regime—essentially paying him to stay in power—the coming of the industry has also given rise to interstitial spaces that before seemed foreclosed by control, surveillance, and paranoia. As Adelaida Caballero (personal communication, 2015) has written, some Equatoguineans (though certainly not those active in any kind of opposition) now joke that the *dictadura* has become a *dictablanda*.[8] At the very least, thousands of international industry personnel come and go every year, loosening (at least in the capital city) the sense of hermetic claustrophobia and isolation that had long enveloped this small country. Citizens of the US (and now of China, also) no longer need visas to visit the country, greatly facilitating the increased entry of journalists and researchers. The Extractive Industries Transparency Initiative (see chapter 6) briefly mandated something close to civil society meetings, in which citizens were invited to talk about governance and oil revenue. But Equatoguineans doubt the potential of these spaces, and with good reason. The memory of indiscriminate violence and death is also its threat (Ávila Laurel 2011).

Guineans remember la triste memoria. They whisper about it. To the extent that little except mass killings or incarceration changed under Obiang, the sadness continues, albeit in different ways for different people. For wealthy, educated Guineans who have returned from lives and educations abroad to pursue their fortunes in this homeland of new and seemingly infinite possibilities, the change is radical and exciting. Familial memories of violence pester, however, and they too whisper about them in hushed voices around the dinner table. But their experience of home is, at least in part, one of renewal and possibility, and they often defend Equatorial Guinea against its critics in one moment, and shake their heads in defeat and disgust in the next. For the poor, most of whom did not leave, or perhaps found themselves in Gabon rather than in Spain or England, opportunity means jobs as security guards or maids, along with new restaurants and cars that they can't afford lining the streets. They are promised public housing in the boggling construction as they are dispossessed of the land on which they lived (Mba 2011; Appel 2018a).

In this time of radical change, it was not only Equatorial Guinea's history that mattered. The US-based transnational oil and gas industry's own histories of violence, subjugation, secrecy, and misinformation had also reached a specific moment in the mid-1990s. From long histories of complicity with and support for repressive regimes (Adunbi 2015; Saro-Wiwa 1992; Yergin 1993; Mitchell 2011; Silverstein 2014; Watts 2004; Vitalis 2007) to the enduring corporate practice of organizing their transnational labor force as "divided, segregated, and paid different wages according to race" (Vitalis 2007, 22; Butler 2015), and from violent dispossession, displacement, and despoliation of communities and ecosystems (Sawyer 2004; Saro-Wiwa 1992; Falola and Genova 2005) to the industry's role as the Angel of the Anthropocene, the turn of the twentieth century was a time of unprecedented exposure and critique of the US-based oil and gas industry. In response, major corporations began to implement a suite of practices—from more aggressive corporate social responsibility agendas to participation in various transparency and accountability programs, including the Extractive Industries Transparency Initiative.[9] These practices were designed to change the growing perception that oil companies were nothing but the necessary evil of modernity.

While corporate social responsibility and transparency programs are largely outward facing, designed to secure oil and gas corporations' increasingly tenuous social license to operate, perhaps the most profound change that occurred in the industry at this moment *seemed* like an inward-facing change and like a feat of engineering: the offshore. As this book will go on to detail, oil's offshore is not merely a response to geologic fact—whether hydrocarbon deposits are located subsoil or subsea—but also an infrastructural *choice* intended to minimize the political risks of visible, accessible production. Equatorial Guinea came on-stream at just the historical moment when—largely in response to the unmitigated disaster in Nigeria—the industry decided that the offshore was useful not only as an organizing principle for industrial operations, but also as a guiding metaphor for its relationship to production sites more broadly. For US companies in Equatorial Guinea during my research, *not to be like Nigeria* was a mantra, shorthand for Shell Nigeria's infamously disastrous presence in the Niger Delta. In particular, the mantra gestured to the robust structures of responsibility that typified corporate involvement in Nigeria, with Shell providing often-unreliable water, light, or education in a tangled relationship with local states (Watts 2004; Zalik 2006, 2009; Saro-Wiwa 1992).[10] The industry setup in Equatorial Guinea was a self-conscious and explicit response to this ongoing disaster. At least on paper, the arrangement between US oil companies and that which is "outside" them in Equatorial Guinea was radically attenuated, with corporate social responsibility subcontracted out and companies separated by multiple layers of liability from that which surrounded them. "Offshore" was shorthand for this shift, and thus it referred not only to mid-ocean production platforms, but also to the guiding metaphor of apparent distance between corporate and national daily life.

It is important to refuse the industry account of the offshore as a technical breakthrough that enabled radically different forms of work, profit-making, or corporate relationships to place. Rather, we might better understand offshore infrastructure as enabling certain forms of *continuity*. Practices that had been met with increasing resistance onshore—unimpeded environmental degradation; labor suppression, including paying workers according to race, and providing separate and strikingly unequal housing facilities; and lack of meaningful training or technology transfer opportunities—can be newly naturalized in offshore work, ostensibly justified by the novel techno-social configuration of the open ocean, the geophysical demands of subsea hydrocarbon, and the forms of infrastructure necessary to respond to those conditions (not to mention the invisibility of the production setup to the general public; Zalik 2009). With onshore communities seemingly disinter-mediated by the offshore production process, and resistance itself presented with new spatial challenges, forms of national rule, regulation, oversight, and state-corporate complicity become increasingly central to the production of oil and gas in the offshore era. In other words, the US industry's long history of active collusion with authoritarian regimes was particularly relevant to its mid-1990s arrival in Equatorial Guinea.

While, as this book will show, the Equatoguinean state is fractious and divided, and far from a homogeneous oppressive force, each and every Equatoguinean I came to know (whether functionary, tycoon, or *a pie*) was afraid of The State in one way or another. That fear facilitated unimpeded oil production without meaningful public participation. Recall again Ed Chow's words from the earlier epigraph: "A strongman president can make all the necessary decisions. It's a lot easier to win support from the top than to build it from the bottom. As long as we want cheap gas, democracy can't exist" (in Silverstein 2014, 7). Press and state-independent media, the possibility of gathering in groups for political debate, and even the willingness to articulate critical ideas and opinions outside the privacy of one's home or close associates were all but absent and often illegal in Equatorial Guinea during my time there. With the few exceptions of citizens openly affiliated with the opposition, who experience regular jailing and other forms of harassment and abuse, "everyone is in their own corner," as one of my friends put it (*Todo el mundo está en su propia esquina*).[11]

This, then, is the historical conjuncture at which Equatorial Guinea and the US oil and gas industry found one another. And it is this conjuncture that the licit life of capitalism is set up to manage. Given these histories of violence and suppression, *how* is Equatorial Guinea converted into just another oil exporting place? How do oil and gas emerge *as if* untouched by these histories? How is the industry so relentlessly able to abdicate responsibility for supply site entanglements? How, in short, is capitalism in its own image possible? These are the questions this book seeks to answer by focusing ethnographically on the *licit life of capitalism*. Rather than use this book to bring critical attention to the scandals that saturate capitalism's daily life, not least in the oil industry and not least in sub-Saharan Africa, I suggest that oil in Equatorial Guinea counterintuitively offers an ideal place in which to explore what we might take to be the opposite of scandal. Contracts and corporate enclaves, offshore rigs and economic theory are the assemblages of liberalism and racialized labor, expertise and technology, gender and spatialized domesticity, which seem to make an industry operating on the edge of legitimacy and legality formally legitimate, legal, and productive of extraordinary profit.[12] This approach to capitalism echoes Saidiya V. Hartman's

(1997) approach to the routinized violence of slavery, in which she focuses *not* on invocations of the shocking and the terrible, but on "those scenes in which terror can hardly be discerned" (4). This attention to the licit undertakes an anthropology of capitalism that proceeds not from a sociology of error, but from the question of how what currently exists has been stabilized (Roitman 2014, 78; see also de Goede 2005). Rather than a (mis)representation to be deconstructed, capitalism here is understood as a constant construction project to be followed through research. Each chapter—The Offshore, The Enclave, The Contract, The Subcontract, The Economy, The Political—focuses on one site where the licit is made.

What I referred to earlier as the as ifs of capitalism that so many of my interlocutors were at pains to approximate—the labor-intensive processes of abstraction and standardization, and the practices of spatial and sociopolitical distancing—are the conditions and ends of the licit. In the section that follows, I explore this relationship through country–company entanglements, the embodied work of disentanglement, and the forms and processes I have come to refer to as the licit life of capitalism.

ENTANGLEMENT AND DISENTANGLEMENT

> The closer we look at the commodity chain, the more every step— even transportation—can be seen as an arena of cultural production . . . yet the commodity must emerge as if untouched by this friction.
> —Anna Tsing, *Friction: An Ethnography of Global Connection*

The oil and gas industry seeps into every corner of Equatorial Guinea's daily life, from keeping a regime in power to the ways in which children are educated; from staggeringly vast infrastructural projects and reconfigured modes of property adjudication to mid-ocean employment. And yet, the industry creates and inhabits an eerie distance from its supply site. How is this distance made and maintained? As Tsing's words above suggest, the technology, labor, contracts, and imaginaries that move hydrocarbons from subsea to futures markets are full of the messy friction of cultural production, deeply and often illicitly entangled with lives and landscapes in Equatorial Guinea. Even so, the commodity emerges "as if untouched" by this friction. Again, how?[13] Methodologically, this *how* asks us to start from what anthropologists have become so good at recognizing—the complex entanglements, histories, and multiplicities of daily life—and then trace the processes by which that complexity and contingency are often so effectively mustered into capitalist projects, as well as the accumulation, dispossession, and retrenchment of intersubjective differences that reliably accompany them (Bear et al. 2015). This *how* asks us to start from the particular histories of both Equatorial Guinea and the transnational oil and gas industry, and then watch the processes through which those histories are sublated into something called "global capitalism." Part of this process, of course, is to understand how the types of distancing and social disembedding that we're often taught are intrinsic to capitalism are, in fact, *made* through daily bodily, affective, and technical practices. Capitalism's distance was something many of my interlocutors aspired to and fought for.

To illustrate, early in my fieldwork, I sat down with the Canadian human resources (HR) manager of a major US oil and gas firm. As I pattered awkwardly about my emerging project, he smiled and nodded from his chair, at one point leaning off to the side to leaf through a desk drawer. I stumbled to a stop, and he passed a document across the table. It was a recently published article from *African Studies Quarterly* titled "The Political Economy of Oil in Equatorial Guinea" (McSherry 2006). "Everything you need to know is in this article," the HR manager said to me. "If the government here doesn't get its act together, this is what's going to happen." I was familiar with the article, written by a political science graduate student who, having never visited the country, had applied resource curse theory to Equatorial Guinea, positing that "oil has exacerbated already present pathologies in Equatorial Guinea's political economy, paving the way for future problems of underdevelopment, instability, and authoritarian rule" (McSherry 2006, 24). My field notes from that day recount my surprise that an expatriate HR manager read *African Studies Quarterly*, my nascent anxiety about the ubiquity of resource curse theory (on which there is more below), and not much more. But as my research stretched to fourteen months over the next two and a half years, that early encounter often came back to haunt me. While driving through a rural area with a British corporate social responsibility manager from another company, for instance, we passed innumerable small fires burning in front or back of people's homes. "Another program we have," this manager remarked, "is a garbage program that helps locals learn how to dispose of waste properly, so that these fires can stop contributing to local air pollution." The haunted feeling returned. In the private compound where this man lived and worked, a towering gas flare burning crude's gaseous by-products lit the

Figure Intro.1. Gas flare on Endurance compound.

sky twenty-four hours per day, a practice so toxic that it is known to create its own microclimate.[14]

It was only through multiple encounters like this second one—in which I was told by an oil industry manager that locals needed to learn about waste disposal and environmental protection despite the arguably unparalleled global environmental wreckage that his industry has wrought—that I began to understand what haunted me about the first. Each of these moments shows a startling habitation of distance between those running US oil firms in Equatorial Guinea and the country in which they happen to find themselves. For the HR manager (white, Canadian, male, living temporarily in central Africa), the resource curse as a ubiquitous and traveling form of economic theory enabled a particular kind of postcolonial common sense: it provided an authoritative, causal narrative that located the pathological effects of oil extraction squarely within "the African state," while his own work for the corporation disappeared from view (Appel 2017). From this habitation of distance, Chakrabarty's (2002, 66) "particular way of seeing," a felicitous and consequential "Africa" emerges, an Africa of pathological states and aggrieved citizens. And, an implied other also emerges in the form of the benevolent corporation that proffers *liberalism*: market rationality in the face of corruption, standardization in the face of irregularity, and universal environmental standards in the face of ignorance. How are these aporias made? How do those whose work is so intimately entangled with life in Equatorial Guinea find themselves looking out at the country as if from afar? The boundary-making projects that stretch from the construction of racialized identities to the construction of physical walls chronicled in the chapters of this book allow companies to bemoan poverty, pollution, and kleptocracy "out there," as if they have nothing to do with them, while they work furiously to disentangle their operations, residential footprints, corporate practices, legal presence, shareholder value, and moral identity from life "outside their walls."

At issue here is the oil industry's *intentional, aspirational* disentanglement from sociopolitical membership in Equatorial Guinea. Anthropology has long used grounded ethnographic research to show how failures to engage with the specificities of place, people, politics, and history have impaired innumerable projects—developmental or humanitarian, activist or capitalist. Encountering economic theory in the field with the HR manager, for instance, could prompt me to dwell on the problem of misrepresentation: how naïve that all resource-rich nation-states could be conflated in resource curse analyses; how illogical to think that the same theory could be applied everywhere. To take this approach, however, would be to miss the productivity of economic theory in shaping the world (Callon 1998; MacKenzie et al. 2007; Miyazaki 2013; Holmes 2014; Appel 2017), and it would also miss the ways in which the HR manager was himself profitably at work in the world that these (mis)representations help to organize (Mitchell 2002). Thus, my analysis moves in the opposite direction. I follow the work of the oil companies themselves, for whom disengagement from Equatorial Guinea's specificity was not a mistaken starting point (ready to be "exposed" by the anthropologist), but an always-unfinished project they worked daily to build.

Within the oil and gas industry in Equatorial Guinea, the cosmology of profit disentangled from place was an explicit *goal* of many industry people with whom I worked, not a flaw, a mistake, or something of which they were ashamed. The techniques, subjectivities, and discourses through which this disentanglement was partially realized—whether offshore accounts for tax "planning," or workers paid differently according to their nations of origin— were not controversies or scandals, but "best practices" buttressed by powerful legal regimes, moralities, and naturalized understandings of capitalism itself. Thus, to persist in "uncovering" local complexity beneath the smooth surface that the industry was laboring to create would be to ignore the intentionality of its disengagement, its partiality and felicitous "as if" qualities, not to mention the spectacular accumulation and near-total abdication of liability that this work produced. Capitalism in its own image, then, becomes a project, a constant ongoing experiment, a desire, a haunted hope. Take for instance oil industry guidelines from McKinsey & Company, widely considered the world's most prestigious management consulting firm, which advise oil companies to "Go Modular":

> To be able to move to modular standardization, oil and gas companies need to make changes in two main areas. The first is project design, where they must adopt modular architecture and reuse standardized modules across multiple major oil and gas plants. The second is organizational: most oil and gas companies come from a tradition of building stand-alone projects designed to specific geological conditions. Going forward, they must make the reuse of existing modules the norm in their organizations. . . . Clear guidelines, including a sound business case, direct which modules or submodules can be standardized and which must be customized. In the end, each will have its own standardization strategy, ranging from identical design to a set of discrete options to fully customized. (Hart et al. 2013)

This business advice speaks directly to the *desire* and *aspiration* I chronicle in this book.[15] Standardization, replicability, and disembedding from social context do not inhere in something called "capitalism." Rather, they are aspirational—work-intensive projects many of my interlocutors worked daily to build, chronicled in the chapters that follow.

The infrastructures, the contracts, and the economic theory are *the stuff* of the licit life of capitalism, *the stuff* that promises to create the modular distance toward which McKinsey & Company counsel oil and gas firms. The promise in each of these forms is the performative appearance of compliance, legal and economic liberalism, and "transparency" in an industry increasingly notorious as the Angel of the Anthropocene. These technological and legal forms are also fundamentally social forms. They rely on gender, race (whiteness in particular), and ideologies of liberalism for their felicity. The making and maintenance of the enclave or the subcontract aim to build intersubjective and semiotic distance and oppositions—US oil companies : Africa :: West : non-West :: global : local :: standard : corrupt :: licit : illicit :: liberal : illiberal—and centrally, I will argue, white : nonwhite. To look at what, precisely, the US oil industry brings with it from place to place is to look not only at the mobility of technical, legal, and infrastructural forms, but also at the mobility of segregation, white supremacy,[16] gendered domesticity, and what Chatterjee (1993) has called "the rule of colonial difference," wherein the industry asserts the universality of post-Enlightenment legal liberalism, while simultaneously constituting Equatorial Guinea as an exception to that universality (Mehta 1997, 1999). The mobility of segregated colonial difference (for instance, the residential enclaves spatially regulated by nation of origin, "skill level," and kinship structures that I chronicle in chapter 2) and of technical, legal, and infrastructural forms rely on one another, and they require one another for their licitness and performativity. In short, the forms of racial segregation that have long characterized the industry—the global mobility of Jim Crow across the long twentieth century (Vitalis 2007, see also Butler 2015), and the technical forms that the industry also carries from place to place (Barry 2006; Appel 2012c)—amplify one another as "Western," "global," and "standard."

In *America's Kingdom*, political scientist Robert Vitalis (2007) presents a history of Saudi ARAMCO, but more pointedly, he offers "a history of the long, unbroken legacy of [racial] hierarchy across the world's mineral frontiers." As he explains in the book:

> Texaco, Chevron, Exxon, and Mobil—ARAMCO's owners—accumulated decades of experience in dozens of locales: Beaumont, Bakersfield, Coalinga, Maracaibo, Oilville, and Tampico. And they laid out each field and camp everywhere the same way, decade after decade, with the labor force divided, segregated, and paid different wages according to race. . . . The incontrovertible fact is that it was a purposeful strategy deployed consistently and unaltered across most of a century. (22)

I hadn't read Vitalis's work before leaving for the field and, consequently, hadn't looked for this "unbroken legacy" of residential and workplace segregation when I arrived in Equatorial Guinea. Reading *America's Kingdom* upon my return, I was stunned by the consonance of his historical description of the US oil industry in Saudi Arabia in the 1930s with the contemporary situation in Equatorial Guinea, where it seemed that little in the way of transnational corporate practice had changed. From different ethnographic sites—subcontracts, enclaves, daily life on offshore rigs—several chapters in the book chronicle how the industry's careful segregation of gendered whiteness from "others," its cordoning-off, and its selective engagements via corporate social responsibility or philanthropy sanctify and indeed domesticate the power and sovereignty that US oil companies wield internationally. Conversely, the felicity of what geographer Andrew Barry (2006) has called the "technological zone"—a space within which differences among technical practices, procedures, and forms have been reduced, or common standards have been established—aids and abets segregation and racialized inequality. The ability to appeal to "standardized" contracts, or mobile offshore infrastructures that ostensibly separate hydrocarbon production from local life, or economic theories like the resource curse offers white supremacy what Cheryl Harris (1993, 1795) describes as "the legal legitimation of expectations of power and control that enshrine the status quo as a neutral baseline, while masking the maintenance of white privilege and domination." Segregation, paradoxically, is used to heighten standardization, repetition, and universality, and to buttress select postcolonial meanings attributed to whiteness, including expertise, technology, power, money, hard work, meritocracy, and philanthropy.

SYSTEMATICITY AND MULTIPLICITY, IN AND BEYOND CAPITALISM

The ethnographic discovery of the work required to unevenly animate the licit life of capitalism—the bundled and repeating set of infrastructures, contracts, and forms of expertise, and the mobility of the workers required to realize these—brought me to the understanding of capitalism as a project, not a context. Moving away from totalizing theories that attribute to capitalism a singular, intrinsic systematicity on the one hand, or an endlessly varied, specific, and fractured form on the other, following the work required to instantiate Equatorial Guinea as an oil exporting place allows us to account for the relationship between capitalism's seeming coherence and power and the radically heterogeneous sites through which those qualities are made (Bear et al. 2015). Modular infrastructures, contracting, and labor regimes do not possess an inherent logic, rationality, or sameness. Rather, their intended standardization must be brought into being through the work required to build and maintain them, work that is technical, legal, and expertise-laden at the same time as it is social, affective, and crowded with racialized and gendered norms and roles. Capitalism, then, is "constructed" in Latour's (2005) sense—in contradistinction to social constructivism—in that we can account for both its solid, objective reality and its contingency by attending ethnographically to its making processes. "Capitalism can be performative only because of the many means of producing stable repetition which are now available to it and which constitute its routine base" (Thrift 2005, 3). From its quotidian white supremacy to its infrastructure, the global hydrocarbon industry and its spectacular profit are not a structural outcome of something called "capitalism," but the concrete outcome of layers of work and history through which specific and far-reaching coherences—profit, licit business practices, and the abdication of responsibility—eventually become robust and durable, despite the contingencies of their making processes (Çalişkan and Callon 2009).

This approach has theoretical implications for anthropology beyond the study of capitalism. In each chapter, I engage local and historical specificity on the one hand, and the scale-making work through which larger projects (capitalism, but also the state, whiteness, economic theory) continue to do their work in the world on the other (Tsing 2015). This theoretical approach is particularly applicable to anthropological accounts of the African state. In response to the widespread assertion that African states are corrupt and pathological, the anthropologist generally suggests, *"Rather than starting from pathology, let me show you the local, postcolonial logics by which African state X actually works."* This invaluable approach spans critical African studies, from Meyer Fortes and E. Evans-Pritchard's (1940) account of African political systems, through Mbembe's (2001) account of postcolonial governance in Cameroon. As Fortes and Evans-Pritchard (1940) put it, "Political philosophy has chiefly concerned itself with how men ought to live and what form of government they ought to have, rather than with what are their political habits and institutions" (4). Mbembe (2001), half a century later, wrote that extant scholarship on the African state "undermines the very possibility of understanding African economic and political facts" (7) and, instead, produces a situation in which "we know nearly everything that African states, societies, and economies are not, [but] we still know absolutely nothing about what they actually are" (12). The demand from these thinkers is that anthropology can and should offer textured accounts of how African politics actually work, rather than dwelling on their distance from an imagined liberal state. This is indispensable work, and I do much of it in this book. And yet, misunderstandings of Equatoguinean or "African" economic and political norms, often assumptions about legal liberalism as the only basis on which wealth and power can be licitly amassed, are not simply wrong (and thus in need of an anthropologist to redress them); they are also incredibly productive understandings in the world. As described in this book, transnational contracts and economic theory, to name only two examples, gain inordinate power because of the perceived corruption of the Equatoguinean state. Again, to simply end the analysis having shown the complex local logics by which the state "actually" works would fail to account for the effects of "misrepresentation" in the world.

The resource curse as a circulating form of economic theory "in the wild," as Michel Callon might put it (in MacKenzie et al. 2007; see also Appel 2017), is a paradigmatic example of this problem. Emerging from economics and political science, resource curse literature (Ebrahim-Zadeh 2003; Hirschman 1961; Humphreys et al. 2007; McSherry 2006; Sachs and Warner 1995) offers an analysis of the typical oil state and its pathologies. It suggests that Equatorial Guinea will now become a member of a class of states that includes Nigeria, Venezuela, and Kazakhstan, among others, in which the influx of oil money fuels a distinctive form of pathological development, with a concomitant set of economic and political problems, including corruption, inflation, armed conflict, antidemocratic tendencies, and the misdistribution of oil revenue. While Mitchell (2009) correctly asserts that resource curse scholarship focuses too narrowly on oil as money, thus failing to account for the oil itself, this is a failure of analysis but not of effort. Resource curse theory, in fact, contends that it is the "natural" properties of hydrocarbon that shape its political outcomes. Pathological outcomes occur repeatedly in oil-producing states, according to resource curse theory, in part because they emerge from qualities "inherent" in oil as a natural resource: it is extracted, not produced; its extraction requirements are technology intensive, though not labor intensive; and sector activities and wealth generation are enclaved, or disconnected from domestic economic and political processes, because of the affordances of the resource itself. While I do not deny the effects of hydrocarbon's materiality and geology, my ethnographic research shows that choices about labor practices, enclaving, and contracting are fundamentally political projects, so far from "natural" to the resource that I watched the tremendous work done on their behalf on a daily basis. By attributing this disconnection or enclave-nature of the sector to the natural properties of hydrocarbon, rather than to the repeating forms of capitalist extraction assembled to commodify it, resource curse theory overlooks the tremendous amount of work oil and gas companies do to produce and maintain selective disentanglements from local contexts.

Scholars in anthropology and geography have also critiqued the resource curse, until quite recently, they also focused too narrowly on oil as money. (See Shever 2012; Bond 2013; Barry 2013; Appel et al. 2015; Appel 2017; and Weszkalnys 2015 for new directions.) In the more canonical literature, scholars resisted blunt resource curse analysis in part for its commodity fetishism (Watts 2004), and in part for the ways in which it imposes a universalizing model on radically different places around the world (Coronil 1997; Okonta and Oronto 2001; Sawyer 2004; Watts 2004). Accounts that explored extraction in terms of the sociocultural effects of oil money show how oil wealth takes on mythic and spectacular qualities, at times miraculous and at times cursed. Unlike more permanent or "earned" wealth, oil money seems to burst from the ground—slippery, ephemeral, tainted, debasing (Coronil 1997; Taussig 1980; Peet and Watts 2004). In both the earlier anthropological literature and in the resource curse literature of which anthropology is so critical, the industry recedes. All but invisible as an object of ethnographic study or sustained analysis, it becomes merely a revenue-producing machine, a black box with predictable effects (Appel 2012c). Once the industry has disappeared from analysis, the well-documented pathologies of oil-exporting places then appear to reside only in state mismanagement of oil money, rather than at many different points within the carbon network, including the United States, where the politics of oil has repeatedly escaped democratic control (Mitchell 2009). As Mitchell (2009) writes, "Accounts of the oil curse diagnose it as a malady located within only one set of nodes of the networks through which oil flows and is converted into energy, profits, and political power—in the decision-making organs of individual states" (400). And Watts (2004) writes: "What is striking in all of this resource-politics scholarship is the almost total invisibility of both transnational oil companies (which typically work in joint ventures with the state) and the forms of capitalism that oil or enclave extraction engenders" (53). This book picks up their call.

I offer this critique of work that analyzes oil too narrowly as money not as mere theoretical quibble. On the contrary, I dwell on it because, as the anecdote with the HR manager I recounted earlier illustrates, powerful theories have social lives, or as Simpson (2014, 100) put it, "teeth that bite through time." The resource curse, in particular, became an ethnographic object in the field; specifically, it offered an authoritative and mobile academic architecture on which the disappearing company phenomenon could rest. If, as I have argued, the industry's mobile offshore infrastructures and labor regimes, or contracting and subcontracting setups, are intended to attenuate corporate liability, to create the effect of a separation between the industry and Equatorial Guinea, then resource curse theory has fallen for it, letting the industry off the hook by allowing it to recede from meaningful analysis. Consider British Petroleum's (BP) description of their Baku-Tbilisi-Ceyhan pipeline, which, they claim, will produce "no rusting pipes or rubbish visible on the ground, while the material consequences of the pipeline would flow indirectly through state budgets and community investment programs" (Barry 2013, 116). In other words, BP actively asserts that oil will only be visible or tangibly experienced as money in state budgets and investment programs. As long as oil only remains important, visible, or contestable in its incarnation as money, the industry's work can remain invisible. Conveniently, in this account, it is *state apparatuses* that are ostensibly responsible for the wise investment of oil rents. Thus, insofar as resource curse theory allows industry entanglements with local sociopolitical, economic, and environmental life to recede, "the African state" can be diagnosed, once again, as cursed.

What can we see differently when we attend to processes that take place long before oil is transubstantiated into money: exploration, seismic studies, discovery, investment, contract negotiation, labor regimes, construction of infrastructures and enclaves both on land and in the middle of the ocean, and joint ventures between national and transnational companies? All of these processes are entangled with the histories of colonialism, racial segregation, and global forms of gender differentiation and oppression that accompany any large-scale project. This broader concept of the carbon network, what we might think of as the construction projects of hydrocarbon capitalism, was eminently visible during my research in Equatorial Guinea. During the summer of 2006, when I first arrived for preliminary fieldwork, an enormous liquid natural gas plant was under construction, which would eventuate in a $1.4 billion facility from which the first gas cargo was exported in May 2007. Luba, a fishing town thirty minutes south of the capital, was in the ever-deferred process of becoming the region's oil logistics center, including land for industry infrastructure storage, fabrication, rig and vessel repair, deep-water docking facilities, a fuel bunkering facility, and more. The hydrocarbon legislation still extant as of 2016 was enacted in November 2006. In short, it was not only Hiltons and lavish residential complexes that were under construction during my time in Equatorial Guinea. It was the oil industry itself—its infrastructures, legal frameworks, socialities, and conditions of possibility. Once we offer equal ethnographic attention to the industry, then new approaches can emerge to analyze oil exporting states and the porous governance and power borders between states and corporations.

LAST WORDS: A DIFFERENT DANGER

The analytic process of showing something apparently singular or binary (capitalism, race, gender, the state, neoliberalism) to be heterogeneous—full of historical contingency, complexity, and multiplicity—is vital work, theoretically and politically. And yet, as I have repeated in this Introduction and will explore throughout this book, analysis can't rest there—Aha! Capitalism doesn't work as it claims to!—lest we entertain the fantasy that in the process of showing capitalism to be multiply and contingently constituted, we have somehow undone its power. At the same time, however, in tracing the work through which contingency and complexity are so often mustered into durable capitalist projects, here through ethnographic attention to the oil industry in Equatorial Guinea, there is a twin danger in making capitalism seem more coherent and hermetic than it is in reality (Banta 1993; Gibson-Graham 1996, 2006; see also Povinelli 2006 on liberalism). Ethnography, as a methodology, helps us through this apparent analytic impasse.

Extended attention to the daily lives of those working to build and participate in petro-capitalism in Equatorial Guinea (a great many of whom had participated in similar projects elsewhere in the world) offers new insights into capitalism as an embodied practice. First, it is haunted—by failure, by controversy, and by never quite being all that it claims to be. In Equatorial Guinea, this was the memory of Nigeria as a model failure, of company towns riven by strikes and worker and community organizing through nearly a century of global extraction, of visible and disruptable onshore infrastructures, of untiring economic theory that lingers in development plans and in the desk drawers of expatriate managers. These hauntings reveal capitalism *itself* as spectral. As AbdouMaliq Simone (2012) writes, "the spectral rests in . . . the conceit that particular kinds of things can be built anywhere regardless of the specificities of setting or the practicality of use." The spectral haunts Equatorial Guinea because this desire to build without regard to specificity— capitalism in its own image of scalability, efficiency, and disembeddedness— is only that: desire, aspiration, failure. As I hope to show in the pages that follow, this is not an insight of critical theory. Rather, this was very much the starting point for many of the US and British migrant managers with whom I worked, who knew full well about the complexities of producing oil and gas in any place, and were merely trying their hardest to manage those complexities and to frame out those they could. Then, for the wide-ranging overflow that remained, they relied on everything from ostensibly standardized contracts, to whiteness as property (Harris 1993), to the invocation of "best practices" that were neither codified nor enforced to enact the licit life of capitalism.

To invoke capitalism as a project, not a context, as I did at the beginning of this Introduction, is already to point to the contingencies of its making, to suggest that it is neither hermetic nor coherent. Because of oil's newness when I arrived in Equatorial Guinea, the project of hydrocarbon capitalism—its building and maintenance—was very much available for ethnographic study. As Latour (2005) has written: "When you are guided to any construction site you are experiencing the troubling and exhilarating feeling that things *could be different*, or at least that they *could still fail*—a feeling never so deep when faced with the final product, no matter how beautiful or impressive it might be" (89). The oil industry is a particularly generative site in which to follow the work of construction, because oil too often has in common with capitalism the analytic privilege of being deemed more suitable as explanatory referent than ethnographic object. Like capitalism, oil is often conceived metonymically—oil is money; oil is geopolitics; oil is modernity—rather than as an ongoing project available for ethnographic inquiry (Appel et al. 2015). This shared quality folds over on itself when oil and capitalism are theorized as co-constitutive. Analysts of all stripes will point to oil's $6 trillion annual market value to pose it as the world's most important commodity or to note that, because the global oil trade is denominated in dollars, its spectacular profits provide liquidity to financial markets, and, reciprocally, finance capital enables contemporary extraction in that enormous amounts of up-front capital are required to access increasingly inaccessible hydrocarbon deposits. Here oil and capitalism are conflated, both empirically as constitutive of one another and analytically as objects that are not also subjects. This conflation works to encircle and enclose our economic imaginations: in both oil and capitalism, and in oil capitalism, we fear that there is no alternative, or that alternatives are constantly deferred. Here I insist that both oil and capitalism are projects, not contexts; that neither oil nor capitalism can give a mirrored account of the other; and that both have immanent alternatives.

And yet, that capitalism is a project does not mean that it can be undone simply. As this book has emerged from various iterations as articles and talks, I have constantly been asked: *Where is the resistance? Where is the otherwise?* As an activist, I am both deeply empathetic to these questions and skeptical of what they are asking. Bringing capitalism's "otherwises" into being is a profound challenge that requires much more than simply calling it a project. Among other strategies, this task requires an intimate knowledge of capitalism itself (Thrift 2005), a strategy to which this book aims to contribute across its six chapters.

Chapter 1, "The Offshore," continues from the helicopter ride that took me, and aims to take the reader, from the shores of Equatorial Guinea out above the ocean, to the rig, to the FPSO, to the tanker that takes oil to market. Starting from the provocation that the production chain of Equatorial Guinea's oil was clearest to me by helicopter, far off the country's shores, I show that the view from the helicopter is redolent with qualities often presumed intrinsic to capitalism: standardization, technical mastery, and the disembedding of economic interaction from social context (Granovetter 1985; Polanyi 2001; Simmel 2011). In contrast, the view from fourteen months of fieldwork made clear the extraordinary amount of work required to produce glimmering approximations of those (ostensibly intrinsic) qualities. From the helicopter ride, chapter 1 moves to working life on the rig, the imbrication of the financial and industrial offshores, the question of infrastructure, and the lives of offshore oil workers through questions of race, labor, and risk.

Chapter 2, "The Enclave," brings us back onshore to look at the domestic and corporate enclaves of US oil firms in Equatorial Guinea. Drawing on feminist theory, which has long held that marriage and kinship are political and economic affairs constitutive of capitalist practice (Wynter 1982, 2003; Enloe 1990; Federici 1998; Davis 1983; Yanagisako 2002; Stoler 1995, 2010; Bear et al. 2015; Hoang 2015), this chapter analyzes domesticity and daily corporate life together. Putting Vitalis's (2007) work on segregation in the world's mineral enclaves in conversation with Barry's (2006) concept of technological zones, I show how the whiteness of the compounds comes to signify a certain kind of licit practice in which white : nonwhite is semiotically mapped onto standard : corrupt :: global : local.

In chapters 3 and 4, "The Contract" and "The Subcontract," I turn ethnographic attention to the contract form. In chapter 3, I explore the production sharing contracts between US oil companies and the Equatoguinean state; and in chapter 4, the subcontracts between companies and their workers. At the intersection of capitalism and legal liberalism, I am interested in the long life of imperial debris (Stoler 2008), and in the "shielding" work of powerful contracts: the effort to create legal spaces in which companies claim sovereignty over everything from environmental law to national taxation policies. Following the work of "body shops" (the industry term for labor brokers) and their licit discrimination based on racialized nationality, these chapters examine Pateman and Mills's (2007) contention that "the global racial contract underpins the stark disparities of the contemporary world" (2).

In the book's final two chapters, I track the work of two hegemonic ideas/ ideals in the liberal imagination—national economies and transparency.

Chapter 5, "The Economy," looks at Equatorial Guinea's national economy in the wake of oil. Denaturalizing an economy as something that grows or shrinks, is liberalized or closed, this chapter starts from a national economic conference to reflect on the performativity of economics (in particular, its racialized uses in the postcolonial era), futurity and desire, and fantasies about the private sector. The second half of the chapter draws on ethnographic material from the Ministry of Finance and Budgets to discuss national accounting, bureaucracy, and the magical realism of budgets. Finally, chapter 6, "The Political," draws on my participant observation position with the Extractive Industries Transparency Initiative to return to the question of political possibility, to critique liberal theologies of social and political change, and to argue that our epistemologies of capitalism—how we come to know it—affect how we might seek to change it.

These six chapters move in a progression that begins with the refusal of the oil-as-money approach that has typified so much scholarship on oil to date, instead drawing attention to the transnational oil companies themselves and the forms of capitalism that oil engenders (Watts 2004). Thus, chapters 1 and 2 start with the corporate form and the making of space, race, and gender, processes that shape the rhythm of work and the industry's domestic spaces in Equatorial Guinea long before oil transubstantiates into money. Like the gendered and raced production of space, the contracts that I turn to in chapters 3 and 4 also long precede (and, in fact, structure the flows of) oil's specie transubstantiation. It is only in the two final chapters on the national economy form and epistemologies of transparency that I finally address oil as money, now able to analyze it as only part of a much more capacious and consequential capitalist project.

CHAPTER ONE

THE *Offshore*

The helicopter touched down gingerly on the rig, and João, the rig's safety coordinator, immediately whisked me to the radio room for a safety training mini-course on video. After administering an exam that tested my comprehension, João had me sign a liability waiver and then put me in my required personal protective equipment (PPE) of hardhat, safety glasses, gloves, earplugs, coveralls, and steel toe boots. A gregarious Brazilian *capoeirista* and vegetarian in his late forties, João had been in the offshore oil and gas business for twenty-eight years, and had been on this particular drilling rig—which I'll call the FIPCO 330—through a series of contracts that had taken him and the rig from the Irish Sea to Turkey, then Angola, the Congo, Gabon, Cameroon, and Nigeria, and now to Equatorial Guinea. Built in 1973 in a Texas shipyard, and owned by offshore drilling contractor SeaTrekker,[1] the FIPCO 330, and many of the men on board, moved around the world from contract to contract under the Liberian flag, a mobile technosocial assemblage at work today in Equatorial Guinea's offshore waters as they were in Turkey's, and as they will be in Ghana's. Operating companies—the Exxon-Mobils, Chevrons, and BPs of the world—contract with offshore drilling contractors, including SeaTrekker, for the FIPCO and rigs like it, paying up to $1 million *per day* for offshore rig rental. With many of the workers like

Figure 1.1. Deepwater offshore, Nigerian waters, 2006. Photo by Ed Kashi/VII. Used by permission.

João already on board, contracted rigs move into position to begin the grueling twenty-four-hour workdays that will eventually bring hydrocarbons to the surface.

On this particular March day, there were 115 men working on board the rig. They came from twenty different nations: Australia, Brazil, Britain, Cameroon, Canada, Colombia, Croatia, Equatorial Guinea, France, India, Nigeria, Norway, the Philippines, Romania, Russia, Serbia, South Africa, Ukraine, the United States, and Venezuela. Among the workers on board, only four worked directly for the operating company, which I will call Smith, and only twenty-five worked directly for SeaTrekker. The remaining eighty-six men were hired from fifteen different subcontracting companies, which provided everything from directional drilling experts to on-board cooks, radio operators to mud engineers. In total, there were seventeen different companies at work on the rig.

The iconic image of a "lone oil rig floating in an endless sea, [seemingly] detached from any sociohistorical or political-economic referent" (Sawyer 2012, 710) asks us to picture petroleum production as a feat defined exclusively by technological prowess and evermore sophisticated engineering.

But João and his fellow rig workers, together with the complex contracting and subcontracting arrangements that organize both their transnational working lives and the intricacy of the corporate form itself, begin to indicate some of the effort required to make that iconic view from the helicopter possible. In this chapter, through an ethnographic account of the FIPCO 330—as infrastructure, as worksite, as node in a vast corporate archipelago, and as a repository of capitalist desire, fantasy, racialized inequality, and tension—I trace the work that allows Equatorial Guinea to recede into the distance. I trace the work required to define this production process—subsea hydrocarbon deposit, to rig, to Floating Production, Storage, and Offloading (FPSO) vessel, to tanker, to market—as "offshore."

While the rig visit that frames this chapter was merely twelve hours long, the broader question of "the offshore" was ubiquitous in my fieldwork, the great majority of which was on land.[2] That Equatorial Guinea's entire industry can be described as "offshore," despite the expansive terrestrial investments and transformations the oil and gas industry brings in its train, speaks to the flexibility and productivity of the category. Specifically, on the FIPCO 330, in both Equatorial Guinea and in the global oil and gas industry more widely, there is a capacious overlap between petroleum's *industrial* offshore and its financial offshore. Rather than imagining the financial offshore as "a significant spatial metaphor" in economic globalization, and the petroleum offshore, in contrast, as "historically embedded within socially defined space" (Zalik 2009, 558), ethnographic attention to the FIPCO shows that the financial and industrial offshores are imbricated zones of fantasy and techniques of profit generation, both of them productive metaphors *and* historically embedded social phenomena.

Offshore oil operations have in common with offshore financial setups the idea (the desire, the design, the intention) that there are spaces where the production of profit can evade or minimize contestation and oversight. Just as offshore financial arrangements are designed to minimize regulation, taxation, and accessibility, oil's offshore is not merely a response to geologic fact—whether hydrocarbon deposits are located subsoil or subsea—but also an infrastructural *choice* intended to minimize the political risks of visible, accessible infrastructure. Indeed, one of the themes of this chapter is the extent to which offshore infrastructure carries far more than the crude, seawater, or liquid natural gas for which it is, at least in part, designed. To the extent that infrastructures can participate in the materialization of certain political goals and fantasies, this chapter asks: What social and political worlds are made when oil is drawn out of the earth from a depth of 35,000 feet? What are the effects of the offshore as a capacious category? I suggest that it is, in part, the overlap between the industrial and financial offshores that makes the view from the helicopter possible, that makes it seem *as if* Equatorial Guinea has disappeared.

The first section of this chapter briefly locates the FIPCO in the archipelagic corporate form of transnational oil and gas companies, before turning in the second section to the shared history of the industrial and financial offshores to animate the ethnographic material that follows: infrastructure, capitalist desire and fantasy, labor, race, and the *habitation* of the industrial and financial offshores for Equatoguinean rig workers. In each of the chapter's sections, the offshore yields questions about responsibility and sovereignty. In a geographic and legal space where attenuation of liability is part of the project, who is responsible for offshore lives and outcomes? Workers' lives? Environmental outcomes? What is the relationship between the Equatoguinean state and US-based oil companies offshore? How is power exercised and abdicated here and to what effect? How is it that practices that are increasingly untenable onshore—organizing and paying labor differently according to race—are given new licitness offshore?

Finally, in each section I try to hold the productive simultaneity of capitalism in tension—contingent and obdurate; full of gaps, yet spectacularly performative. Thus, on the one hand, the chapter engages the familiar anthropological approach of using ethnographic intimacy to show how an ostensibly frictionless space like the offshore is, in fact, a teeming and tense social world. On the other hand, while insisting on the centrality of that approach, I also insist on a key shortcoming: that it can leave us without the tools to understand the persistence and performativity of the offshore itself. In Nigeria alone, more oil seeps out of scantly regulated offshore platforms *every year* than escaped in the entire Deepwater Horizon catastrophe.[3] And, more than $140 billion per year leaves Africa for tax havens and offshore financial centers or via transfer pricing, which is close to four times the annual amount of development aid to the continent (Bond and Sharife 2009).[4] In other words, we have to take seriously the simultaneity of the offshore's teeming sociality *and* its felicitous performativity. The point, then, is not to expose the offshore as mere metaphor or misconception, but to show how that metaphor is made and maintained, and importantly, made *real* through its consequential effects in the world.

The company that owns the FIPCO 330 is incorporated in Switzerland, with additional headquarters in Houston, Texas, and the Cayman Islands. The vessel itself is registered and sails under the Liberian flag, ostensibly operating under Liberian law. In Liberia (as well as in Panama and the Marshall Islands, among others), flags of convenience (FOCs), or "open registries," generate significant state revenue by offering vessels protection from income taxes, labor demands, and other regulations, often advertised as "nonbureaucratic" or more "efficient" maritime administration. Today, over two-thirds of vessels in the shipping industry worldwide sail under open registries. Now ubiquitous, the practice has its origins in the oil industry. As Standard Oil boomed in the 1930s, Esso—Standard's shipping subsidiary—transferred twenty-five ships to Panama's fledgling registry. The colonial geographies of global capitalism's "offshore" are immediately apparent here. Panama, Liberia, and the Marshall Islands become more "efficient" sites of maritime administration not because of their strategic insertion in a market of equals (as "comparative advantage" might suggest) but because of their *subordinated* entanglements in ongoing relations of dispossession and coloniality. In other words, what Stoler (2008) has called imperial debris—in this case, national juridical systems that are broadly permissive toward corporate interests offshore—is the terrain on which comparative advantage is made.

Opening out from the FIPCO to the operating company that contracts it—itself a limited liability local subsidiary of a parent corporation we too simply refer to as "an oil company"—and out still further to the diffuse geographies of the parent corporation itself, we begin to get a sense of the intimate interconnections between financial offshore practices, including tax havens and transfer pricing, and industrial offshore practices that move the FIPCO, João, and his coworkers from Equatorial Guinea to the Congo. It was through conversations with Donald that I came to understand the centrality of corporate geographies to the daily life of US oil companies "in" Equatorial Guinea.

By the time I came to know Donald in 2007, he had been in the oil and gas industry for twenty-eight years. In his twenties, he moved with his wife Cheryl from Utah to Houston to work in the industry and raise their children. Over years of work in Texas, Donald climbed through the management ranks, the kids grew up, and eventually he and Cheryl began the peripatetic lifestyle of so many of the migrant managerial industry workers I came to know in Equatorial Guinea. Together, Cheryl and Donald had lived and worked in the transnational oil industry in Russia for five years, Ireland for two years, Japan for one year, and Indonesia for six months. When I asked him, "Where's home?" Donald replied, "Equatorial Guinea, but my tax residency is in Houston."

Donald and Cheryl were thoughtful and kind interlocutors in the field, a generosity openly informed by their devout Christianity. Cheryl had only recently joined Donald in Malabo, and she readily admitted to feeling isolated living in the company's compound, yet uncomfortable beyond its walls. In order to grow more comfortable in her new surroundings (as she had eventually in Russia, Japan, and Ireland), Cheryl would occasionally accompany me in out in the city—to the market, to pay my electricity bill, to eat lunch. She would ask me about my life in Malabo, and I would ask her about life as a nomadic manager's spouse in Equatorial Guinea and beyond. Donald and Cheryl invited me for meals in their home on the Endurance compound, and they were perhaps the only two migrant oil personnel who saw the inside of my Malabo apartment. Over my fourteen months in Equatorial Guinea, Donald graciously hosted me for hours of interviewing in his office; invited me to accompany him on various facilities-related work excursions; and shared otherwise unattainable information, including lists of Equatoguinean laws relevant to the petroleum sector (compiled by the company's legal team) and insight into the corporate structure of his company, which I will call Endurance. In my effort to make sense of our long conversations, I initially sketched a conventional organogram relating the parts of the Endurance Corporation hierarchically to each other.

In figure 1.2, the aesthetics of legible hierarchy and alignment suggest a clear relationship between corporate subsidiaries in Equatorial Guinea (at the bottom) and the "parent" corporation (at the top), and by extension, clear paths of responsibility and redress. But as I paid closer to attention to my interview notes with Donald and continued pressing him on the corporate form over fourteen months, and as I worked more closely with Equatoguinean rig workers (discussed later in the chapter), I began to understand the intended *effects* of this kind of corporate archipelago. Thus, it became analytically important to subvert the organogram.[5] Dizzying, densely layered, and circuitous, figure 1.3, and the description below, more accurately depict the attenuation and convolution of responsibility and liability offshore.

The headquarters of Endurance Equatorial Guinea Production Limited (EEGPL) sits on an isthmus of Bioko Island. A subsidiary of the US-based Endurance Corporation, EEGPL is a petroleum producer that shares this isthmus with two other companies: one producing methanol and one liquid

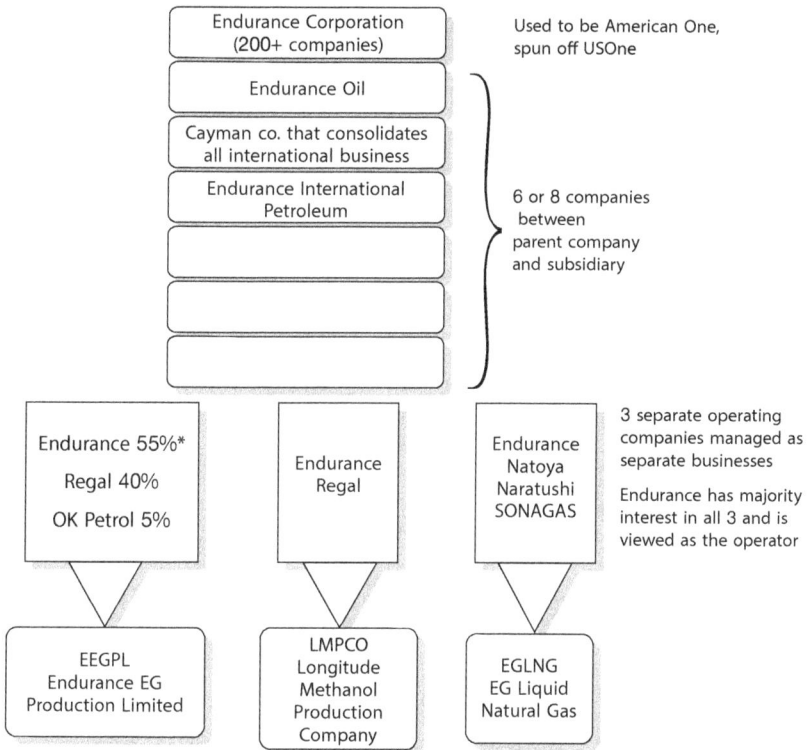

Figure 1.2. Conventional organogram of the Endurance Corporation.

natural gas. These three companies are managed as separate businesses, but the Endurance Corporation is the largest among a consortium of investors in all three, and it acts as the operator. Leaving Bioko Island and moving into Endurance's archipelagic corporate geography, we come first to Japan, where we find the consortium of companies that funds the liquid natural gas project in Equatorial Guinea. From there we move to the Cayman Islands, where Endurance International Petroleum is only one of over two hundred registered Endurance corporate subsidiaries. From there we move through eight other subsidiary levels of the company until we get to Endurance Oil, headquartered in Houston, and Endurance Petroleum, headquartered in Findley, Ohio, both publicly traded, with shareholders unevenly distributed around the world.

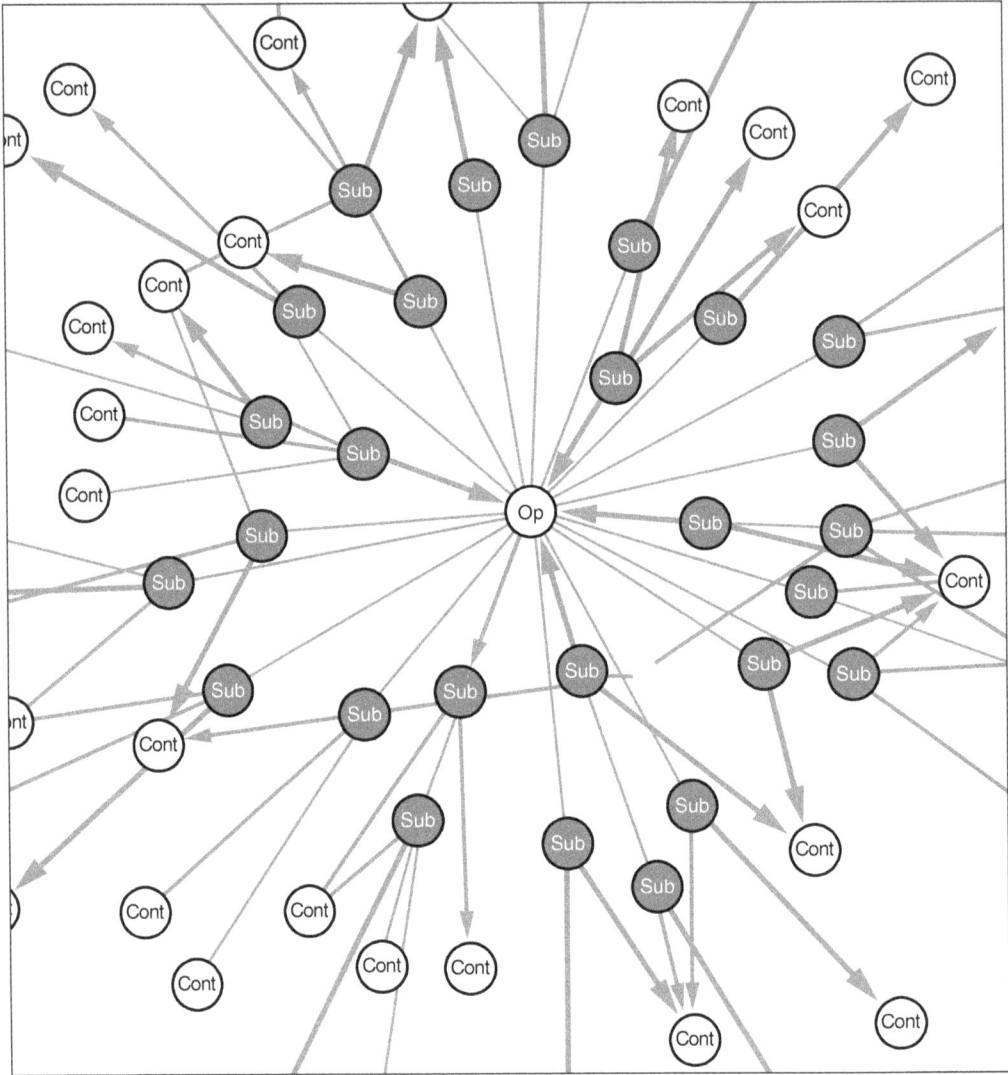

Figure 1.3. Corporate Archipelago I: Corporate Form. *Op* indicates *operating company*, *Sub* indicates *subcontractor*, and *Cont* indicates *direct contractual relationships*.

As Donald explained it to me, "Most of these structures are set up for tax 'planning' and liability." Donald used his fingers in an exaggerated quote/unquote gesture around the word "planning," implying quite plainly that *planning* was a euphemism for legal tax evasion. "This arrangement shields our parent company from liability," Donald continued. The finance manager of a separate company told me a similar story: "The Company consists of numerous business units divided geographically. This is [the] Africa Unit. Our Equatorial Guinea operation is set up in the Cayman Islands for tax advantages. [Other locations in the Africa unit, including] Algeria, Libya, Egypt, Ghana—each has separate tax laws. Africa is one business unit but the tax regimes are different." And still a third manager from a third company explained: "Here in Equatorial Guinea we are a separate company, Regal Energy Equatorial Guinea LTD, a Cayman Island–incorporated company, and a wholly owned subsidiary of Regal Energy Inc. In every country we go into, we set up a second account. There are advantages to keeping the revenues outside the US if you have operations outside the US." The corporate geographies Donald and other managers describe here, including parent companies; multiple subsidiaries; separate headquarters, operations, and finance locations; and consortia investment—all separated by various levels of juridical independence—are practices of risk mitigation and liability dispersal. They allow corporations to enjoy tax loopholes in a variety of jurisdictions; facilitate practices of transfer pricing;[6] and spread the "corporeality" of corporate personhood into a dispersed and legally slippery archipelago (Maurer 2008).

Note, however, that the corporate archipelago in figure 1.3, by itself, cannot make Equatorial Guinea's subsea hydrocarbon deposits into either shareholder value or gas in your tank. Instead, this arrangement relies on an overlaid series of proliferating corporate forms called "body shops"—the industry term for labor brokers detailed in chapter 4—which are paid handsomely to bring mostly men, mostly from the Philippines, into Equatorial Guinea, as well as workers like those I introduced at the beginning of this chapter from across the world's oil diaspora: Ecuador and Scotland, Kazakhstan and the US Gulf States. Body shops are also publicly traded companies working to provide value to disperse shareholders. As I go on to detail in this chapter, the laborers they bring to Equatorial Guinea work rotating shifts on mobile offshore infrastructures that are themselves subcontracted from major oil service firms. These firms—Halliburton, Bechtel, Schlumberger—are also among the world's largest and most profitable publicly traded companies. Notice that this overlaid corporate geography rigorously reins in not only tax

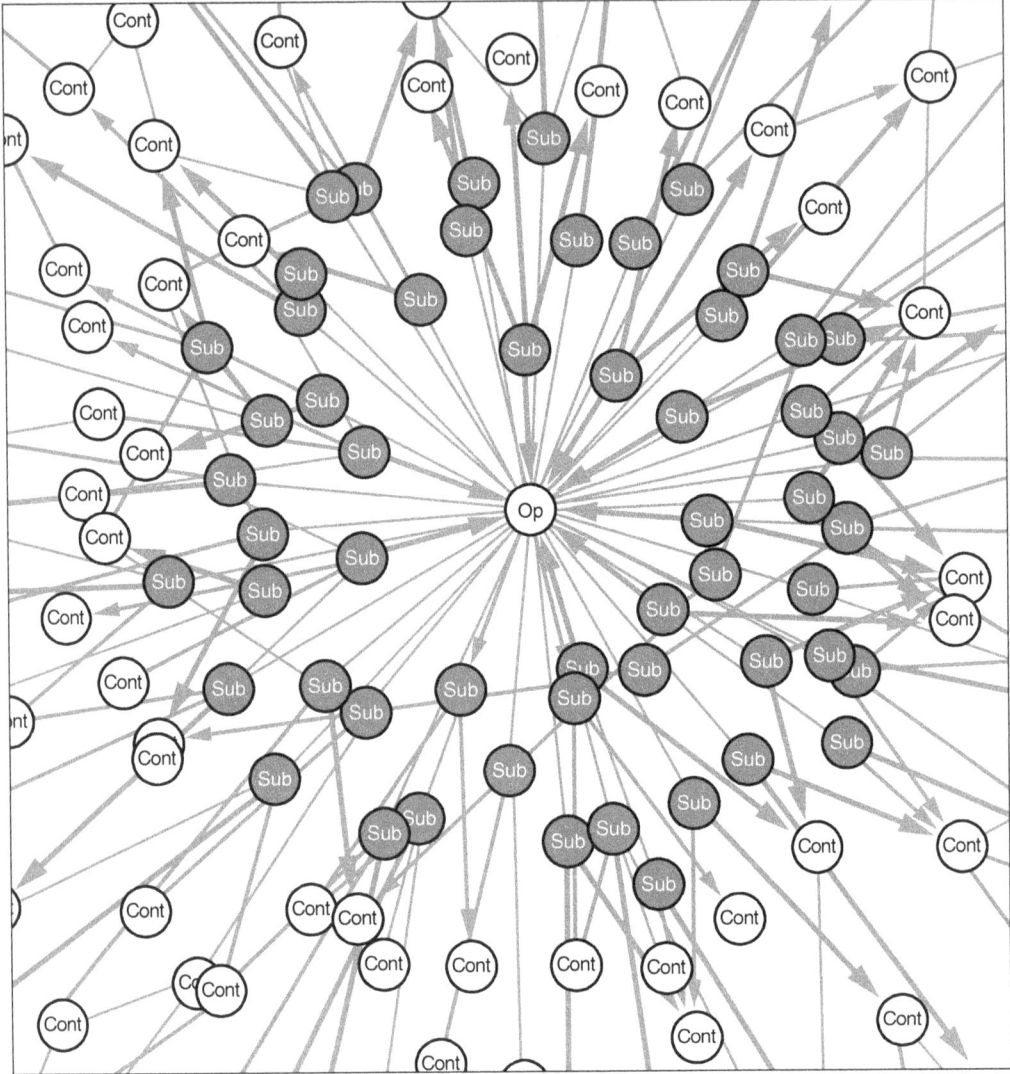

Figure 1.4. Corporate Archipelago II: Subcontracts and Labor.

liability, but also local investment in Equatorial Guinea, with payroll, banking, insurance, labor, transport, technical inputs, and other industry-related services all based elsewhere.

I want to emphasize here that Donald and I worked *together* on this account of his company's corporate form. In other words, these forms of liability dispersal are neither secret nor illicit, but open strategies for profit maximization, risk management, and the attenuation or spreading of liability for everything from offshore platform pollution to labor contestation. While Donald expressed ambivalence about tax "planning," at least acknowledging via gesture that the practice is contested or, for some, unethical, this collaborative sketch of Endurance's corporate form "in" Equatorial Guinea offers a clear look at the licit life of capitalism as legally sanctioned, widely replicated, contested, and yet, ordinary. The open, licit character of these risk-dispersal and profit-maximization practices is precisely what this book draws attention to: the legal, repeating, and flexible ways in which the oil and gas industry sets up in supply sites, and the grounded effects of these arrangements—strategic sovereignty, attenuation of liability, and profit relentlessly escaping both taxation in the United States and investment in Equatorial Guinea.

People in and close to the industry in Equatorial Guinea naturalized the effects of these arrangements, narrativized them and made sense of them, through ideas about the offshore, and offshore infrastructure in particular. Offshore rigs like the FIPCO 330 are industrial atolls in this vast corporate archipelago, and many of their sociotechnical qualities—mobility, technical complexity, transience, and the (in)visibility of their mid-ocean operations to onshore populations—amplify the promises of the offshore corporate form. To illustrate, when I wrote João to schedule the follow-up overnight visit he and I had discussed during my initial rig sojourn, he responded with the following email: "Surprise! We no longer work with SMITH but with another client named 'Regal Energy.' Regal Energy also has an office in Malabo as we are nearby Malabo now. . . . You have to rush as we will soon leave to Congo . . . Around the beginning of August." Within three months, this rig had not only switched operating companies (and consequently, locations at sea), but had also received sailing orders for the Congo, with João and others in tow. The fitful and unpredictable temporalities, contract relationships, and geographies indicated in João's email make even these massive infrastructures— some of the largest mobile infrastructures in the world—seem fleeting and spectral. This apparent ephemerality of massive infrastructures further adds to the slipperiness of the corporate form offshore. The next section attempts to think through the offshore theoretically as a capacious metaphor, practice, and historical spatialization, before turning back to ethnographic material on offshore infrastructure and those who manage and labor on it.

THE OFFSHORE

> There is a historical and structural relation between concepts of the free sea, free trade, free world economy, and those of free competition and free exploitation.
> —Carl Schmitt and G. L. Ulmen, *The Nomos of the Earth in the International Law of the Jus Publicum Europaeum*

The offshore is a broad category "that consists of all sorts of sovereign spaces essentially defined by their relative lack of regulation and taxation compared to nation states" (Cameron and Palan 2004, 91). Think tax havens, export processing zones, or special economic zones. To be offshore, in other words, is to be outside of a set of relations that do not necessarily correspond to literal shores, including regulation, taxation, "inefficiencies," and even conventional conceptions of citizenship. Many offshore arrangements offer different rules for residents and nonresidents, a disaggregated, pay-for-the-privileges-you-want and the-strictures-you-don't citizenship. Maurer (2008) summarizes this quality of offshore designations as "the disaggregation or unbundling of citizenship, jurisdiction, and nationality" (158). (See also Picciotto and Haines 1999; Ong 2006.) Offshore rigs are no exception in this regard. Working under flags of convenience, like the FIPCO's Liberian flag, which facilitate the most permissive shipping and labor standards, the subcontracting arrangements that proliferate in offshore oil effectively deregulate everything from labor guidelines, to environmental assessment, to the provision of security. This is true not only in Equatorial Guinea, but in offshore oil arrangements around the world. Referring to the outer continental shelf of the US Gulf of Mexico, where oversight and regulation were "streamlined" to accelerate exploration in the wake of the 1970s oil crisis, legal scholar Meg Caldwell, geophysicist Mark Zoback, and energy engineer Roland Horne together refer to the Gulf as "the sacrifice zone" (Caldwell et al. 2010). Zalik (2009) has also written that Mexico's offshore is "environmentally deregulated" (293) insofar as restrictions on maritime movement around platforms render spills and other practices largely invisible (see also Reed 2009; Woolfson et al. 1996; Kashi and Watts 2008; Bond 2011, 2013).

In his introduction to *The Offshore World*, Palan (2006) suggests that despite its evocative reputation for placeless economic interaction, the financial offshore is in fact deeply terrestrial. "Offshore evokes images of the high seas and exotic locations. But often this is not the case: offshore economic activity does not take place in some barge floating in the middle of the ocean" (2). Rather, financial transactions originate or pass through London, New York, or Tokyo; offshore manufacturing, including *maquiladoras* or export processing zones, are to be found in enclaves of developing countries; and tax havens have a geographic presence in Cyprus, Luxembourg, or the Cayman Islands. In each case, *shores*, or territorial sovereignty, is indeed central to the making of offshore spaces. "Offshore is sustained by the very principles of sovereignty that it is claimed to have undermined: export processing zones are territorial enclaves produced by the state; tax havens are taking advantage of the right to write law and grant legal title" (Palan 2006, 102).

Offshore, in other words, relies on sovereignty to abdicate sovereignty; sovereign states make political choices to revoke regulatory oversight in specific portions of their territory. But in considering offshores (and finance more broadly) in the Caribbean (Maurer 2008, 2010, 2013; Hudson 2017a) or Africa, where histories of colonialism and underdevelopment are everywhere on the surface, we have to think about sovereignty differently, as something closer to what Cattelino (2008) has called "sovereign interdependency." Small island economies in the Caribbean, for instance, become tax havens in a process entangled with histories of imperialism, colonialism, and slavery, the aftermath of which makes them more attractive/vulnerable repositories for fleeing capital. As with flags of convenience in Panama, Liberia, and the Marshall Islands, the long wake of colonialism is a more apt description of tax haven status than is participation in a market of equals (Maurer 2008). So too in Equatorial Guinea, where state prerogatives to give US companies free rein on platforms are less about sovereign power and more about imperial debris, histories of colonialism, and profound inequality in the global economy.

The petroleum offshore is particularly generative for thinking about sovereignty because it pushes the offshore into arguably its most literal instantiation. In the petroleum offshore, economic activity *does* take place, at least in part, "on some barge floating in the middle of the ocean" (Palan 2003, 2).The FIPCO 330, along with the FPSO and the tankers that transport oil to market, are all (broadly speaking) barges floating in the middle of the ocean in which offshore economic activity—though not the kind Palan has in mind—takes place. In the oil and gas industry in particular, linking the industrial and financial offshores focuses our attention on the centrality and proliferation of multiple offshores as intertwined techniques for the promises and seductions of capitalism, including the production of profit and the abdication of responsibility across the unequal geographies of the postcolonial world. Tax havens, transfer pricing, and flags of convenience are central to quotidian spectacular accumulation in the oil industry, as is the lack of external regulation and oversight on offshore platforms and throughout the hydrocarbon commodity chain. The commodity chain moves alongside another set of practices in which oil companies hedge the risks of notoriously volatile oil prices by investing in oil futures markets (themselves scantly regulated), which then become central not only to the price of gas or the outcome of pension fund investments, but also to the funding required for the industry's ever-deeper-water explorations (Guyer 2009; Sawyer 2012; Johnson 2015). In other words, petroleum's industrial offshore and various financial offshores are intimately intertwined in everyday practice, a relationship with a specific juridical history in the Law of the Sea.

Both the financial and industrial offshores find juridical identity in an ongoing legal conversation around the Law of the Sea, a body of laws governing the relationships among sovereignty, land, and oceans, which has evolved from fourteenth-century jurists through the United Nations Convention on the Law of the Sea. Due to the historical need to defend territory against naval threats, the legal boundaries of a state had to extend beyond strict territorial boundaries into the surrounding seas, as did political authority (Hampton and Abbott 1999). The resulting discontinuity between physical shores and legal sovereignty opened the door for juridical spaces defined in *other than* territorial terms—spaces still organized by state sovereignty, but subject to different sets of regulations. Borrowing from Prescott (1975), Palan (2006) writes that "the earliest legal provisions for coastal waters involved the *distribution of favors by rulers*, such as exclusive rights to shallow fishing grounds and salt deposits in tidal marshes, *exemption from port or harbor dues, and unhindered transit through straits*" (23; emphasis added). The offshore in Equatorial Guinea and elsewhere is a contemporary site of Prescott's "distribution of favors by rulers." Sovereignty is both intentionally and inevitably stretched thin offshore, allowing for exemptions and "unhindered transits" of all kinds. In Equatorial Guinea, foreign oil companies do not pay import duties; they are permitted to set up proprietary ports of entry; and "industry standards," rather than Equatoguinean law, exercise quotidian governing power over daily life on the rigs.

According to the Law of the Sea, states have domain over minerals and other resources within their Exclusive Economic Zones (EEZs) up to two hundred nautical miles off their shores (see figure 1.6). The first paragraph of Equatorial Guinea's hydrocarbon law, translated from Spanish, makes direct reference to their EEZ: "The fundamental Law of the Republic of Equatorial Guinea consecrates and designates as the property of the people of Equatorial Guinea all resources found in our national territory, including the subsoil, continental shelf, islands, and the Exclusive Economic Zone of our seas" (República de Guinea Ecuatorial 2006a, 1) Because offshore oil platforms are still technologically limited to the continental shelf, they remain within the physical boundaries of the Equatoguinean state. However, the daily life of regulation and its absence on platforms, explored here and again in chapter 4, introduce a series of issues around state and corporate sovereignty offshore where, in practice, sovereignty is distributed and (often strategically) fluid between the state and the innumerable contractors and subcontractors that we see on the FIPCO. The Equatoguinean state and foreign companies use and struggle over sovereignty strategically. It is not something that one or the other has or has lost in any straightforward way. By willfully signing contracts that compartmentalize their territory, the Equatoguinean government is given the opportunity "to support unfettered capitalism while denouncing it: to bemoan their loss of power and sovereignty while contributing to that very loss" (Palan 2006, 190). Oil companies manipulate the concept too, "happy to invoke national sovereignty when pressures are placed on them to improve their human rights or social responsibility records; and yet only too happy to operate in an environment in which they could get away with just about anything" (Kashi and Watts 2008, 46).

The archipelagic corporate form I discussed earlier facilitates the slipperiness of corporate sovereignty offshore. This disaggregation or dispersion is, in effect, the legal (licit, intentional) thinning of liability, accountability, and responsibility, such that what seems clearly to be the singular exercise of corporate power—global companies in contract with governments around the world, maneuvering the world's largest mobile infrastructures and reaping spectacular profit—in practice fractures rapidly into a legally slippery tangle of subsidiaries and consortia and subcontractors.[7] The next section tries to arrest some of that slipperiness and materialize offshore infrastructure by asking the questions: What is it? What does it do? What do managers and Equatoguinean government personnel imagine it to be doing? I pay special attention to the forms of desire and fantasy offshore infrastructures produce, and the kinds of distancing effects they enable.

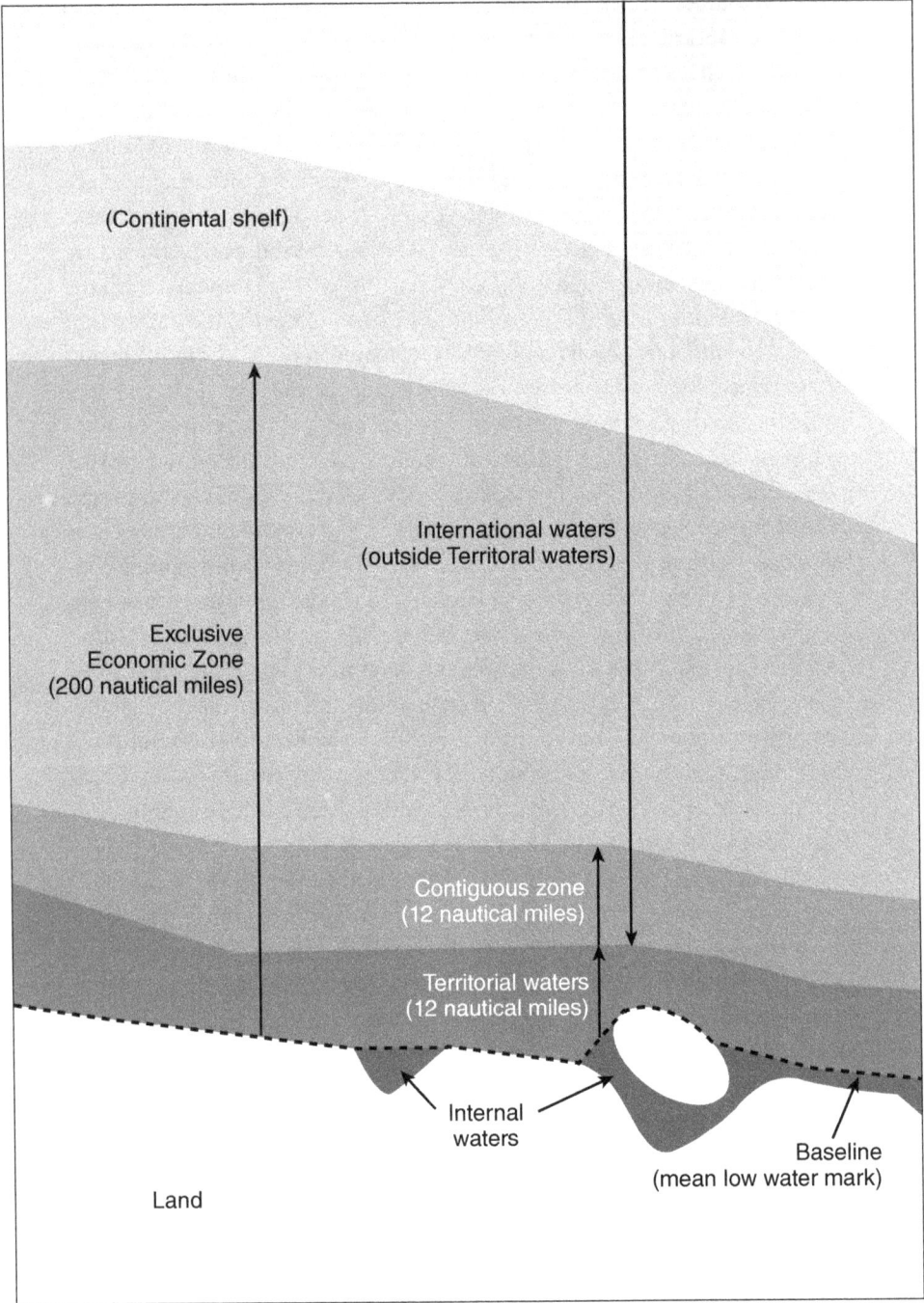

Figure 1.5. Exclusive Economic Zone (2).

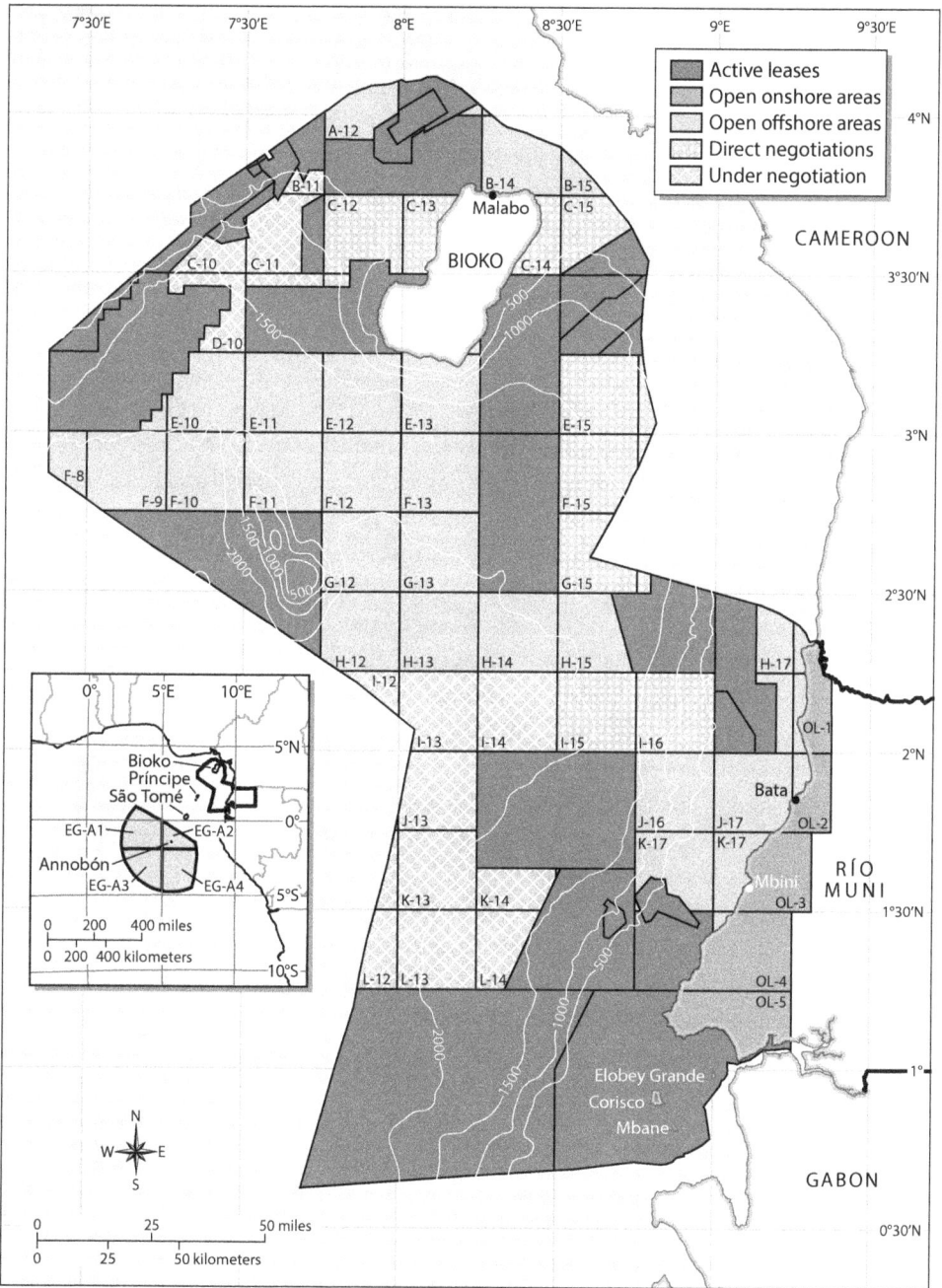

Figure 1.6. Exclusive Economic Zone (1). Note how in this concession map, Bioko is visually swallowed by "the offshore."

Infrastructure exerts a force—not simply in the materials and energies it avails, but also the way it attracts people, draws them in, coalesces and expends their capacities. Thus, the distinction between infrastructure and sociality is fluid and pragmatic rather than definitive. People work on things to work on each other, as these things work on them.

—AbdouMaliq Simone, "Infrastructure: Introductory Commentary by AbdouMaliq Simone"

Offshore infrastructure in Equatorial Guinea's waters includes drilling rigs, platforms, and FPSO vessels. Rigs are used for drilling, completion, and production; platforms are put to use postcompletion to process oil and gas and prepare them for export. Essentially offshore plants, platforms are among the largest movable, human-made infrastructures in the world. A typical platform may have up to thirty wellheads, with directional drilling allowing subsea reservoirs to be accessed at different depths and remote positions up to five miles off of the platform (Conaway 1999). Once reservoirs have been accessed and drilled and the oil is brought to the surface, the FPSOs store the crude until it is offloaded onto tankers for transport to market. While they vary in size, FPSOs have tremendous technical capacities: they are able to process up to 80,000 barrels of oil per day, reinjecting 150,000 barrels per day of seawater, handling 170 million cubic feet of gas, and storing 1.6 million barrels of crude and condensate.

While each platform is commissioned and built by individual operating companies, designed to last the lifetime of the field (after which they will undergo some combination of disassembly, repurposing, and intentional sinking), rigs like the FIPCO and FPSOs are rented and are moved around the world from contract to contract. During my fieldwork, with oil prices at or above $150 per barrel, rigs were in high demand worldwide, and operating companies at work in Equatorial Guinea were waiting up to three years for a drilling rig to move into position. While the FIPCO was built in a Texas shipyard, many newer rigs are built in Korea, purchased by an offshore drilling contractor like SeaTrekker, subcontracted by SeaTrekker to operating companies to drill offshore, and then floated from Korea to wherever the contract happens to be. On the FIPCO, I was told repeatedly how archaic the facility was, including the redolent explanation that Africa, after all, is where rigs go to die.

As Simone (2012) intimates in this section's opening epigraph, the relationship between these infrastructures and the people who animate them is fluid and mutually productive. People work on things to work on each other. Here, I pay particular attention to the forms of desire and fantasy, inequality and segregation that these infrastructures facilitate, both for those who manage them and for those who work on them. Infrastructures, Brian Larkin (2013) writes, "emerge out of and store within them forms of desire and fantasy and can take on fetish-like aspects that sometimes can be wholly autonomous from their technical function (329). . . . The sense of fascination they stimulate is an important part of their political effect" (334). Indeed, offshore petroleum infrastructures are multivalent sociopolitical projects—not merely a response to geologic fact, but a *choice* about visibility, accessibility, and political risk. To illustrate: On both the continental and island landmasses of Equatorial Guinea, liquid hydrocarbons seep to the surface. Locals note the seeps. Migrant geologists working for US companies know about them as well, and yet there is no onshore exploration in Equatorial Guinea. Why? Because the industry's cost-benefit analyses of prospective infrastructure projects consider not only the quality of crude, the distance to market, and the anticipated cost of extraction, but also, centrally, the perceived risk. As one geologist working in Equatorial Guinea explained to me, "An onshore well in Algeria costs $3 million. Offshore here wells are from $15 to $80 million. If you do find a large onshore reservoir, it's very economic, but there are associated political risks. If this country were to go through a civil war, our structures out in the water are safe. But look at Nigeria; nothing's to stop people from coming onto your facility, stopping production, blowing up the facility." The geologist articulates a particular form of capitalist desire common to both the industrial and financial offshores—a space where "associated political risks," whether taxation or militant resistance, attenuate.

Infrastructure, in other words, is envisioned, built, and operated to solve political problems simultaneous and even paramount to its material capacities—politics by other means (Mitchell 2011). Part of infrastructure's affective power emanates from the enfolding of political projects into concrete, steel, copper wire, and asphalt. But lest we take this as an insight of critical theory, somehow a radical rejoinder to naturalized assumptions about infrastructure, industry managers and Equatoguinean state functionaries with whom I worked openly discussed offshore infrastructure as a means of (hoped-for) control.

Offshore facilities promise distance from political entanglements, community entitlements, visible pollution in inhabited areas, and militant attacks and bunkering focused on accessible pipelines. That much of the commodity chain can take place in the middle of the ocean, from exploration to processing to exporting, seems to at least partially remove oil and gas companies from the ugliest, most visible, and most publicized negative effects of their industry. My discussions with migrant managers and Equatoguinean government appointees often substantiated these ideas of offshore disentanglement. Consider the comments of Mauricio, a recently appointed Equatoguinean government official:

> Onshore/offshore is an operation question. Socially there is more positive than negative to it. An oil infrastructure has a lot of environmental problems. When you build that onshore next to a community, there is more potential for problems for that community. Environmentally, the safest way to have an oil facility is to have it removed from social settings. It's an advantage for offshore operations. Having an onshore operation involves a lot of piping, infrastructure, that for some people may not be beautiful architecturally. It may not be impressive for environmentalists and people who care for trees.

While Mauricio starts his reflection by suggesting that the difference between onshore and offshore is an operations issue—merely a spatial question about where the work of exploration and extraction gets done—he segues immediately into social and environmental visibility and impact concerns. He suggests that, at the community level, the visible infrastructure of onshore production is aesthetically disruptive (as it may not be beautiful or architecturally impressive). But he also acknowledges that aesthetic objections are related to environmental objections; since "an oil infrastructure has a lot of environmental problems," whether it is onshore or off, it's best to get those problems away from people.

In addition to noting community-level environmental problems, this government appointee flags a set of supra-local issues, including regulation and visibility. Often when "environmentalists and those who care for trees" are not "impressed" by what they see (and the ability to *see* onshore infrastructure is a crucial difference), there is a call for action or regulation. By removing the visuals of oil infrastructure and operations to the middle of the ocean, there is a noticeable attenuation of public and government attention, facilitating unimpeded production. Eugenio, one of the few Equatoguinean petroleum engineers working in-country, explained the on-platform results of this attenuation:

> Normally, in the United States for example, the more petroleum you extract, the harder it is to clean the water. In other sites the amount [of oil] taken is conditional on water quality. Here there is not this conditionality. Here there is no outside testing. On the platform from time to time people are told to prepare for an environmental assessment, but it's always someone from within the company, and the results are always good.

Both Mauricio and Eugenio are overtly concerned with the environmental impact of offshore production, claiming either that this impact is minimized when infrastructure is moved away from communities or that it is only regulation and assessment that are minimized. Both men, however, framed their understandings of the offshore with concern for local outcomes, whether pollution or community protection from the petro-project.

Migrant managers, on the other hand, voiced markedly different concerns when discussing the advantages of offshore infrastructure. They focused not on threats to Equatoguinean environments and communities, but on the threats those communities posed to their operations. Here, three different managers (two from the US and one British) offer remarkably consonant offshore aspirations:

> Offshore makes it easier. Reduces investment risk. . . . You're not exposed— you're shielded from the masses, shielded from interaction. You can control the interaction. You can contain the asset, it's a clean containment. If a boat drifts in our area, we call the Navy. There's a lot less opportunity for negative interaction and distraction. In Nigeria people steal oil. We don't have that. It's clean. Less leakage, shrinkage. Controls are tight.

> It's expensive offshore, but it's clean. No laying pipelines through jungles, uprooting villages. There's nobody out there. We lay pipelines in the seabed, and it doesn't bother anyone.

> Offshore has made it easier. You're on an island, if you know what I'm saying. Diamond mines in Angola are an absolute nightmare. Armies get to you. Pirates get to you. [You have to] have massive South African war vets to secure the places. When you're out there on an oil rig, you've got a huge moat around you. That has made it easier. It's more expensive to get the oil out of the ground, but you don't have to worry about onshore issues, which are massive expenses.

These managers make a series of claims about the control and containment facilitated by the offshore—of potentially volatile sociopolitical situations (note the comparisons with Nigerian oil and Angolan diamond mining), of profit margins (less "leakage" and "shrinkage"). The offshore setup at least forestalls the risks attached to visible spills and to attack by armies, by oil bunkerers, and by MEND, who was often rumored to be planning an attack on neighboring Equatorial Guinea installations.[8] Avoided too in the offshore setup are the unpleasant tasks of relocating entire villages and negotiating the attendant set of community claims that come with visible onshore extraction—for employment, reparations, and development projects—or of hiring "massive South African war vets" to secure onshore installations. Again, the idealized industrial offshore these managers describe coincides with the evocative promises of the financial offshore, sites "of disinterested and placeless economic interaction" (Cameron and Palan 2004, 105). In avoiding the risks people bring, these managers envision the offshore setup as reducing contestation, even if it cannot be avoided completely.

It is interesting to wonder to what extent these managers' fantasies of the offshore are multifaceted. They clearly hope for a controllable environment in which to make a profit. But do they also hope, from within an industry infamous for disastrous environmental and community consequences, for different environmental and social outcomes? Is there, in these statements, either an acknowledgment or a desire to resolve what one might call the contradictions of capitalism?[9] In the field, I was stunned again and again by the extent to which US and British management failed to acknowledge the imbrication of their industry with the environmental, social, and political problems by which we were all surrounded. They loved to talk about these problems—how corrupt Equatorial Guinea is; how it's so hard to see so much wealth next to so much poverty; how much litter there is. But, as with the Canadian human resources (HR) manager, or the corporate social responsibility (CSR) manager touting environmental education, there was a near total disconnect between the deep imbrication of their industry and the daily lives they critiqued. While Venezuelan, Brazilian, and Nigerian oil workers in Equatorial Guinea often talked about the industry as a historically problematic endeavor in which they were choosing to participate, the most highly placed workers (nearly always from the US or the UK) rarely acknowledged the widely criticized effects of their industry.

Rather than a hoped-for reconciliation of capitalism's contradictions, what I heard repeatedly from managers was more akin to haunting. The accessibility of infrastructure in other places and times across hydrocarbon history haunted their descriptions of the offshore in Equatorial Guinea. Nigeria repeatedly oozed to the surface in my informants' collective memories (both personal stories and anecdotal accounts of industry history), as did histories of uprooting villages in Latin America and laying pipelines through inhabited forests in Indonesia, as well as resistance to those actions. Indeed, from Saudi Arabia to Mexico, early oil work was typified by foreign companies bringing in massive onshore infrastructures, what William Reno (2001) has called "BYOI" or "Bring Your Own Infrastructure," for extraction, production, and export. Local labor at these sites, brought in initially as construction workers, cooks, and technical trainees, lived in tents, cheek by jowl with American and European workers in the most comfortable enclaved conditions. Robert Vitalis (2007) and novelist Abdel Rahman Munif (1989), among others, have chronicled how this intimate segregation of domestic infrastructures led to worker organizing, strikes, and eventually nationalization across the twentieth century.

The emergence of offshore infrastructure changed the course of that trend, and Equatorial Guinea came on-stream just at the historical moment when, largely in response to the unmitigated disaster of Nigeria, the industry decided that the offshore was useful not only as a organizing principle for industrial operations, but also as a guiding metaphor for its relationship to production sites more broadly. In Equatorial Guinea, one repeating US industry mantra was *not to be like Nigeria,* shorthand for Shell Nigeria's infamously violent presence in the Niger Delta and the robust structures of responsibility—with the company providing often-unreliable water, light, and education in a tangled relationship with local states—that typified their involvement. In Equatorial Guinea, the arrangement between the industry and that which was "outside" it was, on paper, radically attenuated, with corporate social responsibility subcontracted out and companies separated by multiple layers of liability from that which surrounded them. "Offshore" was shorthand for this shift.

And yet, it is important to resist the temptation to think of offshore infrastructure as a "clean break"—in other words, a technical change that somehow enabled radically different forms of work, profit making, or corporate relationships to take place. Rather, we might better understand offshore infrastructure as enabling certain forms of *continuity.* Practices that had been met with increasing resistance onshore—paying workers according to race, providing separate and strikingly unequal housing facilities, lack of meaningful training or technology transfer opportunities—can be newly naturalized in offshore work, ostensibly justified by the novel technosocial configuration of the open ocean, the geophysical demands of subsea hydrocarbon, and the forms of infrastructure necessary to respond to those conditions. But of course, the desired distance from the haunting does not inhere in the infrastructure itself; instead, it has to be made and remade with tremendous effort. Nowhere is that effort clearer than in the daily life of rig labor.

"THERE'S NOBODY OUT THERE"

Offshore production may not involve the direct displacement of towns or villages; running pipelines through people's backyards and drinking water supplies; or generally placing the considerable technology, logistics, and humans required to get hydrocarbons out of the ground in visible or disruptive spaces. This does not mean, however, that "there's nobody out there," as one manager claimed above. On the contrary, given the extraordinary requirements of constructing, moving, operating, and maintaining the largest mobile infrastructures in the world, oil and gas companies put people out there in large numbers.

The 115 men from twenty different nations on the FIPCO 330 (and on rigs all over the world) live "rotating" lives—spending a few weeks working and sleeping on the rig, and then a few weeks at home in Equatorial Guinea, the Philippines, Scotland, or wherever home may be. In this arrangement, each employee has his "back to back": when one man leaves the rig to spend his twenty-eight days off, another man with the same job description comes to take over the constant work for his twenty-eight days on.[10] On the March day I spent on the FIPCO, twenty-one workers left the rig by helicopter, and twenty workers arrived to take their places. Referred to in the industry's (English) lingua franca as a "hitch," each of these three-to-six-week shifts on the rig is characterized by grueling twenty-four-hour workdays. While schedules require twelve hours of work, the rig does not "stop" at night, but rather is operational twenty-four hours per day, year round. Hence, everyone on the rig is on call around the clock, and people routinely have to get up in the middle of the night to address problems. The Equatoguinean rig workers I came to know called each hitch a *marea*, the Spanish word for tide, but also evocative of seasickness, *mareado*. Workers, both international and Equatoguinean, described their rotating lives as surreal temporal experiences. One man from Houston explained that he felt like he had two parallel lives, each of which stopped when he wasn't there and started again when he arrived, "as if on DVD." As they fly on and off of the FIPCO and rigs like it, most of the foreign workers barely set foot on Equatoguinean soil. Although they fly into the Malabo or Bata airports, they often then fly immediately out to platforms via helicopter, or spend one night on private company compounds before leaving for the rig the next morning. As one man put it, "I live here like I did in Angola: from the airport to the rig."

The 115 men and their labor twins from twenty different nations and seventeen different companies, flying in and out of Equatorial Guinea to the FIPCO rig alone, is clearly a logistically and financially intensive arrangement. Companies justify the considerable work and expense of recruiting these men—negotiating their visas, moving them between rig and home, paying and insuring them—by the extent to which the wide-ranging subcontracting of production processes diffuses liability for everything from employee health insurance to explosions, away from the operating companies and into the loose web of overlapping subcontracts and legal relationships diagrammed above (see figures 1.3 and 1.4). While I discuss the quotidian details and implications of extensive (offshore and onshore) labor subcontracting in chapter 4, here I am interested in the daily life of rig work and workers, saturated with incessant comings and goings, helicopter rides, transnational plane flights, and personal mareado temporalities, where the desire and fantasy of the offshore for some becomes the lived inequality and exploitation of others.

Among the men on board the FIPCO, contracted by multiple firms and recruited from multiple countries, there were unambiguous working hierarchies. The Offshore Installation Manager (OIM) was at the top, with three supervisors under him and four "leads" under them; and then a series of workers organized by "levels" two through five, with five being the lowest. On the FIPCO, as well as on the other rigs I studied, white Americans, Brits, and Canadians held top positions almost without exception, with nationalities diversifying through the middle, and Equatoguinean workers at the bottom, with a few in level three. The two highest positions on the FIPCO— Company Man and OIM—were held by a white North American and a white Australian, respectively. White Norwegian, English, and Canadian workers held all of the supervisory and technician positions, with Russian, French, Filipino, US, Colombian, Venezuelan, and Serbian workers in positions that included Operator, Field Engineer, Mud Logger, Crane Operator, Mechanic, and Directional Driller, among others. With few exceptions, Equatoguineans occupied the bottom positions on the rig: eleven of the fourteen catering positions, Painter, Welder, Roustabout (an industry term for unskilled labor), Floorhand, Derrickhand, and Assistant Driller, among others. Men were assigned sleeping quarters on the rig according to their level, with those at the top in private rooms, and those at the bottom in bunks that hold four or more people.

Nationality is central to this form of labor organization. Subcontracting companies (discussed at length in chapter 4) calculate each worker's pay and rotation schedule according to his nation of origin, using cost-of-living indices from international ratings agency Standard & Poor's (S&P) Financial Services. According to this system of differentiation, American and UK laborers work a "28/28," that is, twenty-eight days on in Equatorial Guinea and twenty-eight days off at home, considered the best schedule. Filipino workers have the least desirable schedule: eleven weeks on and three weeks off (an "11/3"). Vitalis (2007) has argued that these racialized hierarchies, structured under the rubric of skill differentiation, have come to characterize oil operations around the world (see also McKay 2007 and chapters 2 and 4). Firms, including those in my research, have long argued that wage, schedule, and facilities segregation is not a question of racism, but of skill level. Indeed, increasingly specialized methods of oil extraction (offshore, oil sands, shale oil) require increasingly specialized labor and minimal unskilled labor. These requirements then map onto geographic inequalities in the production and dissemination of technical knowledge and expertise. And yet, even the few locals who occupied semi-technical positions—radio or crane operator, and certainly Eugenio, the Equatoguinean petroleum engineer I quoted earlier—complained repeatedly that they were kept indefinitely at the level of "trainee": "When they bring a [white] South African . . . I have to guide him but I'm the 'trainee.' I spend six months showing him the work, and once these six months are finished he becomes [my boss], and I'm still the 'trainee.'" In other words, unequal opportunities for training and education aside, even when two workers *do the same job*, they are often categorized, paid, and scheduled according to their nationality, a categorization that often maps too neatly onto race (a discussion I continue in chapters 2 and 4).

Dividing labor by nationality and race, of course, has been a strategy to keep wages low and inhibit worker solidarity and organizing far beyond the world's mineral frontiers. Offshore, that strategy is given new lived empirics: work comes in short, intense bursts of mere weeks at a time; workers never know with whom they will be working on any given rotation, or what language(s) coworkers will speak; and they are always under the watchful eye of management, quite literally, in a claustrophobic floating metal atoll. You cannot steal away to a bar or join one another for a home-cooked meal, much less hold an organizing meeting at a local church in your off-hours, if some of you are in the Philippines, others in Venezuela, and still others in Equatorial Guinea. Equatoguineans, in theory, could meet one another for labor organizing over dinner, and indeed many rig workers with whom I worked were friends or family members who socialized in their time off. But, as discussed in the introduction, Equatorial Guinea is a paranoid and arbitrarily violent dictatorship, with a particular kind of authoritarianism and labor history (Campos Serrano and Micó Abogo 2006) that has proved highly productive for US companies.

Again, offshore infrastructure in Equatorial Guinea does not enable a novel configuration of racist labor practices. Rather, it enables the continuity of practices that grew increasingly untenable onshore, but that go back over one hundred years across the world's oil frontiers. In extraction's offshore age, the kinds of social worlds produced on mobile rigs and platforms enable these practices of segregation to continue, facilitated by the *seeming* disappearance of the infrastructure on which they take place. Infrastructure here carries far more than the crude, seawater, or liquid natural gas for which it is, at least in part, designed. To paraphrase Larkin (2013), while infrastructure is matter that enables the movement of other matter, it is also matter that enables the movement, literally the mobility (in the case of rigs), of certain forms of politics—here, inequality, racism, and the disempowerment of workers. This allows us to expand on Larkin's point about affect in that it is not only awe and fascination with infrastructure from afar that stimulates political effects; but also, as infrastructure comes to intimately shape people's daily lives, the rhythms of their labor, and their relationships to one another, it becomes central to what Mazzarella (2012) has called the professional coordination of affect or affect management. Infrastructure has a synaesthetic effect wherein racialized difference becomes sensory and tactile—how long you can rest at home, and how long your hitches on the rig last; who you sleep with, and in how big a room; the relative heft of an s&p-calculated paycheck reflecting the fungibility of Filipinos and Colombians, Brits and Nigerians offshore. As Massumi (2002) writes, "The ability of affect to produce an economic effect more swiftly and surely than economics itself means that affect is itself a real condition . . . as infrastructural as a factory" (45).

Within these racialized hierarchies, working life on the rig is rigid and exacting. *Risk* is the specter under which workers' arduous, wearying daily schedules take shape; this is risk not only in terms of accidents and safety precautions, but also in terms of shareholder value and company reputation. Through "typical day" narratives of rig work, I end this chapter with a discussion of Equatoguinean workers' embodied experiences of rig life. I explore the highly ritualized safety practices that saturate their working days and situate these practices historically within an industry sea change that took place after the Exxon Valdez and Enron debacles, drawing these experiences into dialog with the productivity of risk in oil futures markets. The chapter closes with a broad look at risk that puts rig work, futures markets, and these workers' daily onshore lives in Malabo into the same frame. This is the *lived* experience, the *habitation* of the industrial and financial offshores.

TYPICAL DAY

Daily work on rigs is intense, regimented, and exhausting. The low-skilled, labor-intensive positions of floorhand, roustabout, pumphand, derrickhand, motor operator, and crane operator that Equatoguinean men tend to occupy are closely controlled, occasionally high-pressure, and function within inflexible schedules. At home in Malabo, local rig workers inundated me with talk of strict schedules and elaborate rituals of control and safety. Two men from different companies and platforms explain:

> We work from 6:00 a.m. to 6:00 p.m., with two half-hour breaks and one lunch break of an hour. We work for ten hours and have two hours rest.... I have to get permission to do any type of work. There is a procedure for getting permission.

> They call me at five in the morning to eat breakfast. Then there is a pre-work meeting, to see if there are any conflicts, to see if there are any problems in any jobs, to discuss. [This meeting is from] 6:00–6:15 a.m. At 6:30 we sign the permits to work. You cannot work on anything without the permission of a supervisor. The risks have to be analyzed. At 7:15 you begin to prepare your tools, survey the work to be done, and begin to work. My work is complicated. If I make a mistake—if [I allow] the pressure [in the system] to rise to 100 percent—I will shut down the whole plant. On any given day we have between one and three permits to work, depending on the day. After completing three permits to work, you have to go back to the Offshore Installation Manager, and then you can continue working. At 5:00 p.m., in my case, I stop working to fill out a report which I send to the supervisor about corrections or equipment that has failed. At 5:45 or 6:00 p.m. it's dinner, and to sleep. If something happens during the night, they come to call me in my room. After dinner I shower, and get in bed at 9:00 p.m.

I talked to men who worked on five different rigs across all three major operating companies and they all had similar "typical day" narratives. Everyone started at 6:00 a.m. after having breakfast (available starting at 5:00 a.m.). Some had two fifteen-minute breaks before and after lunch; others had two thirty-minute breaks. Lunch for everyone was one hour. Everyone stopped working at 6:00 p.m. to eat dinner. With few exceptions, each man made the point that he was exhausted at the end of the day and would fall into bed. Equatoguinean rotation schedules were either "2/2s"—two weeks on and two weeks off—or "28/28s," working and living on the platform for a month at a time, with the following month off and home on land.

Labor schedules and tasks were monitored closely, explicitly for the purposes of risk avoidance and safety. (Remember that my own rig visit started immediately with a safety training mini-course on video, a test to ensure I had paid attention, and a liability waiver. And João, my guide, was the rig's safety coordinator.) Petroleum production is widely acknowledged to be a risky industry, rife with explosion and fire hazards, limb-threatening equipment, and noxious chemicals, among other perilous potentialities. The open ocean of the offshore and the helicopter rides required to get there concentrate and exacerbate these dangers.[11] My own rig visit felt like a protracted exercise in both embodied and ornamental safety rituals—from what I was given to wear; to what I was told to take off; to the ways I was trained to walk, climb, and descend stairs; to the safety videos, manuals, and waivers I was required to study and sign, literally from the moment I stepped off the helicopter. João instructed me to take off all rings, earrings, necklaces, the hair band around my wrist, and anything else that could snag or catch. None of these items were allowed on the rig. While walking on the rig, if a railing was available I was instructed to hold it at all times, in particular on stairs, which, depending on their pitch, one had to descend facing backward.

For Equatoguinean rig workers, this exacting fixation on safety stretched beyond bodily adornment and comportment to the system of Permits to Work, mentioned by both men quoted above. Essentially a job permission slip, each permit details the job to be done, which tools will be included, whether the work is hot (welding) or cold (lifting, relocating equipment), the possible risks involved, and how those risks will be avoided. Tasks cannot proceed without a Permit to Work, secured from supervisors at the morning meeting. If a task arises during the course of the day for which one does not have a permit, a worker cannot proceed with the task until he secures a new permit. Permits to Work must be posted visibly at the job site on the rig, so that as others circulate they can discern what is going on at any given location. Each permit must be removed and cleared with an authority when the

Figure 1.7. Author in personal protective equipment (PPE).

job is completed. On the FIPCO, if jobs were considered complicated—those involving multiple personnel and tasks—they required an additional THINK drill and a special permit. Also known as the THINK Process for incident prevention, a THINK drill is a five-step, pre-job meeting:

1 Plan—Develop a step-by-step plan.
2 Inspect—Inspect all equipment to be used.
3 Identify—Identify all potential hazards.
4 Communicate—Communicate all relevant information.
5 Control—Control the operation.

While these details are mundane in and of themselves, they offer a sense of the intensely regimented, hyper-scheduled, and monitored working day on the rig. All of the men with whom I had in-depth interviews at home in Malabo repeatedly mentioned their respective companies' overt and constant

CHAPTER ONE

attention to safety. Talking in amazement about helicopters—a notoriously dangerous offshore technology—Antonio explained:

> I am telling you that perhaps for you this wouldn't be so incredible, but in our environment we had never seen things like this; maybe on television. We have never had equipment like this. It makes me say, where am I? How do you control so much technology? For the helicopter we watch a safety video, [that covers] emergency landings, what to do; if the helicopter falls how you can escape; where is the emergency equipment; where are the escape boats. [We wear] double auditory protection and life jackets. The pilot asks you if there has been anything that you didn't understand. There has not been a single helicopter accident.

For Antonio and others, the complexity, power, and potential hazard of the equipment with which they worked often generated a reverence that naturalized the rhythms of their working days. Antonio's words recall Larkin's (2013) insistence on infrastructure's capacity to "generate desire and awe in autonomy of its technical function . . . the sense of fascination [infrastructures] stimulate is an important part of their political effect" (333). Rogelio, in a safety soliloquy that would make his employer proud, explained that "for [this company] safety is *first*. It is worth more to finish the day without an accident than to complete the work that you have been given. [We count the] days that we are able to go without an accident." As Antonio and Rogelio make clear, this safety-saturated atmosphere was not only experientially definitive of working offshore, but also a welcome characteristic of rig work. While this gratitude in the face of serious hazard is readily understandable, it was an important ethnographic discovery for me, because it so directly contravened my own visceral response to safety measures in the Equatoguinean industry more generally.

In the eight months of fieldwork that preceded my visit to the FIPCO, among my strongest impressions of the oil industry in Equatorial Guinea was a corporate culture so safety-saturated that it bordered on the comedic. On one occasion, I was in a company vehicle with an British industry employee who was driving painfully slowly, and an apparatus beeped loudly any time he hit forty miles per hour. Cars whizzed past us. When I chided him, he told me that every time the apparatus beeped a report was sent to Houston headquarters. On a walk through another company's compound with the wife of a migrant manager, I bent to tie my shoe, and she joked that her husband would need three signatures to secure permission to do what I had just done. Industry offices were wallpapered with safety achievements—how many "incident-free" days, how many months without a "lost-time" incident. Acronyms abounded—keep it SIMPLE, THINK drills, START. In Malabo, the T-shirts that give safety acronyms their public lives could be seen on the backs of many local men, women, and children, having found reincarnation in the used clothing markets.

These elaborately choreographed and audited safety rituals are the outcomes of earlier offshore fantasies run aground, the hauntings of the Exxon Valdez spill, the Piper Alpha disaster, and the Enron/Arthur Andersen scandal, a list to which the Deepwater Horizon is now certainly to be added (Bond 2013). As many management informants explained to me, this series of disasters and their nightmarish human, environmental, public relations, and shareholder consequences motivated a corporate culture sea change to newly confront these risks, affecting everything from accounting practices to rituals of rig safety. Where I scoffed at beeping, speed-monitoring apparatuses and ridiculous acronyms, for my migrant management informants, these were the procedures through which they controlled and monitored working environments, and the audited outcomes (incident-free days) could be used in shareholder reports to reassure investors that times had changed.

For Equatoguinean rig workers, the micromanagement of time and tasks that defined their labor environment was exhausting but welcome. Yet for most of them, the Enron debacle or the Exxon Valdez spill that instigated the safety regime for which Permits to Work are a synecdoche were snippets on television, if they were anything at all. An analysis of the Enron/Exxon Valdez outcomes complicates workers' faith in the safety rituals they carried out. The shift in corporate practices in the wake of these disasters, in part, actually rescinded responsibility for workers and replaced it with internal, self-regulated safety procedures intended to keep "recorded" or "lost-time" incidents down, and stock prices up. As one Equatoguinean man who lost a finger in a rig accident found out, neither the operating company to which his work provided oil, nor the subcontracting company to which his salary provided profit, could help when he could no longer work on a rig. And, as two other workers reported, their OIM's act of throwing broken radial saws into the ocean on not one, but two, occasions, in order that the incidents would go unreported, prompts us to ask: For whom is the offshore arrangement simpler or safer? For whom does it redistribute risk and where does that redistributed risk *go*?

Alfredo was an Equatoguinean economist who had long lived abroad and completed his postgraduate studies in London before moving home to work first for the Major Corporation, and later for Regal Energy. When I asked him what he did as an economist at Regal, he responded: "Controls: audit, corporate governance, business ethics, Foreign Corrupt Practices Act. I design and implement processes and procedures for control and compliance." When I admitted that I had no idea what "controls" meant, he offered an example, explaining that he had recently been working to implement a system that would allow company vehicles to pay tolls without having to stop at the toll booths that separate central Malabo from the airport road on which company headquarters were located. When I remarked on the apparent triviality of that project in relation to corporate governance and business ethics, he continued that he handled anything that had to do with "control and safety," from the crucial to the humdrum. "These have been the key words," he emphasized, "*safety* since the Exxon Valdez and *control* since Enron/Arthur Andersen; i.e., look for what the company is trying to avoid."

Alfredo explained that before Enron, audits looked only at financial statements, but the Arthur Andersen scandal exposed glossy financial statements as mere surfaces prepared to encourage shareholders to trust company finances. These statements revealed little about what was actually going on in the company, let alone about the processes that led to the figures therein. In 2002, in the wake of the Enron scandal (and others, including Tyco and WorldCom), the US Congress passed the Sarbanes-Oxley Act, a federal law intended to strengthen corporate accounting controls. In practice in Equatorial Guinea, Alfredo remarked: "SARBOX, or SOX 404, as we lovingly refer to it, guides what I do. [There are] procedures for absolutely everything, and the procedures are standardized in almost all affiliates. The company maintains them everywhere they go. If I was to work in an accounting department anywhere in the world, I would already know the procedures." Alfredo was not my only informant to mention the aftermath of Enron and the seemingly arcane 2002 legislation. (I certainly had never heard of Sarbanes-Oxley before arriving in Central Africa.) David, the manager of a major oil services company, explained that after Sarbanes-Oxley, "you and I can't do business on a paper napkin. . . . But before Enron that was different. When I was in South America we did a lot of dodgy things. It used to be that the ends justified the means. [The attitude was] go ahead and do it and we'll fix it later."

When I asked David about the potential ramifications of paper napkin business in the post-Enron era, he responded that, "the oil industry is small, and that kind of behavior is not admired. It's quite regulated. You have one scandal and they blacklist people. One scandal, and can you imagine the impact on the NYMEX stock price?"

While I discuss the proceduralism and self-regulation described by both Alfredo and David at greater length in chapter 3, what interests me here is threefold. First, both the Exxon Valdez disaster and the later Enron scandal ushered in new "keywords" in the oil and gas industry: safety and control. Second, these keywords have fundamentally changed practices on the ground—from how many signatures one needs to tie one's shoe, to audit procedures following new US laws, to elaborate risk-avoidance rituals and their tabulation in the recording of days without a lost-time incident (Radial saw? What radial saw?). Third, the grounded practices to control risk—both financial and industrial—are *primarily* indexed to shareholder value, secondarily indexed to human safety, and not at all to labor rights. "One scandal, and can you imagine the impact on the NYMEX stock price?" I want to dwell for a moment longer on risk where we find it here: at the intersection of accounting procedures and Permits to Work, or of financial and industrial practice.

The packaging of risk as a commodity is among the most profitable of contemporary financial endeavors (LiPuma and Lee 2004; Thrift 2005; Zaloom 2004; Poon 2009; Power 1997, 2007). Indeed, Guyer (2009) suggests that risk be added to land, labor, and money as a fourth category in Polanyi's famous list of commodity fictions. Although the oil and gas industry is, of course, traditionally reliant on the production and sale of a tangible industrial commodity, it is also deeply invested in what Zaloom (2004) calls "the productive life of risk" (365). Of the approximately two hundred million barrels of oil traded per day on NYMEX, much of it is "paper oil," or the buying and selling of futures contracts. Futures markets are technologies for distributing risk: oil companies and others who need a constant supply buy futures, essentially a contract on future delivery at a price agreed-upon now. Investment banks and US pension funds are also heavily invested in the oil futures markets, not because they need oil, but because the investments can be extremely profitable (Guyer 2009). Based on her involvement in the Chad–Cameroon pipeline project, Guyer also points out that financial instruments are inserted at multiple stages in these brick-and-mortar oil industry projects, from debt servicing to actuarial calculations (2009, 209;

see also Leonard 2016). This intercalation of industrial and financial productivity involves multiple moments in which to trace "risk as a problem of practice" (Zaloom 2004, 368) or "the actual ways in which risk instruments intervene in the world" (Guyer 2009, 218; Johnson 2015).

On the one hand, in futures markets, risk signifies an opening; it conjures opportunities for increased profit in a zone of possibility and chance (LiPuma and Lee 2004; Miyazaki 2003; Riles 2004; Maurer 2005; Thrift 2005; Zaloom 2004). These are risks one should take, because the yields they promise outstrip and are, in fact, increased by their dangers. On the other hand, as I will detail below, the risks addressed by Permits to Work and helicopter safety videos in the lives of temporarily subcontracted and semiskilled Equatoguinean employees evoke volatility and fear, conjuring the narrowing of opportunities and prospects. These are risks one *should*, but probably cannot, avoid (see also Simone 2004; Guyer 1995, 2004; Ferguson 1999). (See Peterson 2014 for this dual nature of risk in Nigerian pharmaceutical markets.) And yet, these are not simply different moments or places of risk, but rather comprise an ethnographic scene that shows profitable risk and exploitative risk to be mutually dependent. To what extent is productive, profitable, and voluntary risk enabled by, and *funded* by, the destructive and seemingly intractable risk shouldered by (racialized, classed, gendered) others? As Karen Ho (2009) has pointed out, in the mortgage crisis of 2007–2008, Wall Street's professional risk-takers relied on the income streams of middle- and working-class homeowners (see also Appel et al. 2019; Roitman 2014; Poon 2009). For professional risk-takers, the repackaged mortgages were short-term securitization opportunities, whereas for the homeowners they were both *homes* and long-term investments. The professional risk-takers required the risk of the homeowners. Dick Bryan and Michael Rafferty (2011, 2018) have also argued that the working and middle classes are central to the profitability of financial risk. Their pension funds, mortgages, auto loans, and health insurance payments are securitized, packaged, and sold as commodities. Households, and one might even say labor, are the stable source of investment; so too, I want to suggest, with subcontracted rig labor. To rephrase the questions I asked above—For whom does the offshore redistribute risk, and where does that redistributed risk *go?*—using different language: "What is risk as a transacted thing? From whom and to whom is risk transferred? Since mitigation can only ever be partial, where is the excess located in relation to a theory of ownership?" (Guyer 2009, 215; Maurer 1998).

My visceral response to the oil industry's comically relentless safety practices was intensified by contrasting them with daily life in Malabo, a city which was essentially without the safety and risk-prevention provisions of many other urban environments. Common sights in the city included construction or road improvement projects haphazardly set up in the middle of everything, with day laborers performing welding; swinging metal beams; using jackhammers and bulldozers; and creating huge ditches that dropped into the bowels of the old colonial undercity, without safety glasses, hard hats, or safety equipment of any kind, let alone a sign or brightly colored tape alerting pedestrians to walk elsewhere. As a foreigner, navigating scenes like this was definitive of living in the city, *especially* given the overwhelming quantity of hydrocarbon and construction industry-related heavy machinery, equipment, and materials in constant circulation and use (Appel 2018a). In the claustrophobically small city, pedestrians routinely walked directly through these sites on their way here or there, hoping not to get sprayed by welding spatter or fall into a ditch. It was also common to see huge flatbed trucks careening though the city streets holding stacks of unsecured metal tubes or rebar piled high in pyramid shapes, with day laborers perched precariously (to me), yet apparently comfortably (to them), on top. One day I did hear that one of these trucks took a roundabout too fast, and all the tubing fell off, along with the men, one of whom was killed in the accident. In a country where the public hospital was known as a place where people went to die, not to be treated for broken necks, the risk was a serious one.

Official work for the US-based oil and gas companies that form the subject of this book would never be performed under these conditions, but rather, under their comic opposite of the beeping and reporting-to-base speed-limit car monitor. Indeed, I laughed when I saw the photo in figure 1.8 for the first time. I had intended to capture a migrant manager mansion on a private residential oil compound, but instead, I unintentionally captured perhaps the only twenty square feet in Equatorial Guinea that contained both a fire hydrant *and* a speed limit sign.

Taking the contrast of needing three signatures to tie your shoe with overburdened trucks careening around corners with workers perched on top into consideration, one can hear the rig workers all the more clearly when they marvel at helicopter safety videos and recite industry slogans. They are understandably thankful for this industry-specific, transplanted conception

Figure 1.8. Fire hydrant and speed limit sign on the Smith compound.

of safety in what they know to be a highly technical and often dangerous environment. But this conception of safety that allows Equatoguineans working on rigs to potentially survive a helicopter crash cannot remove them from the larger insecurities and risks of their lives. In fact, these workers' very removal onto offshore rigs for up to one month at a time—despite the acronyms of THINK and START—actually exacerbates the most pressing and dangerous insecurities they face. While management can work furiously toward the disentanglement promised in the offshore setup, subcontracted local workers cannot simplify the risks they negotiate on a daily basis as they try to reconcile the promises of Permits to Work with the promises of security for their own lives and those of their family members. As one Equatoguinean rig worker put it forcefully to me: "How are you going to talk to me about safety, if you know that your child has no water or no light and no medical care, not to mention your wife? They don't have anything to eat today and you're talking to me about *safety*!?"

During my interviews with Equatoguinean offshore workers, work-related risk on the platforms was not the locus of their concern. Seatbelts, hard hats, and safety glasses, while welcome, did not begin to address the main locus of risk that they faced on a daily basis. In Malabo and its surrounding residential communities of Ela Nguema, Lamper, and Campo Yaounde, these men and their families lived with sporadic electricity, no running water, and inadequate healthcare. Malaria and typhoid were rampant, and child mortality from afflictions as basic as diarrhea was common. Thus, while the risk of a helicopter falling out of the sky was indeed grave, as was the risk of cutting off a finger at work, when compared with the deep, daily insecurities of these men's lives, those risks and the elaborate rituals set up to avoid them seemed as trivial as the acronyms used to remember them. Two workers explain:

> [Working on the platform] is very risky, difficult. To have to be there for twenty-nine days is very difficult. It could be that something happens to my child and the [agency that subcontracts me] will not help me with anything. [I am] very worried about my family and everyone else.

> In my particular case to live on the platform is difficult. My family is far away. . . . The most difficult is that we have sixty minutes of communication every week. This isn't enough because they calculate it in a distinct way. If the person doesn't pick up the phone they cut minutes. You can't pass your limit. . . . Ultimately when you have a problem onshore and you leave the platform, those days you don't get paid. For example, if you're on the platform for ten days and you have a problem onshore and you leave for two days, they cut those days. Being [on the platform] sometimes my head hurts because of the pressure. I think of my family, sick children. I can't leave the platform. If I leave there isn't any money. What will it have been worth?

The first worker talks about the risk of platform work, but does not have helicopters or fire hazards in mind. The risk is that, while away from his family for twenty-nine days, something could happen to his child, and not only would he not be there, but he worries (correctly) that the agency employing him would do nothing. The second man brings up the issue of communication with home. The only way to know if everyone is okay, or more accurately, what is not okay on any particular day, is to call home. However, the company allows less than ten minutes per day of phone time, with time counted off for incomplete calls. Imagine negotiating in six minutes per day what to do if a child or uncle has malaria, if a relative has died, or if there's a fire in the neighborhood, all common occurrences. Should he leave the platform to take care of it? He doesn't want to leave because then he won't get paid and "What will it have been worth?" As one rig worker's wife put it:

"It seems bad when they leave for so long and I'm home alone suffering with the children. There are [six] of us in the house and my husband is the only one that has work." The *cost* of living in Equatorial Guinea—sporadic water, electricity, healthcare—is not factored into offshore work. As Mbembe (2001), Simone (2004), and Roitman (2005) have pointed out in other postcolonial African contexts, the calculus of compensation is radically divorced from actual labor value. Although neither unusual nor specific to postcolonial Africa, this divorce takes on a particular severity in contexts where insurance and social welfare have long been provided by networks of personal relations, as is the case in Equatorial Guinea, not by contractual obligations won through labor rights or citizenship entitlements.

The details of rotation schedules and phone-time allowance—seemingly trivial—take on serious weight for Equatoguinean laborers:

> Now our shifts are two weeks on and two weeks off. Before we had to work and live on the rigs for four weeks with four weeks off but because of our families, because they are home without electricity, because we cannot communicate with them, we had to change that schedule. After three years, we complained to the company and asked to have a 2/2 rotation. At first the company didn't accept, but eventually they did.

> Many are leaving the company and [the company] knows. You are only given two minutes per day to talk to your wife. We begged, please give us time to talk to our families. But they forget that [the] French can talk to their wives on the internet, or the phone cards that let you talk for hours. But the rates here only let you talk for fourteen minutes per week. They say they understand our condition, but the company really doesn't. You are cut off from your family completely. With all this difficulty you prefer to be with your family alone.

Where foreign rig laborers could count on internet connections and international phone cards to keep them in touch with home, international inequality in the spread of technology in homes (let alone electricity provision in Malabo) guaranteed that for local labor, fourteen minutes per week on the phone with one's family was simply insufficient. The subcontracted conditions under which these men worked intentionally abdicated responsibility for the precarity of their onshore living conditions, a precarity best stabilized by the presence of *people*. The more able bodies in the house, the better to manage life's daily challenges. With healthy men gone, even for short periods of time, vulnerability and worry increased exponentially (Moodie and Ndatshe 1994;

Meillassoux 1981). One man captured this grave misunderstanding of conditions in what became a productive phrase to think with: "We are working like Americans but being paid like Africans." As he explains:

> The cost of living is so high here. There's no water. There's no electricity. You go to the hospital, you die there. Here this money isn't acceptable. When you tell us this is a lot, we ask, for whom? We are working like Americans but being paid like Africans. . . . You can't have it both ways. Either make us work like Guineans and treat us like Guineans, or make us work like Americans and treat us like Americans.

In this man's narrative, to "work like an American" is to work long, hard hours in a safe environment; to be able to talk to your wife for free and endlessly over the internet that she has in her home, enabled by the electricity she also has; and to be compensated accordingly. To "work like an African," on the contrary, is to work fewer hours and be compensated less, with the idea that the remaining time spent with the family putting out literal and figurative fires is compensation in and of itself. To "have it both ways" is to make these men work as if they were Americans, but to compensate them as Africans.

CONCLUSION: OFFSHORE FUTURES

This chapter has traced some of the work that allows Equatorial Guinea to recede from view, as the hydrocarbons from its Exclusive Economic Zone are extracted, produced, and exported from an offshore in which oversight, liability, and ties to terrestrial communities are intentionally stretched thin. I have suggested that a broadening of the category "offshore," to encompass financial and industrial relations at the same time, helps us to understand how it is that profit can be disentangled from the places in which it is made, at once increasing risk for some and making it more profitable for others. Whether through disperse corporate geographies for tax "planning," in which all of the major companies in Equatorial Guinea operate through Cayman Island subsidiaries, or through mobile technosocial assemblages of colossal infrastructure with racialized labor already on board, the offshore becomes a space of deregulation in terms of finance, labor, and the environment. Central to this project of disentanglement is the diffusion of liability in the form of managing risk. Oil futures markets and Permits to Work both participate in this project, and risk is at once enormously profitable and an unbearable burden, radically unequal in its qualities and distribution.

Writing of traders in Chicago's Board of Trade Futures market, Zaloom (2004) notes that the daily practices of their working lives "encourage the production of subjects who can sustain themselves under high-stakes conditions and thereby draw profit from economic risk" (366). Compare this to the situations of Antonio, Rogelio, and the other workers quoted above, who were at once thankful for the risk-avoidance practices in which they had been trained, and utterly marginalized from the spectacular profit their risks were reaping elsewhere. Like Zaloom's traders, risk here has arguably produced subjects who can sustain themselves under high-stakes conditions, but their futures hardly profit from their practices. Late one Malabo afternoon, I was sitting in my apartment with a group of four Equatoguinean rig workers home from their marea. An impassioned friend they had brought along interrupted one as he responded somewhat listlessly to my questions about their working lives, salary, and home. "My friends do not have a future. They can't even build a house. . . . Neither my friends nor anyone who has worked in the industry has any type of guarantee for life having worked this many years. The cost of life here—food, school fees, hospital, medications—it's all gone. They don't have social security; they can't buy a house, nothing. As older people, they will be poor." Risk of oil bunkering or community entitlements is certainly minimized in the offshore arrangement. Risk's foil— liability—is also stretched thin as companies operate in tax loopholes and regulatory paradises. Who profits from these practices, whose risks they hedge and whose they exacerbate, whose futures they guarantee and whose they foreclose, is also clear as the open ocean, seen from above.

Ethnographic attention to the FIPCO 330 and the larger corporate archipelago of which it is a part offers the opportunity to follow how certain qualities often understood as intrinsic to capitalism—standardization, replicability, indifference to local context—are *built*. Here, petroleum's industrial offshore not only concretizes a metaphor, but also shows how desire-filled fantasies of capitalism itself—as placeless, as frictionless—come into being in tenuous and work-intensive ways. In other words, the offshore is real. Its effects are real. It is not without friction; it is not the capitalist utopia of placeless economic interaction. It is full of specters of political risk: men from twenty different countries and seventeen different companies consequentially divided by nationality and race; Equatoguineans underpaid and held indefinitely at the level of trainee; and a corporate form so multiple and attenuated that, paradoxically, it can seem to disintegrate altogether. But nor is the power of the offshore, its effects, undone by attention to this teeming and contentious sociality.

In chapter 2, the ethnographic material seemingly moves from offshore to onshore—from rigs to the companies' gated residential and corporate enclaves. Yet, as a general industry descriptor, "the offshore" still encompasses these spaces, as companies work to distance themselves from life outside their walls. Racial segregation remains central to this work, and raced and gendered domesticity and intimacy emerges along with it.

CHAPTER TWO

THE *Enclave*

Figure 2.1. "True to Texas" Miller Lite sign on Endurance compound.

Every Wednesday since roughly 1999, "the wives" have gathered for two hours of food, drink, and a card game called Continental in the well-appointed home of whoever agreed to hostess that week.[1] These women—the wives of migrant oil and gas company management assigned to Equatorial Guinea— socialize, plan charity projects, trade coffee beans for Crystal Light, and re- lax over cards, casseroles, nachos, crudités, cakes, and Diet Coke. Although far from the only scheduled weekly activity these women shared, which in- cluded mahjong, stitching, water aerobics, tennis, and calligraphy among others, cards was the most popular and the longest standing. As one woman who had been living in and out of Malabo since 1999 remarked, "We always played cards on Wednesdays. And unbelievably, every time I've come back here, Continental is still held on Wednesdays. That has continued. And many times some women have wanted to change that and do other things, but it's always come back to playing Continental." Continental on Wednesday afternoons in this suburb of Texas just outside Malabo was, by the time I arrived in 2006, a native ritual.

Women took turns hosting Wednesday games in their homes, located in the private residential/industrial compounds owned and operated by US oil and gas companies. A trip to one of these compounds for those living outside their walls meant lengthy stops at gates and guardhouses, where security personnel checked your requisite invitation against a logbook; where you offered license plate numbers, phone numbers, and other identifying information; and where you had to leave official identification at the security office for the duration of your stay on the compound.

One Wednesday in September 2008, I was on my way to cards at the largest compound just outside Malabo with a few wives from another compound. (I tried to carpool to cards whenever possible because Malabo's public transportation—informal taxis—was prohibited on compound grounds.) As had become our Wednesday ritual, my carpool mates and I stood together for almost thirty minutes in the security checkpoint building while the young Equatoguinean man at the desk searched in vain for our identification information in a database that never seemed to be updated, despite our weekly visits. As each of us gave him our full identification information again, and he painstakingly entered it again, the waiting wives began to talk about home. Only days before, Hurricane Ike had badly damaged much of Houston and eastern Texas. These women and many others in Equatorial Guinea had homes, family, and friends in the area who, in the wake of the hurricane, had been living without electricity or running water for several days. One woman expressed worry about her twenty-four-year-old daughter in Houston who had told her that there was no more gas, and ice was sold out at local stores, as were ice chests. She wondered aloud, "How will she eat?" But in the next breath she noted the incongruity, the strangeness, that people in Equatorial Guinea live without electricity and running water every day.

Indeed, people in Equatorial Guinea, including those residing in the capital city, live with endemic typhoid and malaria, and largely without running water and reliable electricity in their homes. In affluent areas, the lack of public infrastructure gives way to private provision, and those with the means to do so buy generators for electricity and/or put tanks on their roofs for water. But even my next-door neighbor—the Ministry of Finance and Budgets—was routinely dark for days and even weeks at a time. The water that occasionally ran for twenty minutes each morning in my Malabo apartment flowed untreated through corroded, colonial-era pipes. Education and healthcare systems were similarly erratic, as were food staples in local markets. With no industrialized agriculture, when the border closed with Cameroon (which it often did), fresh vegetables rapidly disappeared from Malabo's markets and prices skyrocketed on the dwindling piles that remained.

Meanwhile, not four miles away from the capital's public infrastructure woes, these wives and I stood at a checkpoint waiting to enter something very different. Once past the guard, the gates and walls of the Endurance compound opened onto manicured lawns, towering ceiba trees with landscaped hedges and flowers, paved roads with speed limits and fire hydrants, stately suburban homes with SUVs in garages, and a sprawling office/industrial complex (Appel 2012d). The sheer concentration of resources within this compound and others like it in Equatorial Guinea is difficult to overstate. The Endurance compound alone generated enough energy to power the entire country's electricity needs twenty-four hours per day, every day. Food to feed foreign employees was shipped in from Europe and the United States. The luxurious mansions in which migrant management lived were serviced by their own sewage and septic systems, and appointed with flat screen televisions, wireless Internet, and landline phone service with Houston area codes. Each house had an Equatoguinean maid; the office complexes had Equatoguinean janitorial services; and Equatoguinean gardeners and landscapers (employed by a Spanish-owned company with roots in the colonial era) maintained each compound's pristine grounds. These compounds also included pools, gyms, basketball and tennis courts, restaurants, bars, and at one location, a movie theater and small golf course. Malaria, endemic to the Equatorial Guinea just on the other side of the wall, had been all but eradicated within the compounds. In one particularly evocative description, an older Equatoguinean man who had never been inside one of these compounds, but knew them through local lore, described what was behind the walls as *una limpieza terrible;* literally, "a terrible cleanliness," but perhaps more accurately a fearsome cleanliness, where fearsome can mean both terrifying and awesome.

Management-level migrant oil workers and their wives living in these lavish "suburbs of Houston" received up to a 75 percent salary increase for working in what was known in the industry as a "hardship post." Equatoguineans, regardless of their class position, governmental authority, or employment with the company, were prohibited from living on the compounds. While locals came and went as security guards, maids, gardeners, secretaries, local content managers, or government liaisons, public transport was prohibited from entering the compounds, and all nonresident personnel had to wear an employee or visitor's badge at all times while inside.

This chapter is about these compounds, domestic and corporate life within them, and the *lived* disjuncture from life outside their walls. Inspired, on the one hand, by the spatialization of the enclaves themselves, in which domestic and corporate life cohabit, and on the other by feminist anthropology, which has long held that domestic arrangements of marriage and kinship are political and economic affairs constitutive of capitalist practice (Wynter 1982, 2003; Enloe 1990; Federici 1998; Yanagisako 2002; Stoler 2010; Bear et al. 2015; Hoang 2015), this chapter analyzes domesticity and corporate daily life together. In particular, I am interested in how the segregation of the enclaves, not only from life "outside their walls," but perhaps more importantly, the raced and classed segregation *within* their walls, is evidence of the cultural work required to enliven the licit life of capitalism. I am interested in how, for instance, reproducing the nuclear family structure (if only for the highest managers) mobilizes white heteronormative marriage to communicate industry morality and even standardization, in contrast to "local" corruption, or how segregating foreign worker housing by "skill level" maps neatly onto racial and national categories, and in turn onto rotation schedules and salaries.

Here, the scalability of global capitalism, so clearly seen in the offshore, is always a question of domesticity and gendered, racialized subjectivities; in this chapter, white womanhood in particular. "Race [is] a primary and protean category for colonial capitalism and . . . managing the domestic [is] crucial to it" (Stoler 2010, 13). As I argued in the introduction, in relating Vitalis's (2007) work on segregation in the world's mineral enclaves to Barry's (2006) work on technological zones, the whiteness of the compounds, their racial segregation, actually comes to signify a certain kind of licit practice, in which white : nonwhite is semiotically mapped onto tenuous infrastructural and technical distinctions between West : non-West :: standard : corrupt :: global : local. Segregation and intimate life, paradoxically, come to signify heightened standardization, repetition, and universality, drawing on select postcolonial meanings attributed to heteronormative whiteness, including expertise, virtue, technology, meritocracy, and philanthropy. The white women with whom I waited at the gate that day—the caretakers of white male physical well-being, and the guardians of racialized morality and privilege (Stoler 2010)—are a *constitutive piece* of the licit life of capitalism in Equatorial Guinea.

Rendering capitalism licit is, in part, about the ways we ethically partition responsibility for "others," and how those partitions are at once individually embodied *and* materialized in corporate, residential, and urban planning. This chapter focuses on the work of making tenuous separations and disentanglements on behalf of the licit life of capitalism. We might think of the material in this chapter as *the domestic offshore,* to tie the enclaves to the offshoring practices I discussed in chapter 1. Rather than simply showing how apparently separate experiences on either side of the enclave walls are, in fact, part of the same economic and ethical picture (which, of course, they are), I ask, instead: By what processes are "the proliferation of connections" between the enclaves and wider Equatoguinean life framed or bracketed into *convincingly lived separations*? (Callon 1998, 4). What work went into the imagining of these enclaves? What kinds of unmapping and frontier-making (Tsing 2005)? What kinds of white masculinity and femininity work? If racialization is always a relationship, this white gender work in relation to what ideas about "Africa," Equatorial Guinea, and Blackness? To expose connections that were intentionally severed leaves intact the processes by which that separation came to be; it overlooks the considerable *work* required to produce the effect of separation itself. In the enclave, disentanglement is crafted in architecture, infrastructure, exceptional fiscal statuses, food consumption, telephone area codes, heteronormative white domesticity, and residential labor segregation—an assemblage that allows those who might otherwise experience the troubled intimacy of oil extraction and Equatoguinean life to exist in what *quite reasonably* feels like a separate ethical picture. Stoler (2010) describes this separation in a different time and place: "A cordon sanitaire surrounded European enclaves, was wrapped around mind and body, around each European man and his home. White prestige became redefined by the conventions that would safeguard the moral, cultural, and physical wellbeing of its agents, with which European women were charged. Colonial politics locked European men and women into routinized protection of their physical health and social space in ways that bound gender prescriptions to the racial cleavages between 'us' and 'them'" (77).

This chapter is divided into five sections. Section 1 discusses the "lay of the land": a descriptive mapping of the major enclave settlements in Equatorial Guinea focused on the relationships among race, labor, and the built environment. Section 2 is an effort to contextualize what I experienced as the exoticness of Equatorial Guinea's gated enclaves through their historical continuity with colonial settlements, company towns, and more contemporary forms of zonal capitalism (Winters 1996), including special economic zones, free zones, and *maquiladoras*. Section 3 traces the decision-making processes that led oil and gas companies to build enclaves in Equatorial Guinea, introducing the idea of ring-fencing—"the separation of non-resident corporate persons from domestic economies and taxes and [the denial] to resident corporate persons the same privilege granted to foreign ones" (Maurer 2005, 479). Section 4 explores practices of corporeal disentanglement: corporate and domestic ring-fencing in the enclaves as an attempted circumscription of health, sexuality, gender, and race (Enloe 1990; Stoler 2010; Vitalis 2007). Finally, through an exploration of the wives in their "golden cages" (as many of them described compound life), Section 5 thinks through the gendered inhabitance and effects of multiple forms of isolation and segregation.

THE LAY OF THE LAND

Methodologically, the enclaves were the most traditionally "ethnographic" of my research sites. Consider Bonvillain's (2009) textbook description of ethnographic research:

> To conduct ethnographic research, anthropologists do "fieldwork," that is, they live among the people they are studying to compile a full record of their activities. They learn about people's behaviors, beliefs, and attitudes. They study how they make their living, obtain their food, and supply themselves with tools, equipment, and other products. They study how families and communities are organized, and how people form clubs and associations, discuss common interests, and resolve disputes. And they investigate the relationship between the people and their larger social institutions—the nations they are part of and their place in local, regional, and global economies." (6)

The familiarly problematic anthropological conceit that a researcher can arrive in an apparently static place where people live and come to know their behaviors, beliefs, and attitudes; how they make their living, obtain their food, and supply themselves with tools; the organization of their families and communities, clubs, and associations; and their relationships to the nations and wider economies of which they are a part, is uncannily close to the information I present in this chapter. Compound walls seemed to circumscribe everything from food provision to housing, kinship to citizenship, giving the enclaves the feeling of anthropology's timeless, bounded community. *And this was precisely their intended effect—that the world inside compound walls could seem to be completely separate from the world outside, despite its deep entanglements with Equatorial Guinea's broader political, infrastructural, and social realities.*

Each of the three largest US oil and gas companies in Equatorial Guinea—Major, Endurance, and Smith—has its own residential/industrial enclave, and all share these broad features of apparent boundedness. All three are private, in that both residents and nonresidents cannot come and go at will, but must stop at guard gates to offer identification, declare the purpose of their visit, and register their presence on the compound. You cannot enter the compound unless you have an invitation, and no public transport from Malabo or Bata is allowed past the gates. Residents often have curfews, or "lock-downs" that respond to regional security concerns, and all three compounds have a badge system where visitors and employees alike must display badges on their persons at all times.

The Endurance and Major compounds sit on old Malabo's outskirts. The Smith compound, in Río Muni, is twenty-five minutes by car from the main continental city of Bata. Among the three compounds, Endurance's is by far the largest due to its substantial onshore industrial footprint consisting of a Liquid Natural Gas (LNG) plant, a methanol plant, and storage and refining facilities. Because the Major Corporation and the Smith Corporation do all of their processing, storage, and offloading offshore, and do not have LNG or methanol plants, they have much smaller onshore footprints. All three compounds contain company headquarters within the gates, offices where both migrants and nationals work on a daily basis, as well as residential facilities for migrant work forces. I spent considerable time in each of the three largest compounds, and the ethnographic data I use in this chapter are drawn from all three spaces. I do my best, however, not to conflate the enclaves, creating a "typical" space. Informants in each company would often go to great lengths to differentiate their corporate culture and compound from the others. Rather than either creating a "typical" compound or dutifully following tales of exceptionalism, I use data primarily from the Endurance Company and compound, the largest of the three both spatially and in terms of personnel. The Endurance compound is also where I spent the most time, and where I had the most interviews and interested interlocutors.[2] Before turning to Endurance, however, I'll briefly describe Major and Smith.

Once inside the secured Major compound—the oldest and smallest of the three—you pass a small helicopter landing pad on manicured grass and come immediately to an office complex and employee parking lot. The office complex is a series of large buildings housing the country manager's office and various departments, including accounting, public and government relations, corporate social responsibility, human resources, procurement, drilling, and other offices. Past the office complexes is a second automatic iron gate through which you have to be buzzed again to access the residential area, which includes three parallel streets of two- and three-bedroom Tudor-style suburban townhouses. Each home has a front and back yard, with manicured grass, flowers, and trees tended by Equatoguinean gardeners. The Major compound also includes a pool, a gym, a large multipurpose room (for church services, parties, and other large functions), and a bar/restaurant.

The Smith compound is the only large compound on the mainland, and of the three is located the farthest from the main city (in this case, Bata). One can feel this distance in the layout and aesthetics of the camp. It is by far the most beautiful, surrounded by lush greenery on three sides, and on the fourth it faces a seemingly endless white sand beach that stretches from Gabon to Cameroon, with Equatorial Guinea and the Smith compound in the middle. Catamarans, kayaks, and other water sports equipment linger behind the fence that closes the compound off from the beach, to be checked out from the guard who will also take your name and the time of your departure if you choose to go for a beach walk. (One could easily see this compound turning into vacation homes for the elite, or even something like a ClubMed, when the industry leaves. The all-inclusive aspect is already there!) Like the Major compound, Smith is divided into the office area, the food and recreation area, and the housing area. The luxurious houses border a circular road that encloses a small golf course, and some—including the largest home reserved for the country manager—have ocean views (see figure 1.8). All resident employees are given golf carts to drive around the complex to reach not only their offices, but also the restaurant, pool, game room, and small movie theater.

Because both the Major and Smith Corporations do all their extraction, production, storage, and offloading offshore, the only personnel living in these compounds are the high-level migrant managers who spend most months of the year in Equatorial Guinea, although only for two- to three-year stints. In terms of layout and architecture then, the compounds are relatively small, housing roughly thirty expatriate workers at a given time. The Equatoguineans who work in these compounds—whether as maids, gardeners, or security guards, or in higher positions such as government and community relations—are forbidden to live on the compounds.

The Endurance Corporation has substantial onshore industrial facili-

Figure 2.2. Golf carts on the Smith compound, with management houses in the background.

ties that entail increased onshore labor; thus, the scene that unfolds once past their gates is quite different from the other two. First, one need not even reach the security gate to see one of the two constantly burning, towering gas flares (see figure 1.1). Although this flare was said to burn only when the system was "disrupted," the flame was constantly burning during my fourteen months of fieldwork in Equatorial Guinea. I never saw it unlit. This flare and other signs of industrial production take up considerable amounts of space within the compound, including a large electricity plant and colossal LNG infrastructure. Thus, rather than a ClubMed, Endurance feels far closer to Porteous's (1974) description of a company town, with "the general dominance of the settlement by the physical expressions of economic enterprise, brought about by a close juxtaposition of town and plant" (411). The most significant difference between the Endurance compound and the other two is its housing. While all three companies have extensive subcontracts with firms that stock their offshore operations with employees from around the world, Major and Smith (given their offshore setup) do not have to house these workers in their compounds. Rather, they rotate on and off rigs as discussed in chapter 1. By contrast, the Endurance group of companies, with hundreds of contracted and subcontracted employees from over thirty countries staffing the methanol and LNG plants, has extensive housing obligations not only for migrant management, but also for mid-level and lower employees who also rotate, but in this case, to and from the onshore plant.

Once past security in the Endurance compound, the entrance to the industrial complex of offices and plant infrastructure sits straight ahead. A left turn at a second guardhouse brings you to another gate where you have to show identification to the guards, who then electronically admit you into the compound's residential area. Up and over a large hill with majestic ceiba trees and ever-mowed grass, the residential zones partially spread out before you. To the left are the large suburban homes also found within the other compounds, housing upper management, with the largest as always reserved for the country manager and his wife (see figure 2.3). During my fieldwork, it was white Americans, white Brits, and white South Africans who occupied these houses (and job positions). I knew one Brazilian woman, married to an Argentinean man, who lived in this level of the compound. While she considered herself to be white and was considered white by others in Brazil, she made it clear to me that in the compound, the US- and UK-dominant culture made her feel distinctly nonwhite. Two other women—one Filipina and one Thai—married to white American men, rounded out the nonwhite presence that I was aware of in either the large houses or the condominiums for mid-level management. The condominiums are located straight down the hill to the right, overlooking a small man-made lake. Referred to as the "townhouses," they comprise a smaller set of buildings, equally well-appointed with flat screen televisions, state-of-the-art stackable washer-driers, and new appliances. Most of the mid-level management living in the townhouses were stationed in Equatorial Guinea alone, although some men lived there with their wives as well. While these townhomes were absolutely luxurious and spanking new inside, the wives who lived in these buildings lacked expansive common areas, and hence did not host cards. Housing in the Endurance compound was contentious. The larger houses were all occupied, so some couples who "deserved" larger housing were placed in the condominiums. There was also a feeling that the three subsidiary Endurance companies housed within the compound did not get equal treatment in housing. I explore embodiments and effects of these tensions in the final section of this chapter.

I spent considerable time with "the wives" in their weekly activities and thus came to know their housing facilities well. My own whiteness and (at that time) young womanhood produced a form of white racial sorority, fictive kinship where these women looked at me and often said or seemed to

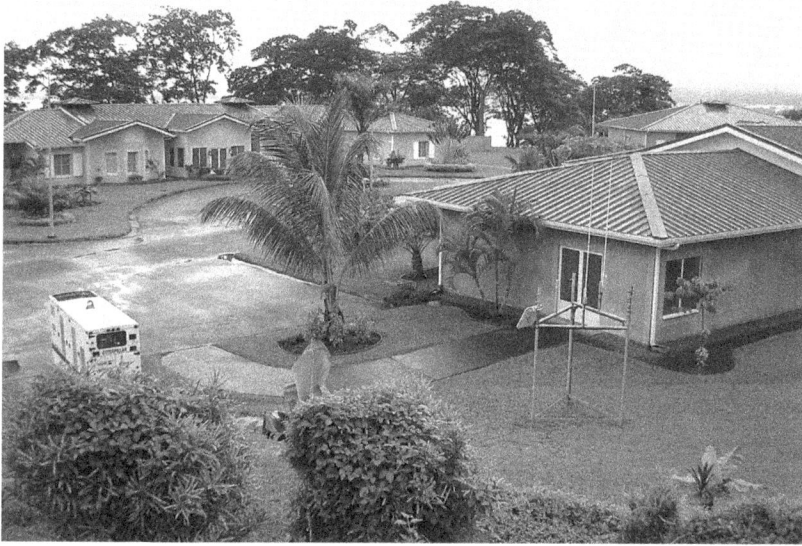

Figure 2.3. Upper management homes on the Endurance compound.

think, "You could be my daughter." This raced and gendered kinship opened their homes to me and got me through the considerable layers of security with which the compound is guarded.[3] For a variety of reasons, including my raced and gendered identity, I did not have this level of access to the rotator camp (figure 2.4) or the barracks (figure 2.5). The men who lived in this housing were rotators; thus, when they were not working, they were "home" in the Philippines, Kazakhstan, or Mexico, and not in Equatorial Guinea. When in Equatorial Guinea their work schedules were outrageously demanding; therefore, it was difficult for me to find ways to socialize with them or encounter them outside of work. They were also forbidden from bringing visitors into their housing, thus the description I offer below comes from a series of visits guided by white management personnel with whom I had established a relationship.

The rotator camp is one step below the condominiums in the housing/employment hierarchy. Within the camp are large apartment-building structures where men (exclusively) live, two to a suite, with separate bedrooms and bathrooms, and a shared common room. While still nice, the accommodations are sparse, having more of an institutional/college dormitory

Figure 2.4. Rotator camp.

Figure 2.5. The barracks or "Indian Camp."

feel. The country manager who led me on the tour of the rotator camp told me that the majority of workers living there were Filipino, but that it housed the most diverse workforce of all housing levels. Others at this occupational level—mid-level engineers, electricians, and plant operators—included men from India and Pakistan, Mexico, Venezuela and Ecuador, Turkmenistan and Kazakhstan, as well as the US Gulf Coast. (Notice that despite the diversity of nationalities, the men generally hail from oil-producing places.) All men housed in these facilities are subcontracted labor and do not work directly for the Endurance corporation. They are not permitted to bring wives (let alone other partners), despite the fact that some of their rotations are up to one year long. They have no cooking facilities and are expected to eat in the canteen for every meal, although many had hotpots in their common areas.

At the bottom of this residential hierarchy are the barracks, a series of modular dwellings lined up in parallel rows. Before we entered these housing units, the facilities manager explained, "I've never been inside these facilities. I think they're similar to offshore: fairly tight quarters with common areas for relaxation." With an anxiety that clearly anticipated critique, he told me repeatedly that they were shutting down this housing level, although I saw no sign of the decommissioning. The barracks I visited had long, trailer-wide corridors with doors off of either side leading into shared bedrooms and bathrooms. They had no cooking facilities, and men in this level were also prohibited from bringing partners. Originally constructed to house the workers who built the LNG plant, the vast majority of whom were Indian, many still referred to the barracks as "Indian Camp." In 2008, the barracks residents were majority Filipino, Indian, and Pakistani. Although clean, the bottom two levels of housing—essentially for semi-skilled and unskilled and mostly nonwhite labor—were far inferior to the top tiers. When I asked the manager giving me the tour why the barracks were so different even from the rotator camp, he explained, "Western rotators are on 28/28s. Their requirements for housing are a little higher standard." Not only were these housing options obviously inferior to the others, but they also had separate recreational facilities, including a small pool and their own bar. By comparison, the Olympic-size pool in the management residential area was attached to the clubhouse—a more formal dining, recreation, and bar area where mostly white management residents paid for meals should they so choose—and wives could often be found sunning poolside. Nonwhite men away from home for long periods of time were not fully welcome in that recreational environment. One white North American expat wife told me that she had recently seen some Filipinos coming to "her" gym. "They're here alone for a year without a break," she said. "Seeing them in the gym creeped me out."

I ate in the workers' canteen (as opposed to the clubhouse) on two occasions, and on both there were multiple menu options advertised by nationality—one day the "American" option was Sloppy Joe; the "international" option was Beef Hot Pot; the "Vietnamese" option was Squid Pakwis; and the "Indian" option was Aloo ka Bharta. Yet of course, multicultural lists of inclusive menu options belie a situation in which men ostensibly divided by skill level are given unequal housing, facilities, and recreational options, not to mention different pay scales, in a hierarchy that falls too neatly along national and racial lines. At the top of this hierarchy are white managers from the US, Western Europe, and (occasionally) South Africa, permitted to bring their wives and given lavish homes and salaries; Filipinos and other workers from the world's oil-exporting diaspora (Kazakhstan, Mexico, Venezuela) are in the middle, on "rotations" that last up to a year, during which time they are not permitted to bring family, and share dorm-like facilities with fellow workers; at the bottom are Indians, Pakistanis, more Filipinos, and the occasional Nigerian, living in shared trailers. Again, these forms of segregation and differentiation are a clear illustration of the continuity with over a century of oilfield practices, the "long, unbroken legacy of racial hierarchy across the world's mineral frontiers" (Vitalis 2007, 19). Note, for instance, that while the United States, the United Kingdom, and South Africa (among others) are multiracial societies, over two years of research in Equatorial Guinea, everyone from rig workers to migrant management from these countries were nearly homogeneously white. The only African American man I met in Equatorial Guinea was head of corporate social responsibility for one of the transnational corporations. The only Asian American man I met in Equatorial Guinea managed a smaller oil services company attached to the larger companies by subcontract. I did not meet a single Black or Coloured South African in Equatorial Guinea, though white South Africans were common. This is to say that the forms of racialized hierarchy—and white supremacy in particular—that unevenly organize life in these countries traveled to Equatorial Guinea with the US-based oil and gas industry. While Vitalis (2007) points to the continuity of these corporate practices, one significant piece *had* changed: Equatoguineans were not allowed to live within the compound.

Some of my Equatoguinean interlocutors, who had lived abroad for most of their lives, received postgraduate degrees, and recently returned to work in the oil industry, had requested to live in the compounds. They argued that their extended families were still abroad, and that they would be uncomfortable in Malabo and not able to work as efficiently as their foreign coworkers without the quality of life to which they were now accustomed, including reliable electricity, potable running water, and adequate housing. They were all denied. When I asked migrant management *why* Equatoguineans were forbidden to live in the compounds, they responded that allowing Equatoguinean residents would force them to "change the rules." Migrant managers were open about the fact that the "rules" of the compound allowed them to control not only the comings and goings of their employees with curfews and lockdowns, but also to control the comings and goings of outsiders. If Equatoguineans lived in the compound, then they would want their families and friends to visit, and perhaps spend the night, all of which was currently prohibited. Fundamentally, the compounds were set up to separate, to disentangle, to partition, and to control. Separations, including the prohibition of Equatoguinean residents, contributed to the *lived* sense that the enclaves were somehow completely detached from the world outside their walls.

Through fourteen months in Equatorial Guinea, I was consistently astounded by the strangeness of the industry's gated enclaves, by what felt like their horrifically archaic raced and gendered divisions of labor, space, and domestic life. I am obviously at home with running water and cable television, and neither golf carts, Texans, nor gross inequality is new to me. Yet there was something about the sheer concentration of these technosocial arrangements tucked within enclave walls that made me feel ever the anthropologist, constantly wondering what the "locals" would think of next. But my ongoing astonishment was merely historical ignorance. The oil industry, and extractive industries more broadly, have long histories of compound life to which I now turn in order to situate what I have described within longer histories of settler colonialism, company towns, and zonal capitalism.

HISTORY AND PRECEDENTS

In his 1931 memoir, an American oil worker living in Venezuela noted, "If ever a white man could live happily with his wife and children in a hot climate, that was in Mene Grande," a US oil camp in the Maracaibo Basin

(Coronil 1997). Citing this passage, Coronil writes, "As veritable enclaves with private roads, schools, stores, and medical supplies, these camps constituted a State apart from the Venezuelan State" (109). And a 1946 article from Chevron's monthly magazine opens with a phonetic introduction: "Sah-*oo*-dee Ah-*ray*-bee-ah, where part of America has been set down amidst rock and sand" (Vitalis 2007, 35). Of Dhahran in 1945, the wife of an American colonel brokering a deal with Saudi ARAMCO wrote home: "The oil town . . . is just like a bit of the USA—modern air-conditioned houses, swimming pool, movie theater, etc. Six American wives have already arrived and more are on their way" (Vitalis 2007, 80). Since the 1930s, firms have built "the basic infrastructure that we associate with modern municipalities—housing, streets, power, water, security, and so on—in order to bring their commodity to market" (Vitalis 2007, 266). Transnational extractive industries have long operated in what William Reno (2001) has called a "BYOI" or "Bring Your Own Infrastructure" environment.

Paul—a white Zimbabwean brought to Equatorial Guinea to finish, commission, and run the Smith enclave—narrated these long histories in Africa, in particular:

> Within [our company], people say that this compound is one of a kind and that there's no other complex like this. But they're wrong. There are lots of these in Africa. Mining has been a huge industry in Africa—diamond, gold, emerald, platinum, and every single mine has a complex like this—larger and more elaborate . . . private game reserves, eighteen-hole golf courses. Shell Nigeria, their whole outlook is different. They have schools with teachers and headmasters and summer trips, wives and children with dad. [Their] complexes have supermarkets.

Indeed, the partitioned use of space has an extensive history across Africa's resource-rich landscape. Here, through brief thought pieces on colonial settlements, company towns, and more contemporary forms of zonal capitalism, I think through the enclaves' adjacent forms—histories and precedents of the strategic use of domestic and industrial space. Each spatial form I consider is both strategically and partially integrated into the economies, infrastructures, and populations of their respective locations, and each is intentionally *dis*-integrated as well. Each form is highly racialized and gendered insofar as those aspects of personhood are central to the organization of labor, living conditions, and freedom of movement.

Colonial Settlements

> Plantations are self-contained worlds. Workers, managers, and the crops they cultivate live together side by side, but regulated by strict hierarchies, the more blatant because they are carved into the landscape. Male managers and their wives live in comfortable houses with gardens and kitchens maintained by local employees and have access to their own clubs with well-stocked bars and refreshing swimming pools.
> —Cynthia Enloe, *Bananas, Beaches, and Bases: Making Feminist Sense of International Politics*

Enloe's description of South American plantation life could, in many ways, be today's oil and gas enclaves in Equatorial Guinea, with oil production replacing crop cultivation (see also McKittrick 2013). The enclaves' geographies in Equatorial Guinea often trace a direct relationship to the colonial cocoa economy. For instance, to be dropped outside the Endurance compound by taxi, you simply ask the driver to take you to *la planta*. As a cognate with the English word "plant," asking to go to la planta is, in part, asking to be taken to the factory. But in Spanish, la planta is also an abbreviated reference to *la plantación*, or the plantation, and the picturesque spit of land on which the massive Endurance compound now sits was, during the colonial era, a large cocoa plantation.

Commercially, colonial (and imperial) extraction brought with it many of the key features still seen in oil enclaves today: private companies with private security, intimately connected to their states of origin; infrastructure set up exclusively for export; and contractual regimes with local power holders guaranteeing special ports and customs areas, or specifying juridical procedures in disputes. As with petroleum, the historical trades in enslaved humans and palm oil were governed by parastatal contractual relationships (Lovejoy and Falola 2003; Mann 2007, 2013). The slave trade, in particular, was so capital intensive that it could only be run by chartered corporations—which operated by means of chartered monopoly, subsidies, and special rights (including the right to wage war) conferred by the British Monarchy.[4]

Labor setups in colonial and imperial production varied dependent on, among other factors, the requirements of the raw material in question. In the case of labor-intensive crops or extractive processes, such as sugar, rubber, and mining, large labor pools were required; often, large town-like complexes were set up to house workers at all levels, from manual, indentured, or enslaved labor through overseers and plantation managers (Ferguson 1999; Stoler 1989b. See also McKittrick 2013, 2016 for contemporary rereading.)

In Equatorial Guinea, after the slave trade grew dangerous at the beginning of the nineteenth century, trade turned from humans to ivory and palm oil, both needed to lubricate the machines of the industrial revolution.[5] With the end of slavery, a merchant class of English-speaking West Africans from Sierra Leone came to Equatorial Guinea, the first of many waves of people who would come to be called *Fernandinos* (Liniger-Goumaz 1989, 2000; Creus 1997). Later, Liberians and Nigerians joined them, and together they made up an entrepreneurial and social group, communicating in Pidgin English and shunned by the local Bubis. Undoubtedly motivated by the rapid accumulation taking place, Spain gradually began to reassert nominal imperial control of various kinds. The first Spanish governor arrived in 1858, followed by missionaries (Liniger-Goumaz 1989, 2000). Cocoa was introduced for the first time at the end of the nineteenth century, some argue by a Liberian Fernandino, and Bioko (in particular) began a transformation into a latifundiary monoculture (Creus 1997; Liniger-Goumaz 1989). Unlike the export of palm oil or ivory, which required only a comprador class to broker deals with ship merchants, the farming of cocoa was both time and capital intensive. The work to be done and the profits to be had ensured that more European colonists and African Fernandinos began to settle on Bioko island. The Spanish set up small-scale cacao plantations with houses for local workers, mostly those the Spanish labeled *civilizados*, who had been converted to Catholicism and worked and lived with their families on the plantation. Because the local population of Bioko Island was so small, the Spanish brought in Nigerians to supplement the labor force, who then grew to become the demographic majority of colonial labor on Equatorial Guinea's cacao plantations. Nigerians too received houses, although much smaller and of much poorer quality, where they lived as migrant laborers without their families. These graduated domestic arrangements—from white colonial managers and their wives in comfortable housing with luxurious amenities, to mid-level foreign labor in smaller housing, to manual labor in camp-like situations—persist in Equatorial Guinea's enclave arrangements today.

The Uncle Ben's Rice, Crystal Light, and Bull's Eye barbeque sauce that crowded pantry shelves in management housing seem easily analogous to "the inappropriate dress, food, and other markers of European culture that Anthony Burgess and George Orwell caricatured in their novels: the jungle planter sweating through a five-course dinner in formal attire" (Stoler 1989, 149). Yet, there is also something fundamentally different about migrant management lifestyles in Equatorial Guinea's compounds. Burgess's and Orwell's caricature comes across in *the sweat*, exposing the five-course meal and formal attire as maladapted, insisted upon only as markers of race, class, and other forms of distinction. Among migrant management in contemporary Equatorial Guinea, there is little sweat, literally or figuratively. Down to the telephone area codes, the air-conditioned transport of suburban Houston life into these small enclaves can feel so complete that, until one goes outside, it is difficult to caricature flat screen televisions, running water, or mansions with manicured lawns and maids as absurd.

Stoler (1989) writes that "colonial cultures were never direct translations of European societies planted in the colonies, but unique cultural configurations, homespun creations in which European food, dress, housing and morality were given new political meanings in the particular social order of colonial rule" (136). Much of this insight certainly applies to these enclaves, but "homespun" they are not. The intensification of technologies of transport, communication, and infrastructure since the colonial era guarantees that these enclaves are appointed in ways perhaps less homespun even than these people's homes in Texas or elsewhere. The furniture is standardized as are the state-of-the art appliances. The compound even has its own voltage setting to accommodate American electrical current. Perhaps the most significant difference from Stoler's accounts, however, is that in Equatorial Guinea there is no colonial governing apparatus, just the exigencies of the industry. This distinction alone brings us much closer to the model of a company town.

The Company Town

A consideration of company towns highlights other continuities between Equatorial Guinea's gated compounds and different historical forms of corporate and domestic enclaving across the world's mineral frontiers. The company town is defined by employees of a single company or group of companies as inhabitants; ownership of the real estate and the houses by those same companies; deliberate residential zoning by nationality/ethnicity and socioeconomic class, including "a small number of luxury dwellings for the attraction and retention of key personnel"; and an elevated level of employee control stemming from the company acting as both landlord and employer (Porteous 1974). As I explore below, only the Endurance compound in Equatorial Guinea might fruitfully be thought of as a company town. The obligation to house a large number of workers from all over the world makes Endurance unique in Equatorial Guinea, but far more like most other major extractive industry complexes around the world. Specifically, as narrated above, the compound is laid out according to what Porteous (1974) has called "deliberate residential zoning from the outset" (41).

> The occupational hierarchy of the plant traditionally has been imposed upon the town by fiat in residential segregation terms. Common elements include (1) deliberate ethnic and socioeconomic segregation in housing location; (2) creation of a graded series of house styles which are allocated to employees not according to need (i.e. family size) but according to class; and (3) creation of separate institutions for each class. (411)

Porteous's schema is based on his work at three major Anaconda Copper Company towns in Chile's Atacama Desert, established in 1916 and enduring in a certain form until they were nationalized in 1971. A century after Anaconda's establishment of these towns, the same deliberate residential zoning processes are visible in Equatorial Guinea's oil compounds.

Segregated residential zoning is not peculiar to company towns. Most cities and towns in the US and elsewhere are racially segregated, and in turn have drastically different housing styles—from mansions to public housing— "not according to need but according to class," and, we must add, race. However, the processes that produce the daily segregation in which many of us live—long histories of redlining and gentrification (better referred to as "racial banishment" [see Roy et al. 2016]), immigration, white flight, industrial collapse, and zoning legislation—are often slow and cumulative. While their effects are systemic and visible, the processes themselves are often implicit enough to be naturalized. The segregation in company towns, on the other hand, is premeditated and deliberate. Far from emerging slowly over time, graduated residential options are put in all at once, and workers are assigned to each according to their job and the temporality of their labor, which coincide neatly (although not perfectly) with race, class, nationality, and the attendant inequalities of the global nation-state system.

Like settler colonialism, the company town model is not historically static. On the contrary, its long and violent history of labor unrest and rebellion, in particular, has meant that companies continually rethink how best to set up their domestic-industrial space. This ongoing rethinking has led to significant differences between the classic company town model—with schools, post offices, and churches, for instance—and what we find today in

Equatorial Guinea. Take Ferguson's (2006) description of company towns on the colonial Zambian copperbelt, which he deliberately contrasts with more contemporary forms in what he calls the "Angola model" of oil development:

> On the Zambian copperbelt, investment in copper mining brought the construction of vast "company towns" for some 100,000 workers. . . . The mining towns, classic examples of colonial-era corporate paternalism, eventually came to include not only company-provided housing, schools, and hospitals, but even social workers, recreational amenities such as movie theaters and sports clubs and domestic education programs. (197)

Paul, the Zimbabwean in charge of the Smith camp quoted earlier in this chapter, commented that older oil compounds in Nigeria have "schools with teachers and headmasters and summer trips, wives and children with dad. [Their] complexes have supermarkets." Porteous's description of Atacama towns, Ferguson's of Zambia, and Paul's of earlier Nigerian oil camps all betray the dynamism of modes of extraction, management of labor, and, with them, company town design. Towns that grew up around both the Atacama mines and the Zambian copperbelt depended first on the idea that the company would be in the region for a long time and second on the need to employ and house thousands of local workers. On the one hand, the temporality of oil in relatively low-producing supply sites like Equatorial Guinea is such that long-term investment in institutions doesn't make sense from an oil company's perspective. But on the other hand, as Paul's comparison with Nigeria shows, there is also the iterability of the transnational oil industry; that is, US oil companies have learned from the failures of institution-building in Nigeria (Watts 2004; Zalik 2004, 2009; Adunbi 2015), and they are investing differently in places like Angola and Equatorial Guinea (Ferguson 2006). In Equatorial Guinea, again, the locals who work in the industry are prohibited from living in the enclaves, and the migrant workers who live in the enclaves, regardless of level, are prohibited from bringing their children. These, among other regulations, leave a small and transient population as enclave inhabitants, no more connected to living in Equatorial Guinea than they were to their postings in Angola, Indonesia, or Russia before that.

Company towns are perhaps most (in)famous for the tension between proffering comfortable living conditions in extractive settings on the one hand, while intimately controlling labor and domestic life on the other. Starting with the town of Pullman, Illinois, in the late nineteenth century, company towns have long been based on "progressive concepts of management and labor relations administered by trained professionals. In order to deter unionization and reduce labor turnover, the 'new' company town attempted to attract workers by providing significantly better working and living conditions" (Crawford 1995, 3). In other words, invoking the recurring theme of self-regulation, company towns rely on a specific relationship between capital and labor, in which workers' living conditions become a matter not of potential industrial conflict, but of company regulation, as Porteous (1974) explains:

> Company housing has proved an effective means of worker control where an entire company town is constructed. If the industrialist is landlord as well as employer, his relationship with his employees extends beyond the plant and into the workers' homes. Employers may thus exert considerable influence over the social and political, as well as the economic life of company towns, sometimes with dire results. . . . Union organization may be prevented, religious bigotry fostered, and social class structures fossilized; dissenters and "radicals" may be dismissed from their jobs, and consequently from their homes and thus from the company town itself. Such excesses, including a lurid history of strike suppression in the cola and copper towns of the American west, in Chile, and elsewhere, have been the basis for the notoriously poor public image of the company town. (410)

Extraction and production in Equatorial Guinea are happening at a conjunctural moment when, in both industry history and Equatoguinean history, active or intentional union deterrence is superfluous. The structure of labor evidenced in both the offshore and in these enclaves—subcontracting to innumerable companies through "body shops" that draw men from Venezuela to Turkmenistan, Pakistan to the Philippines, rotating in and out on hitches of various lengths—now practically ensures that unionization isn't an option.[6] Moreover, de jure, unions are illegal in Equatorial Guinea (Campos Serrano and Micó Abogo 2006). In other words, since Pullman, Atacama, and even Nigeria's earlier oil boom, much has changed in the setup and temporality of labor, and in how migrant oil workers relate to their "homes." And yet, the ability of companies to intimately shape the lives of employees in company towns has not changed. In Equatorial Guinea's enclaves, residents have curfews, and nonresident circulation is closely monitored; residents cannot paint their homes or tend their own yards; and only a privileged few are allowed to bring their heteronormative spouse to the country.

While colonial settlements and company towns draw attention to the long histories of cohabitation of industrial production and domesticity still to be found today in Equatorial Guinea's oil enclaves, those precedents don't

CHAPTER TWO

capture the privileged fiscal and juridical statuses of these ring-fenced areas. There is a final category that Winters (1996) calls "zonal capitalism" through which to consider the enclaves, to grasp the full extent of their exceptional juridical and fiscal status in relation to the nation-state in which they can be found.

ZONAL CAPITALISM

> Zonal capitalism [is] a special economic zone in a subnational area—sometimes walled but always clearly bounded—in which an intensified effort has been made to create a climate favorable to business. Government policies within and for the zone tend to diverge markedly from those applying generally to the national jurisdiction.
> —Jeffrey A. Winters, *Power in Motion: Capital Mobility and the Indonesian State*

Special economic zones come in various forms: Export Processing Zones (EPZS), Free Zones, maquiladoras, and Special Autonomous Zones, among others. Intimately related to the broadly conceived offshore discussed in chapter 1, within these spaces companies are often "freed" from national laws regarding taxation, labor rights, and environmental regulation. The means of production also are freed from state involvement; for example, companies are permitted the duty-free import of equipment, materials, machinery, supplies, and components, using global procurement chains that stretch around the world (Sklair 1993; MacLachlan and Aguilar 1998; Cameron and Palan 2004). With reference to the ways in which corporate sovereignty supersedes national sovereignty within these spaces, Palan (2006) has called this use of space "sovereign bifurcation," and Ong (2006) "postdevelopmentalism," to refer to "a more dispersed strategy that does not treat the national territory as a uniform public space [but instead favors] the fragmentation of national space into various noncontiguous zones" (Ong 2006, 77).

For both Winters (1996) and Ong, zonal capitalism and postdevelopmentalism are state strategies to attract nomadic capital. Taking Indonesia and China as their respective sites, their theorizations rely on an idea of the state as strong and tactical, intentionally ring-fencing parts of their territory in which they can rescind labor laws and attenuate taxation regimes in an effort to draw foreign investment. In Equatorial Guinea, on the other hand, profitable hydrocarbon discoveries can hardly be said to have been state strategies. As Ferguson (2006) and Reno (2001) have pointed out, despite the axiom that African states, in particular, must demonstrate a certain business climate—such as good governance and transparency—to attract Foreign Direct Investment (FDI), Nigeria, Angola, Equatorial Guinea, and Sudan, among others, show that the geologic presence of hydrocarbon deposits alone is sufficient to attract abundant FDI, even in the midst of civil wars, ethnic cleansings, widespread civilian unrest, and deep kleptocracy. In these contexts, the enclave is hardly a state strategy, but rather an oil company strategy through which the industry attempts to "shield" its practices from what may lie outside its walls. In this way, enclaves in Equatorial Guinea may have less in common with noncontiguous zones in China, and more in common with the relationship between small Caribbean economies and offshore finance (Maurer 2005, 2008; Navarro 2010; Hudson 2017a), where historical hierarchies of rank and the state's marginalization are, in fact, what guide the enclaving process. As with small Caribbean tax havens, the impunity that oil companies enjoy inside Equatorial Guinea's enclaves has less to do with the state's position at the cutting edge of capitalism, and more with its earlier abjection therefrom.

Finally, Ong (2006) has pointed out that "these zones [are a] confluence of economic freedom and political repression" (113). Despite other differences, Equatorial Guinea and China certainly share this effect of variegated sovereignty. Although Equatoguinean life outside the enclave walls—with armed soldiers dotting the streets, tanks parked visibly in strategic locations, and military helicopters often buzzing overhead—feels more militarized than the enclaves themselves, the potential for armed response within the enclaves is palpable, if latent. The compounds were frequently "locked down" in response to perceived geopolitical unrest, whether local, in Nigeria or Cameroon, or along various borders. In addition to these forms of overt control of circulation and the policing of boundaries, the political repression in these spaces suffused the texture of daily life in more mundane ways, as I have described above—from regulations on what you can and cannot do to your house, to the clear backpacks issued to Equatoguinean workers; and from strict guidelines about visitors, children, and partners, to the differential treatment of citizens and noncitizens in terms of rights and privileges within these spaces. "Preexisting ethnoracializing schemes (installed under colonial rule) are reinforced and crosscut by new ways of governing that differentially value populations according to market calculations" (Ong 2006, 79). The relationships among national and racial hierarchies, profit, domesticity, and the built environment come together in novel ways in Equatorial Guinea's oil enclaves. They contain, at once, centuries of continuity with settler colonial patterns, the iterative antipolitics planning of company towns, and newer forms of noncontiguous political life in which nation-states refract into policed communities of variegated, racialized sovereignties and profit potentials.

Equatorial Guinea's enclaves are novel in another way, in that, despite narratives of progressive corporate learning and late capitalist flexibility (from full employees to subcontractors, from company towns to special economic/domestic zones), securitized compounds are not inevitable. Indeed, in major extraction sites around the world, many large US oil companies house workers at all levels "in the community." In the next section, I explore why major US oil companies chose the compound model in Equatorial Guinea, making an argument for the "actuarial" enclave. This is an infrastructural form haunted by past uprisings and built to avoid a set of *potentialities*—illness, graft, political unrest—which, if they were to materialize, would ultimately be more expensive than the enclave built at great expense to avoid them. This is cost-benefit analysis as ghost story, capitalism as a calculative palimpsest of white fear and the making of a frontier.

ACTUARIAL ENCLAVES:
FRONTIER-MAKING AND RING-FENCING

Wendy and I sat on the screened back patio of her home on the Endurance compound. With her large suburban-style house perched on the very edge of Punta Europa, the spacious patio offered an expansive view over the ocean inlet that separated the Endurance compound from Malabo. The city was low and visible through the heated haze just across the water, as if a mirage. Wendy had been coming in and out of Equatorial Guinea with her husband since 1999, by far the longest of any of the migrant wives I came to know. She reminisced that her husband had come to Equatorial Guinea originally in 1998 on a "reci," or reconnaissance mission for his oil company. She came with him in 1999 on a "look-see," to determine if the place was a viable living and working option for them after other lengthy migrant stints (as Brits) in Ecuador, Indonesia, Texas, and Los Angeles. As she explained:

> We came out to where we were going to live, just to take a look at the project. The project was here. It was actually on this spot, because they had a gas flare. They had a gas discovery and ... that flare has been there since 1990, since the

discovery. The original flare was massive. It actually lit up the sky. There were very little lights around here. Nothing existed. This was jungle, total jungle. So we just saw the whole thing develop. This, where we are now, where we're sitting now, on the peninsula, was jungle, and we chose this spot. [My husband] went on a "reci," you know, he actually did a reconnaissance mission. He actually walked through the jungle to see whether it would be suitable to start building houses, and where they would build the plant.

Part of the lore of company towns has long been the idea of remoteness, captured in Wendy's description: "Nothing existed. This was jungle, total jungle." Porteous (1974) writes that "most commonly, the company town comes into being through the overwhelming physical fact of isolation. The pioneering entrepreneur, endeavoring to develop a resource in a region remote from established population centers, is likely to find a dearth of speculative builders, local governments capable of providing housing, or workers with the capital or skills necessary for dwelling construction. The company town is thus typical of remote resource frontier regions" (410). In Wendy's narrative, we hear not only about geographic isolation, but also of Porteous's "pioneering entrepreneur," her husband who "actually walked through the jungle" to determine where they might start building houses. Enloe (1990) has also written about this masculinized and most often white character in military bases and plantation economies around the world, where a "rough and rugged cohort of men [transform] the primeval forest into a civilized and profitable plantation belt" (140). Their bravery and ruggedness are crucial in this narrative, where "privilege and profit are [justified via] character and not on race or class" (141). (See also Stoler 2010.) But as is so often the case, the physical isolation or remoteness that Wendy conjures is not what it seems.

Not only is the Endurance compound separated from the capital city by a mere handful of kilometers, but the land itself, Punta Europa (Europe Point), is an exquisitely beautiful and perfectly situated isthmus that had been a Spanish-owned cacao plantation in the colonial era. Along with most of Equatorial Guinea's plantations, it fell into disuse after Macías expelled the Spanish. Although its Spanish owner attempted to reclaim it under the *bienes abandonados* (abandoned property) act at the beginning of Obiang's rule, he was unable to do so, and Obiang then claimed the property as his own.[7] The first contract for oil-related infrastructure development on Punta Europa—the contract that produced the gas flare "that lit up the sky"—was between Endurance's predecessor company and Obiang, to whom they paid rent as a private property owner. During the second phase of Endurance's major expansion, when the company forced the displacement of small communities living on the isthmus, Obiang made a large and public display that the state was "expropriating him too," as he officially sold his property to the government (of which he was, and is, the president). Rather than a company town established in a remote area, at its beginnings the Punta Europa compound was more accurately Tsing's (2005) "zone of unmapping.... Frontiers aren't just discovered at the edge; they are projects in making geographical and temporal experience" (29). In Punta Europa, there were already long histories of ownership, cultivation, and dispossession underfoot, although Wendy narrates her family's early experiences as reconnaissance missions in empty jungle. At the same time, however, it is difficult to suggest that she might have done otherwise.

The overgrown cacao plantations that dot Bioko Island quite reasonably look like "jungle" to those unfamiliar with the landscape, with second-growth tropical flora creating dense shade. Only those with a practiced eye, like the Equatoguineans who walked me through second-growth elsewhere, might know that the cocoa pods deepening from yellow to orange, or the easily walkable routes through the trees with no machete needed, indicate histories of cultivation and inhabitance over the last century. Obiang's ownership of the land, on the other hand, was a widely known fact, albeit one that made most migrant industry managers somewhat uncomfortable—that their presence was enriching a dictator in the most direct and intimate ways—while most Equatoguineans rolled their eyes in bored resignation, accustomed to the complicity.

While Wendy and other early migrant arrivals waited for the compound to be built, they lived in Malabo, as she describes here:

> We settled into Caracolas [an affluent residential community]. The housing accommodation was basic, [but] very, very nice. It was all Spanish so it was beautiful and it had the facilities, but it needed updating. Everything was really quite old. It had to be rewired and the water was really bad. We had lots of remodeling to do. We settled into the house there, and most of my life was in town. There was a very large cultural community. I played tennis with the French. We used to go out to beaches. We used to do a lot of things, but in a multicultural group—Spanish, French, Belgian; we had people from Brazil.

Indeed, Equatoguineans remembered that when oil first arrived, "Americans were all over the city. Now you hardly see them." Wendy was clearly nostalgic for this time—the freedom of movement and the national (although note, not racial) diversity of the migrant community with whom she socialized.

As Wendy's brief time in Caracolas illustrates, the construction of these enclaves was not inevitable, at least according to my migrant informants, many of whom were closely involved in the decision-making processes that led to this spatial arrangement. As the country manager for Regal Energy noted, "In Ecuador and Vietnam [migrant employees] live in the community. In China they congregate you in expat housing. Tunisia we lived in the community. Israel we lived in the community. Here it's unique, this [enclave] school." Others I spoke with had lived "in the community" in places as disparate as Gabon, Indonesia, Ireland, and Japan; however, "in the community" is a slippery designation. While migrant employees in Gabon or Indonesia weren't as isolated from local life or public infrastructure as they are in Equatorial Guinea, in all places they were offered "western-style" housing with separate water and electricity sources, if necessary, in gated and securitized compounds.

When I talked with people involved in the decision-making that led to compound construction in Equatorial Guinea, they consistently responded with a narrative pair: on the one hand, the difficulty and expense of the endeavor; on the other, a sense of having no alternative. As a finance manager I spoke with explained:

> There were two schools of thought when we first came here: one wanted to live in the community and force the issue. Others said no, we need to build a place where people are willing to come and work. I don't know which way is right. [It's] a business decision: it makes more sense to live in the community. It's cheaper if you can live in the community and shop in the places where local people shop. It makes financial sense from a business standpoint. So the pressure in that regard is internal. You've gotta get your project done; you've gotta get the personnel; you've got the realities on the ground to deal with. If we find some very nice housing for expats in Paraiso [another affluent neighborhood in Malabo], from the outside it looks really nice. But the wiring is really messed up and the residents get shocked. The well has chloroform because it was built too close. . . . It's not to the standard that we're accustomed to.

The man quoted above claimed that "it makes more sense to live in the community" because it's less expensive to rent already-built housing and shop where local people shop. The pressure to enclave then is not simple economic pressure, but what he referred to as "internal" pressure. Enclaving, in his estimation and that of all other managers with whom I spoke, facilitated a more focused work environment, attracted people who might not otherwise be willing to work in Equatorial Guinea, and prevented contact with "realities on the ground," which included everything from chloroform in the drinking water, to unpredictable electricity, to volatile regional politics that had the industry constantly uneasy about "Nigerian" unrest.

Ferguson (2006) has argued that this enclaved model, in which food and other industry needs are imported, has the benefit of "low overhead"; its efficiency is almost frictionless or deterritorialized insofar as it avoids local involvement altogether. On the contrary, as the finance manager above intimates, the complex processes necessary to materialize the disentanglements of this model are far more expensive and logistically involved than the alternatives. But the idea is that this strategy avoids a set of *potentialities*—illness, graft, involvement in political unrest—which, if they were to materialize, would be more expensive to contain than the enclave built at great expense to avoid them. Enclaves are, in this way, actuarial: calculating potential future risks and attempting to insure against them in the present. As one manager put it, "We have people out there working on wells that cost $30, $40, $50 million. If they have an upset stomach and they're not thinking about work, a $30 million problem turns into $100 million. So we try to ensure that everything works, the maid service, everything, because the repercussions, the knock-on effect, is magnified." For enclave administrators, the enclaves are about keeping the focus on the project at hand: drilling for oil. For them, the predictability, health, and control that the "self-contained" compounds offer allow workers to keep their minds on $30 million problems instead of upset stomachs or the frustration of a late local delivery.

Provisioning enclaves to ensure these outcomes is to work incessantly and at great expense toward building separations. Of the Smith facility, Paul explained:

> Opening a facility like this takes twelve months. [We had] teething problems: generators black out, the incinerator wouldn't work. Potable water wasn't really potable. We had to modify filtration systems, massive problems with AC. The camp was designed by an American company in Houston and this isn't Houston, this is Africa. It has its own relative humidity, its own dew point. But now this camp is completely self-contained. Electricity, water, sewage, [we] handle our own garbage, incinerate it all here. The bulk of our food comes from Houston. We ship it all in because it's cheaper and the quality is good, and it eliminated the need to deal with a highly inefficient local economy. We spend $150 million a year [in the local economy], but in a manner

that doesn't negatively impact efficiency offshore. [Our] social development program [is] 100 percent local content.

After a long year in which a Danish construction company, along with Spanish, Portuguese, Icelandic, and Greenlandic journeymen, plus eight hundred temporarily employed Bata residents, built the camp according to a Houston-based design, it is now, according to Paul, "completely self-contained." The provision of electricity, water, sewage, garbage disposal, and even food is systematized to serve those within the walls, and, in his telling, independent from those systems outside the walls. That this provision of resources relies on infrastructure outside the compound walls—roads, seaports, airports—does not, in Paul's telling, detract from the compound's self-sufficiency. The effect, instead, is to "eliminate the need to deal with a highly inefficient local economy." Facilitated by this idea of separation, the only relationship Paul sees with the outside world is the $150 million per year spent "in the local economy" on corporate social responsibility programs. Corporate social responsibility becomes the detached way in which oil companies can intervene, from the other side of the wall, redoubling the effect that they are somehow separate, but willing to "help" those on the "outside" (Shever 2012; Rajak 2011; Welker 2014).

Paul's idea that it is cheaper to import food from Houston (and Europe in some cases) is, at face value, shocking and seemingly untenable. Donald's wife Cheryl explained to me the logistics of their "food drops":

Fruits and vegetables and milk are flown in from Europe so that we're guaranteed that we get nutritious food and are able to eat well. Twice a year, frozen meats are brought in and put in a container. I can place an order for meat items that I might need. I do that once every two months based on how much I get so that I can cook and make my menus. I do go into town. I have not bought any of their fresh produce or milk. But I will purchase cereals and olive oils and tinned things, and the American or European products that I'm familiar with. Their interest on the compound is making sure that we are given food items that are cleaned and unspoiled so that we omit a lot of illness. That's another reason for the compound: continuous work without ill health issues.

The procurement procedures Cheryl described were in place for all three compounds, although practiced differently in each. In the Major compound, for instance, residents did not receive regular shipments of fresh produce, but ate mostly canned vegetables or local (Cameroonian) vegetables purchased by the intrepid few in town. This was a a source of tension and jealousy between the Endurance wives and the Major wives, the latter of whom were resentful of the Endurance wives' produce deliveries. Wives in the two compounds often traded various pantry items for imported produce items, a trade they oddly had to keep secret, as it was—in some arcane version of anticorruption practices—considered against company policy. Wives from the Major compound often used the tidbit that they didn't receive fresh vegetable shipments as proof that they weren't as spoiled as the Endurance wives and that they were "roughing it" slightly more.

The rationale for importing food was one of the questions I routinely asked my informants, not only those in procurement or management positions within oil companies, but also relevant Equatoguinean authorities and farmers as well, including one farmer whom I often accompanied on food drops to the compounds. Their answers displayed a remarkable consonance around the undesirability of importing food, but the impossibility of sourcing it locally. In an early conversation with Donald, I speculated that perhaps food import had to do with the industry being offshore. "It doesn't have anything to do with onshore or offshore industry" he responded:

> People want to know why we're not supporting gardens and farms, why we import food. [Agriculture is] not our business. We could set up a group that helped people to plant, but that takes energy from the organization to do things that aren't core. If we could get milk and juice and fresh vegetables locally we would. In Russia, the only thing we brought in from the outside were highly technical pieces of equipment that we couldn't buy locally.

Equatorial Guinea does not have industrialized agriculture, let alone a meat or dairy industry. Boats come daily from Kribi, Cameroon, into Malabo's port to stock the markets with produce staples, including tomatoes, onions, and peanuts. While companies have secured a handful of "local content" contracts with small farmers to bring in local produce, the quantity and regularity of the deliveries does not meet the US-standard needs of resident employees. Thus, as Donald explains it, in order to source local food, the industry would have to get involved in subsidizing and fomenting local agriculture—"setting up a group that helped people plant"—and they're not interested in doing that. Another manager, clearly aggravated by my question, responded: "There is this song and dance about local content. Where are the people that can do the services we need? We don't take food into Ecuador and Vietnam, or any other place we've ever worked. You purchase it locally. We bring it in here because it's not available." Local farmers readily admitted their inability to fulfill the scale of food needs the US industry presented, and lamented the absence of systematized agricultural support from the government. Equatoguinean state officials I spoke with agreed that importing food was the companies' only option. As one of them explained:

> If your country can produce those things, they will be bought from the local economy. Equatorial Guinea does not produce things consumed in EG. In Nigeria you buy everything [in Nigeria] to put food on the platform. When we get self-sufficient here in food and foodstuff, we will not need to go to Cameroon to buy tomatoes or potatoes. It's not because the platform is offshore. If the industry were onshore, you would still get it from Cameroon. That's what you eat here.

And yet of course, while Equatoguineans eat Cameroonian produce, oil industry personnel, by and large, do not. While importing from Cameroon would be cheaper, it would also entangle companies in ongoing border tensions and disputes, which routinely cut off supplies of fresh vegetables from Cameroon for days and, less frequently, weeks at a time. The use of duty-free imports and private ports ensures that materials shipped from the United States or Europe experience less friction, or at the very least, that the friction they might encounter can more easily be framed as technical supply chain disruptions, rather than entanglements in central African politics. It is to these corporate ring-fencing mechanisms of privileged tax and transport statuses, among others, that I now turn.

CORPORATE RING-FENCING

It was 9:50 a.m. on a Malabo Monday morning. I had been sitting in the office of the Secretary General of the Ministry of Mines, Industry, and Energy (MMIE) for fifty minutes, waiting for a 9:00 a.m. meeting to begin. In front of me, one secretary mouse-clicked at a computer game absentmindedly while another read the bible. In a room to the right, a third secretary clicked away at her mouse at the same rate, perhaps playing the same game. Their desks were empty, save the computers; three desks without a visible piece of paper, file folder, or pen. These women seemed to have nothing to do, at least according to my own understanding of office work. The Secretary General seemed to have no formal schedule (as I'm sure he didn't remember telling me to meet him at 9:00 a.m.), and these women certainly didn't seem to be keeping one for him. As people came in and out of the office, one of the secretaries would simply say that the Secretary General had not arrived. Beyond that, it was unclear what work, if any, these women were entrusted with. Here I was in the all-important Ministry of Mines, and the action at 10:00 a.m. on a Monday morning was three secretaries chatting in Fang and clicking away at computer games. The building itself, a ragged ten-story apartment building constructed by a Lebanese company, was essentially empty. As you climbed the stairs floor by floor, outdated concession maps dotted the walls here and there; but in general, it felt like a big, dark building (electricity was often out) with empty offices and the occasional piece of office furniture, often still wrapped in plastic. Although the ministry was soon to move into a brand-new glass skyscraper in Malabo II, the emptiness was not because the action had already moved. That's how the MMIE had been since I had first seen it two years before. In fact, the plastic plaques on the walls of each floor, listing whose offices were where, were an improvement over the water-stained paper printouts pinned on each landing when I first arrived in 2006.

The confusing emptiness of the MMIE offices was in sharp contrast to the more familiar office environs within the compounds of US oil companies: large buildings with open floor plans, crowded with cubicles and swivel office chairs; walls covered with graphs, charts, and posters, as well as company slogans and achievements; enormous white board calendars and schedules filled in with uncannily legible handwriting; office desks and shelving crowded with stacks of paper, books, labeled binders, file cabinets, computers, printers, and fax machines; the thrum of phones and photocopying; and the hustle and bustle to and fro between personalized work spaces, with photos of families, inspirational quotes, and sports team memorabilia.

And yet, in addition to the Presidency, the Ministry of Mines was considered to be Equatorial Guinea's most powerful and effective Ministry. Locals and foreigners alike often extolled its highly trained personnel and bureaucratic presence in industry processes, often in contrast to GEPetrol (the relatively new national oil company), whose mission was still ill-defined to the point of mystery. I share the contrasting office-life descriptions here to draw attention both to how different the working life of the industry *feels* on either side of the wall and to the kinds of work rhythms enabled by infrastructures like reliable electricity and running water, which could be found only within the US company compounds.

The infrastructures that enable constant electricity, wireless internet, running water, and Houston area codes are clearly not only for the domestic comfort of migrant employees and their spouses, but also enable the movement of oil and gas to market in more legibly business-oriented ways. Considered in their commercial capacity, these enclaves enclose separate business practices, ranging from the use of satellites that allow Houston phone numbers in offices; to differential laws, regulations, and taxation regimes; to variegated citizenship rights and responsibilities; and finally to the infrastructure itself—private, duty-free ports, electricity grids, and telecommunications systems. Justifications for these separate business practices stretch beyond arguments for efficiency and profit maximization. The companies also point to the ostensible ring fence within which they operate as spatial and procedural proof that they are separate from the "corruption" outside their walls. "In effect, a border is expected to be established between the oil industry, which now seeks to demonstrate that it is governed according to global standards, and the local economy and society, which lie outside these borders" (Barry 2006, 246). Corporate processes "within" the enclave cloak themselves in discursive and procedural regimes of the global, the standard, the compliant, and the objective, to be differentiated from the arbitrary, the personalistic, and the incomprehensibly local beyond their walls.

Political scientist William Reno (2001) has gone so far as to say that "the private enclave exploitation of resources is a salutary imposition of market discipline and standards of efficiency on corrupt economies. Foreign firms, especially larger ones, offer short-term prospects of filling in for missing state capacity" (4). To the contrary, I would suggest that the enclave is precisely a *performance* of market discipline and standards of efficiency, crafted through spatial, technical, and embodied differentiations that serve as semiotic proxies (Ho 2016) for "the market" and its ostensible standardizations. Like any felicitous performance, these practices create sociomaterial effects in the world; in this case, oil and gas from Equatorial Guinea reaching global markets with remarkable reliability. At the same time, however, the relationship between the enclave's segregated infrastructure and these effects is not causal, as the industry (or Reno) would suggest, but instead mediated by those thick, sticky entanglements with local life that the enclave claims to avoid. The enclave then is a procedural stage for market terms— efficiency, standardization, depersonalization—while the actual processes that get oil to market have everything to do with socialities, compromises, and often political force. One can hear this layering in a Smith's country manager's explanation of the exceptional status his company enjoyed in Equatorial Guinea:

> There's so much revenue generated. Corruption and inefficiency exist in
> spades in West Africa. The fact that we generate so much revenue, we have

direct contacts in [the Ministry of Mines, Industry and Energy] and tremendous influence. If there are difficulties—[given our] $700,000 per day [rig] rental—negative impact by customs, immigration, [we are] able to make a few phone calls and it gets cleared away.

This man intimates that zonal capitalism is made through deep and personalistic entanglements with that from which it claims to be separate. "We make a few phone calls and it gets cleared away." *We use our global, compliant standards to call our highly placed connections at the Ministry of Mines and ask them to please call the lowly customs official who is holding our needed technology at the port and tell him to snap out of it. This is how we differentiate ourselves from the "corruption that exists in spades in West Africa."* Thus, while Barry (2006) is right that "the formation of technological zones has become critical to the constitution of a distinction between global/Western political and economic forms, and their non-Western others" (250), the work of making that distinction cannot be characterized by separation alone. It must also include the onomatopoetic Spanish word that locals used most often to explain the relationship between the oil industry and local power structures: *compinchados*, or accomplices. As one Equatoguinean lawyer put it:

> Obiang gives the companies free rein, and in turn they protect his regime. [The companies] operate on the margins of local law, but it doesn't affect them. This theme of having their own telecommunications system, it guarantees that the government cannot interfere. This is on the margins of current legality and of the country's interests. They are commercial relations in which the industry closes its eyes to what is obviously illegal according to international law in order to do business with the regime. . . . The government has tacitly renounced control of the activities of these companies. As they renounce control, company activities damage the interests of the population. In environmental protection there is not a single control. The damage that they do—to the coast, the pollution—it reaches our beaches and the government doesn't have a single mechanism, there isn't the political will. . . . The fishermen feel abandoned, and the government does nothing.

This man narrates the deep ties and complicities between oil companies and those in power—from the state's granting of corporate sovereignty within the enclaves, to the companies turning a willful blind eye to blatant illegalities. And yet, the *effects* to which Barry and Reno point are pervasive—the performance of a border between the industry and the economy and society thought to be "outside" it (Barry 2006), and an idea of the imposition of market discipline (Reno 2001). The needed technology rapidly clears customs at the port, and the company can claim that they have sidestepped the corrupt outside yet again. Despite the sticky entanglements through which it is made, this infrastructural separation allows the consequential inhabitance of a partition, wherein companies are "imposing market discipline" from behind the walls, paragons of legal and economic liberalism. The separation is a spatial and procedural *stage* on which companies enact removal from and superiority to the legal, environmental, political, and financial situations in which they are causally and irrevocably implicated.

Another aspect of the ring-fence model—or the story told about it—is the extent to which these companies are "tightly integrated with the head offices of multinational corporations and metropolitan centers but sharply walled off from their own national societies" (Ferguson 2006, 203). While it is difficult to argue that Equatorial Guinea is the "national society" for any of the companies in question, their tight integration with head offices in Houston is unquestionable. Their phone systems, internet, and email are all connected and operated via satellite between Houston and New York. Every lobby and many offices of US firms in Equatorial Guinea have two clocks adjacent on the wall—one set to Houston time and the other to Malabo time. Daily working life for most employees who work in the offices is also Houston-centered. As one migrant IT manager explained, "A lot of the people in this building don't deal outside these walls. The nationals [Guineans] are 50/50. In many cases they just deal with Houston. Our logistics people—drivers, movement of materials—they deal a fair bit with the outside."

It is not only Houston that becomes a central reference. Rather, Equatorial Guinea becomes a node in a much wider oil-producing diaspora connected by supply chain logistics, mobile infrastructures, procurement procedures, and shared technologies (Barry 2006). In discussing his job, Smith's information technology manager explained that he delivered technology to "every company office—Indonesia, Africa, Russia—sixteen different countries." Smith's finance manager told me a similar story about her job. Having worked previously in Dubai and Jakarta, she explained:

> Smith's reporting procedures are standardized worldwide. It's not very different. There are set rules and policies we have to follow. Government reporting is slightly different place to place, but what I do—company accounts—are the same. And the more locations there are, the more things are the same in every single location. The company is really trying to standardize. So EG, in terms of reporting, is the same as any other location.

CHAPTER TWO

Houston-centric working life, including globally standardized IT systems and company financial reporting, indeed serve to ring-fence corporations. But again, I would suggest that the level on which this ring-fencing works is procedural, gaining its effects through the performativity of ritual—a performation (Mol 2002). In other words, that Smith's reporting procedures are standardized worldwide is itself the consequential and performative fact. That the contents of those reports, and even the methods through which they are completed, may vary widely does not dim the effect or felicity of apparent standardization, integral to the disentanglement of the ring-fence model.

The performance and invocation of standardized, "global" procedures work to smooth the lived, messy entanglements that characterize even the most mundane details of ring-fenced work. "We do get [preferential treatment] on our major stuff. [But] there are always other issues—registration of cars, residency permits, licenses—that you struggle through," as one country manager explained.

> We spend $850,000 per year on licenses and residency permits. One of the big issues starting to hamper our work [is the] process for visa attainment: requiring police checks for guys from the UK. [The Equatoguinean state now asks for] a certified letter from the local constable. You can't have outstanding warrants or a history of crime. [So now the] whole process takes longer, but even after the process you have people's visas denied with no explanation. So we just finished a two-week turnaround where we shut down the plant and wanted to bring in specialists. One specialty company that we wanted to bring in, none were given visas.

While nearly $1 million spent on registrations, permits, and working visas was a drop in the bucket to both the oil companies and the Equatoguinean government as a yearly expenditure or receipt, seemingly banal bureaucratic forms like car registrations, residency permits, and licenses are in fact analytically rich. Particular sticking points—visas in this case—reveal much about what worked smoothly in the enclaves, what didn't, and why. The visa issue this informant identified as emerging in 2008 is not simple red tape. Instead, in the wake of a failed 2004 coup attempt in Equatorial Guinea that implicated Mark Thatcher, among other geopolitical luminaries (Roberts 2006), the Equatoguinean government grew increasingly wary of who they let into the country. While Spaniards, as the ex-colonial power, had long struggled to acquire visas, most other European foreigners did not. However, the coup and its aftermath laid bare for the Equatoguinean state the long history of personnel exchange between international militaries, mercenaries, private security companies, and the extractive industries. A brief professional biography of Simon Mann, accused of being the coup plot leader, illustrates the point.

White South African Simon Mann had served in the Gulf War before he returned to work in the oil industry in Canada. In 1993, when the National Union for the Total Independence of Angola (UNITA) fighters closed oil installations in Angola, President dos Santos brought in Executive Outcomes, a private military security firm in South Africa (now defunct), who contracted Mann and others to fight. This circulation between private military contracting companies—soldiers for hire in wars both declared and undeclared—and the extractive industries is common. To sharpen the point, the defense Mann and others offered of their apparent coup attempt in Equatorial Guinea was to say that they were, in fact, flying to the Democratic Republic of Congo to provide security at a diamond mine (Roberts 2006). Thus, with Mann's extradition, trial, and jail time in Equatorial Guinea, and even before, the Equatoguinean government became increasingly savvy about the intimate ties between the extractive industries and the defense/security industries. Many of the foreign men I met in the field shared these biographies, including Paul (the white Zimbabwean discussed above), who was arrested and interrogated in the wake of the coup, and the head of security for Regal Energy, who was recently back from Iraq where he had served as J. Paul Bremer's bodyguard.[8] Once Simon Mann was released from prison in Eqautorial Guinea, Obiang contracted him as a defense and intelligence expert.

Another company logistics manager, quoted below, ran into similar, seemingly trivial entanglements that show the extent to which supply chains are, in fact, chains of or proxies for exchange translations along the rails of geopolitics (Tsing 2015).

> We have LOI [letter of invitation] issues. They're trying to get a technician here from Genoa, Italy, to work on the machine, but we don't get our LOIs nearly as fast as Chinese or Arabs. They present the red envelope. We are prohibited from doing that. [Our employees] know if they do that, they won't get reimbursed, because they're going officially. So the LOIs drag out. Literally one week later Endurance gets the LOI; [the expert] will be here tomorrow. During this whole period I estimate the lost revenues in the $15 million gross range for the government's take alone. I can guarantee you that Endurance has pulled out all stops, putting pressure on their people, but things don't move like that here.

In this man's story, we see significant revenue loss from seemingly banal visa and LOI issues that stem from the geopolitics of mercenaries, coups, and oil. This informant also brings up "the red envelope." He claims that the Chinese and the Arabs (referring to Arab Contractors, an Egyptian parastatal) bribe officials to get letters of invitation and other bureaucratic red tape taken care of more quickly. As we saw in chapter 1, US-based migrant informants who worked for large American companies claimed repeatedly that in the wake of the Foreign Corrupt Practices Act (FCPA), the Enron scandal, and Sarbanes-Oxley legislation, they simply could not offer bribes in that way, even if it previously had been standard practice in the industry. Nevertheless, major transnational Chinese and Egyptian companies—mostly in Equatorial Guinea for construction-, road-, and port-building projects—were widely known to operate under a different set of circumstances and guidelines.

Chinese and Egyptian companies—and indeed, French and Spanish ones—were also heavily invested in Equatorial Guinea's booming hydrocarbon economy. While in-depth ethnographic knowledge of these companies' practices was outside the scope of my research, word on the street and in the ministries—among *Equatoguineans*—was that these companies did business in different ways than the Americans, at least in terms of quotidian interactions. I do not doubt, for example, that many Chinese or Egyptian companies often paid a fee for an expedited letter of invitation. An Egyptian friend living in Equatorial Guinea and working for a major construction firm confirmed for me the regularity of these practices in his company. Nor do I doubt that at this level of *trivial proceduralism*, American companies tended *not* to present red envelopes. In this difference, I would like to suggest, it is not a coincidence that American companies were enclaved, while these other companies were not. The performative separation of the enclave and the refusal to give a red envelope for trivial matters were *enactments* of procedural disentanglement intended to frame out the profound implications of US firms in Equatoguinean life. Akin to needing three signatures to tie one's shoe, or to the regulations that prohibited wives from trading pantry items and produce, the scale and depth of US company entanglements in local life renders pitiful the refusal to pay a $5,000 bribe.

In the remainder of this chapter, I examine how the enclaves' most profound effects arguably emerge *not* from procedures of zonal capitalism traditionally conceived, but instead from the embodied experiences of seclusion, racialized social separation, and control that enclaves produce on both sides of the wall. Starting from the feminist argument that the domestic arrangements are not private matters, but political and economic affairs that act to sharpen or mute specific practices in the search for profit (Wynter 1982, 2003; Enloe 1990; Federici 1998; Yanagisako 2002; Stoler 2002; Bear et al. 2015; Hoang 2015), I offer a brief discussion of lived separation before closing the chapter with an exploration of the ways in which "the wives'" presence shaped company involvement with, separation from, and boundary-making between their compounds and the outside.

ON LIVED SEPARATION

Back in the Ministry of Mines. At 10:20 a.m., a Chinese businessman walked in with his Equatoguinean liaison. The Guinean spoke in Fang to the secretary, saying that they had an appointment. He then spoke in fluent Mandarin, I assume telling the Chinese man that the Secretary General wasn't yet in the office. They sat down next to me to wait, and we started talking. The Equatoguinean man had studied in China for five years, and he read, wrote, and spoke fluent Mandarin. The Chinese businessman gave me his card from the China National Offshore Oil Corporation (CNOOC), and I gave him mine, from Stanford University. Speaking English and Spanish, he explained the Equatoguinean oil industry as high risk but high reward, especially given that prices were so high. We talked about Stanford, and how he would like to study psychology and philosophy. We talked about the Olympic opening ceremonies that had just taken place in Beijing. As we chatted, we both overheard the secretary telling another visitor that the Secretary General would travel to Nigeria that day at 11 or 12; maybe he wouldn't come into the office at all. But at 10:41, we got word that he was on his way.

After meeting briefly with the CNOOC representative and his Guinean liaison, the Secretary General welcomed me into his office. Among other questions, I asked, "What does the average Equatoguinean know about the oil industry?" And "What does the rotator from Texas know about Equatorial Guinea?" I quote his response at some length:

> This question, it could be a theme of professional debate. Unfortunately, I am not a sociologist. We have lamented precisely this situation many times. At the cultural level we (Equatoguineans) are hospitable, peaceful people who know how to integrate among ourselves and with others. It's possible that this Guinean character trait—the ability to integrate—has not been able to be noticed from the cultural point of view, perhaps because we have such a small population. We can look at the typical example—the Major Corporation—in the same city but fortified, and no one leaves. Despite the fact that many jog

in the streets, they don't stop. . . . I would say that a form of business discipline has been imposed, one that requires specific behaviors for those who come to work in the petroleum industry in Equatorial Guinea: singular dedication to work. I refer to the exclusive attention to the [work-related] tasks necessary to achieve success in a relatively unknown environment. Nevertheless, there is no doubt that this discipline imposed by the norms of the company collides with the social values of man, the sociability of man. [This discipline is] creating a space of isolation, of exemption. The people exempt themselves . . . and why not? Those who come learn from those who are already here. There has to be much more human interaction to avoid differences, a consolidation of interpersonal exchange. These people need to open themselves. It could improve the relationships.

The Secretary General's explanation of the enclaved spaces was compellingly resonant with those given by expatriate company managers. The managers suggested that the enclaves—with imported food and stable infrastructure—were necessary to ensure that the work got done, that $10 million wells weren't made into $30 million problems by upset stomachs. The Secretary General states it more plainly: the foreign men who work in Equatorial Guinea have been told to come and work, not to stop and chat with Guineans while jogging. He suggests that companies have decided that this exclusive dedication to work is what will ensure "success in a relatively unknown environment." This success comes at a price, however, as dedication to work forecloses other forms of sociality, and the foreigners isolate themselves from the peaceful, accommodating Guineans.

Responding to a similar question, Mauricio—an Equatoguinean functionary in the Ministry of Finance who had also worked for the Major Corporation for years before transitioning into government—echoed the Secretary General's ideas of isolation and partition, stating of the migrant workers that they don't "pick up social views" very well:

They don't live in Malabo. They live in Houston. In [the Major Corporation] we even used to have rules in terms of when you go out and when you come back home [to the compound]. You can't keep company in compound, guys or girls . . . You're not going to pick up good social views, realistic social views, from the Major Compound or the Endurance Compound. They're never outside of that compound unless they're on the bus to the airport.

While the amount of time that expatriate employees spent outside of their compounds for personal/recreational purposes varied, there was definitely a common refrain among US and European single male rotators I talked to: "I rarely if ever go out in public. From the rig to the office. I've rotated into Equatorial Guinea for six years and I've probably spent thirty nights in Malabo. I've maybe gone out for dinner twelve times. I've not taken advantage of opportunities to mingle with the culture. One of my [Guinean employees] says he's disappointed I've never come to his house for dinner." Or another:

> I live here like I did in Angola. From the airport to the rig. In Angola there was a civil war. Some people like going into town. I want to eat my dinner here and go to sleep. In two years in Bata, I've been out to dinner one time. In Aberdeen [we were] part of the culture. In China, it's not a camp situation. I spent time in the city. But I didn't go out a lot. But that's just me. There have been those that partake of the nightlife in Bata, which is discouraged. [There was] one guy who couldn't ever make it back to camp the same night, and they moved him offshore.

In this man's account, to "partake of the nightlife" is a euphemism for engaging with sex workers. This man talks about a "type" of man he defines himself against, a man who—whether in China, Equatorial Guinea, or Angola—likes to "go out a lot." Certainly, there are many men in the oil industry in Equatorial Guinea who meet Guinean women and offer them money, meals in restaurants, and sometimes even more stable arrangements of housing and a living allowance in exchange for companionship and sex. In an effort to constrain this behavior, the large compounds at issue in this chapter all have curfews, ostensibly for security and to ensure a well-rested worker in the morning; but certainly this is also a way to control alcohol consumption and sex, two of the main attractions outside compound walls. Who can and cannot bring wives to the compounds becomes more salient here. The apparent morality and dignity of white upper management is propped up by their domestic arrangement that either keeps them home at night or encourages them to socialize with other married white couples in luxurious housing. As Povinelli (2006) writes: "The intimate couple is a key transfer point between, on the one hand, liberal imaginaries of contractual economics, politics, and sociality, and, the other, liberal forms of power in the contemporary world. Love, as an intimate event, secures the self-evident good of social institutions, social distributions of life and death, and social responsibilities for these institutions and distributions. If you want to locate the hegemonic home of liberal logics and aspirations, look to love in settler colonies" (17). By contrast, nonmanagement migrant workers—the vast majority of whom are men and not white—away from home, alone, for up to a year at a time, are more likely to leave their claustrophobic compound housing after work. (Remember from the earlier descriptions that they have no kitchens or living room areas, and live in dorm-style or trailer-wide facilities.) Given that many of these workers are also from the global South, Equatorial Guinea "beyond the walls" is far less exotic or threatening to them, and many enjoy the opportunity to frequent bars and nightclubs beyond the gaze of their bosses, whether or not they engage with sex workers or long-term companions. Notice too that for men who can't make it back to camp before the curfew, an offshore posting is a punishment, further racializing questions of work and sexual morality.

The enclave and its outside are intensely gendered spaces in which sexuality and kinship—who was frequenting sex workers; who could or could not have wives with them—became entangled in labor control and hierarchies of privilege. The wives, however, inhabited these regulations and hierarchies differently, and segregation both outside and inside the walls took on different meanings and potentialities for them than it did for their husbands and other male workers.

THE WIVES

The wives of migrant managers were my easiest informal entry into the industry early in my fieldwork. Many pitied me: "You live in the city? Don't you want to take a shower?" They invited me to cards on Wednesdays, mahjong on Tuesdays, and stitching on Thursdays. They also invited me to present a seminar on "Women in Africa," and later to offer a workshop on understanding opera as part of their ongoing series of "master classes" presented by women for women living in the compound. I received invitations as a friend, translator, or mere tagalong to their charity drives and to outings in the city or around the island. They invited me to potlucks to say goodbye to various women—one leaving for a new post with her husband in Kazakhstan, another retiring with her husband to a lake house in Wisconsin, and another heading back to Texas. When I left the field, they threw me a potluck of my own.

I played Continental with them every Wednesday afternoon. Chat over the game would turn to predictable questions. "What did you do before?" At my table of card players on one afternoon, there was one woman who had been a tennis pro, one who had been a banker, and another who had a degree in recreation and had led a YWCA. Another woman's father had died when she was young, and she was left to take care of her mother. She explained softly that she didn't get the chance to earn a college degree, and she was clearly self-conscious about losing at cards. Other women I came to know had degrees in law enforcement and police administration, library science, nursing, education, sociology and psychology, and fashion and interior design. Others had a high school education. Many were gifted artists; one was a jeweler. All of them were in Equatorial Guinea to support their husbands in what was widely acknowledged to be an egregiously work-intensive environment. Reminiscing about her own career, Scarlett explained, "Oh, I've worked in retail; as a secretary; been a mom; wife and professional volunteer. I never got to pursue my dreams. In my late thirties and early forties, I grieved the loss of that. I've had a wonderful life. We've been corporate nomads." Sarah, who had worked as a nurse, explained, "I can get a job anywhere. [My husband] does appreciate that I could be at home working, with an identity more or less. I gave that up, and he never takes it for granted."

We would eat Doritos and M&Ms, salads and sandwiches, meringue and chocolate cake. If one wife ran low on something, she would trade with another, Crystal Light for coffee beans. As various wives prepared to leave Equatorial Guinea permanently, each would sell off the contents of her pantry: Stove Top stuffing, Campbell's soup, Grape-Nuts, marshmallows, graham crackers, Uncle Ben's rice, hot chocolate mix, canned beans and beets, and chili. I bought a box full of groceries at one sale, and the wife who was leaving quoted me a price in US dollars as "the same amount that I paid," to assure me she was not making a profit. Because all three companies forbid expatriate workers from bringing children to Equatorial Guinea (for reasons that ranged from malaria, to lack of adequate schooling, to "security concerns"), the women I met were mostly in their fifties and early sixties, dearly missing children in their twenties and beyond at home. They were from New York City; Savannah, Georgia; Ipswich and London, England; Ogden, Utah; Conroe and Georgetown and Houston, Texas; rural Louisiana; and the Midwest. There was one Thai woman married to an American, one Brazilian married to an Argentine, and one white South African woman who had moved to Equatorial Guinea from Bonnie Island, Nigeria, where she had been posted with her husband in the industry and opened a chicken chain. Nearly all of them had lived in Texas with the industry, and among them they had also been posted in Angola, Argentina, Ecuador, England, Gabon, Indonesia, Ireland, Japan, Nigeria, Russia, Saudi Arabia, and throughout the American Gulf Coast.

They were accustomed to expatriate life and its particular rhythms, its comings and goings.[9] The woman leaving for Kazakhstan went for a "look-see," to find out what she would need there and what types of clothes she should bring. We spoke after she returned to Equatorial Guinea to pack up her house for good and, referring to an earlier conversation we'd had, she said, "You know, it's just like we were talking about. There's a whole subculture there ready to show me where to change money, wives calling me ready to pick me up." Despite all the moving, the constant new faces and places, there was a patchy transnational continuity to migrant management life. The "wives" in any given place had most often lived in many other places and were accustomed not only to the transience of this existence, but also to its peculiar durabilities and continuities, including already-established communities of migrant wives. Chief among these peculiar durabilities are postcolonial meanings attached to whiteness, and the political potency of the heteronormative and deeply gendered nuclear family structure. "The Expat Wives' Prayer" (figure 2.6) is a poignant and troubling artifact of these continuities, given to me by one of the migrant women with whom I worked in Equatorial Guinea. The poem is a satirical reinterpretation of the Lord's Prayer that circulates broadly in expatriate communities the world over, on the internet, and in Equatorial Guinea, although the woman who gave me this copy was clearly ashamed to do so.

The poem is saturated with gendered, raced, and classed relationships, implicitly laid across the global north and south. White property—excess baggage, houses, treasures, duty free; relations of white dominance and nonwhite subordination—cooks, maids, drivers, gardeners; and a collective whiteness seeking "divine guidance in *our* way of doing things" (emphasis, tellingly, in original) are all arrayed against "the natives" who the Lord is called upon to help "love us . . . for what we are, and not for what we appear to be worth." In thinking through this poem, first I want to recognize white supremacy by Ansley's (1989) definition: "a political, economic, and cultural system in which whites overwhelmingly control power and material resources, conscious and unconscious ideas of white superiority are widespread, and relations of white dominance and nonwhite subordination are daily reenacted across a broad array of institutions and social settings" (1024). Second, I want to recognize the centrality of racialized sexuality and propriety to the place of white womanhood in the long history of global capitalist projects (Hoang 2015; Stoler 2010): "Keep our husbands from comparing us to foreign women. Save them from making fools of themselves in nightclubs and please, Lord, do not forgive them their trespasses for they

The Expat Wives' Prayer

Heavenly Father,
look down on us, your blessed and humble expat wives
following our loved ones through their working lives,
travelling this earth to lands unknown.
We beseech You, O Lord,
to see that our plane is not hijacked,
our luggage is not lost or pillaged,
and our over-weight baggage goes unnoticed.
Give us divine guidance
in our selection of houses, cooks, maids, drivers and gardeners.
We pray that the telephone works, the roof doesn't leak,
the power cuts are few, and the rats and cockroaches fewer.
Lead us not into temptation but deliver us from weevils.
Save us this day our daily dread ~ of traffic jams.
Lord, please lead us to good, inexpensive restaurants
where wine is included in the meal ~ not dysentery.
Have mercy on us, Lord, if it be the latter:
make us fleet of foot to make it on time,
and strong of knee in case we have to squat.
Give us wisdom
to tip in currencies not yet understood.
Help the natives love us Lord, for what we are,
and not for what we appear to be worth.
Grant us the strength to smile at our maids
over shrunken laundry and broken treasures,
remembering our own mistakes in menial matters.
Give us divine patience when we again explain our way of doing things.
Almighty Father,
keep our husbands from comparing us to foreign women.
Save them from making fools of themselves in night clubs
and please, Lord,
do not forgive them their trespasses
for they know exactly what they do.
Forgive us our expensive treats at Duty Free, for our flesh is weak.
Dear God, protect us from bargains we do not need or can't afford.

And lastly Lord,
when our expat years are over, grant us the favor
of finding friends who will look at our photographs
and listen to our stories
so our lives as Expat Wives
will not have been in vain.

Amen

Figure 2.6. The Expat Wives' Prayer.

know exactly what they do." The "natives" are in one moment the expat wives' cooks and maids, and in the next, "foreign women" against whom they compare unfavorably, and with whom their husbands should not be forgiven for having extramarital sex.

In the penultimate line of the poem, the author capitalizes "Expat Wives" for the first time, as she hopes for friends at the end of her nomadic life with whom to share her many memories, photos, and stories "so our lives as Expat Wives will not have been in vain." This line suggests that wedded domesticity *itself*—providing emotional and housework support to a husband through his corporate journeys—is not enough to redeem the hardships and sacrifices of this kind of life for women. In the sadness of this line, in the sentiment of lives lived in vain, one can hear Scarlett again: "I never got to pursue my dreams. In my late thirties and early forties I grieved the loss of that." In other words, the de facto white supremacy and the class mobility that can come with these postings—women, many with a high school education, living in luxurious houses with maids and gardeners on a 75 percent uplift salary—create a profoundly ambivalent lived experience, full of contradiction, guilt, and grief. Fear, physical and emotional vulnerability, and conflicted relationships to material possessions and "native" peoples braid with the poem's invocations of quotidian white supremacy. These affective ambivalences saturated women's daily lives in the compound, as the end of this chapter will detail, and they betray how these women wrestle with the availability of preexisting systems of raced and gendered inequality to capitalist projects.

The intimately unequal domesticities of race, class, and gender expressed in the poem, and the repetition of domestic segregation in oil compounds around the world, show us *how* the US-based transnational oil and gas industry draws on preexisting systems of raced and gendered inequality, not simply on norms and technologies of liberal capitalism. Or, rather, raced and gendered inequality *are* norms and technologies of liberal capitalism. From subcontracting and body shops that pay workers according to their nation of origin, to the flags of convenience that allow offshore rigs to sail under the Liberian flag, to the acceptability of residential segregation, many of the quotidian practices that constitute the licit life of capitalism depend on the reliability of already existing inequalities. "No firm has to personally invent patriarchy, colonialism, war, racism, or imprisonment, yet each of these is privileged in supply chain labor mobilization" (Tsing 2009, 151). Consider Nigel Thrift's (2005) assertion that "capitalism can be performative only because of the many means of producing stable repetition which are now available to it and constitute its routine base" (3). While Thrift is nodding toward infrastructure here, the stability of race and gender iniquities—*in all their lived ambivalence*—is also more and less consistently available to capitalist projects and, thus, constitute their routine base (Robinson 1983; Yanagisako 2002; Hoang 2015; Hudson 2017a). However, the affective experience of those iniquities—even for those whose intersectional identities place them at the apex of race and class, if not gender, hierarchies—is deeply troubled. Among the wives, "living in vain" came up often.[10]

In every interview and questionnaire, I asked the women what they would change about their lives in Equatorial Guinea. Nearly without exception, they wished to know more about the country, understand more, learn to speak Spanish. "I would like to have the opportunity to get to know Guinean people. It's as if you come here, you pass through the place, and you don't learn anything. It's like living in vain. Rather than losing this time, I would like to have more Guinean friends, more exchange." As was the case in the poem, here again wedded domesticity is not enough to redeem the sacrifices of gendered migrant life. Where the poem's narrator yearned for friends with whom to share photographs and stories once her world travel ended, the majority of the migrant wives I interviewed and came to know well in Equatorial Guinea yearned for a more fulfilling life *outside* the compound, *outside* the walled boundaries of conjugal conscription.

Philanthropic projects of various types offered the most readily available and acceptable method to access this outside. Indeed, the phrase "men are out for prostitutes and bars, women are out for charity" was a widely circulated platitude in the compounds. While no wife had been in Equatorial Guinea longer than three years at a stretch, there were charity projects that had been going constantly since 2000. New wives coming in picked up where others had left off. As Chris left for Kazakhstan, she could rest assured that her art and English classes, her program with Jet Air and the local Catholic nuns, and her milk program would continue. And when she arrived in the North Sea, she would pick up the projects of others. While their husbands were up to their eyeballs in work, and enjoyed an open bar, deluxe gym facilities, and American cable TV in every home at the end of the day to unwind, the women were bored for the most part and filled their time with all kinds of activities both within and outside the compound. They were often more involved in and knowledgeable about life outside the walls than many of their husbands.

Many of the women living in the Major and Endurance compounds were involved in charitable, philanthropic, and educational projects, which were officially separate from the corporate social responsibility programs of the companies, the largest of which were subcontracted to multinational development contractors. Through a sympathetic airline employee at London's Gatwick Airport, the wives brought in large containers of materials—school supplies, toiletries, toys, clothes—and distributed them in Bioko's small, mostly Bubi villages of Moeri, Basacato, Riaba, and Baney, often working through nuns in churches and schools. They hosted game days for local children, and also taught art, English, and stitching classes, occasionally hosted in their homes on the compound, although more often at local schools. They helped to run a summer school for teenagers in Ela Nguema that focused on hotel management, cooking, and welding classes. They also hosted fundraisers and donated the proceeds to local school building projects or for the purchase of a generator.

Yet, for all this activity, nearly without exception, from the most fervent participants in charitable work to those who chose only to give money occasionally, if at all, the women had profoundly ambivalent feelings about their presence in and relationship to Equatoguinean communities. As one woman told me:

> I will donate things, but I don't want to be center stage. To be perfectly honest, I feel like it's a dog and pony show where you go and pass out one pencil. The pencil-passing embarrasses the hell out of me. I'd rather help them paint a house instead of coming all dressed up like I just had a shower and hand out stuff. I don't know where that comes from. I don't want to feel like the great white savior. For me, I'd rather be in the background. I don't think you can save the world. I think you work one person at a time. And I hate when the camera comes out. Oh lord! To be perfectly honest, sometimes I'm embarrassed. We have so much. I'd rather go into a school on a regular basis and help. Give something of myself, maybe? I've volunteered in one of the schools in Nyu Bili. I volunteered in the [company's] computer lab with trainees teaching computer-based math and science programs. I really enjoyed that.

Madalena, the Brazilian woman living in the Endurance compound, had equally ambivalent feelings that she openly narrated in terms of colonialism and race, which set her descriptions apart. When first coming to Equatorial Guinea she had volunteered in a local clinic as a nurse, but she left due to the resentment she felt from the Equatoguinean nurses, which she attributed to their "racist" attitudes toward white people and to a perceived threat that she would take over their jobs. After quitting in frustration, she preferred to stay at home in the compound. As she described it:

> The [charity] projects? No. I'll donate money, but orphanages? No. I'll donate, but not go. "Here come the white people who take our petroleum money and come to give us pencils!" It's like colonialism—they take all the land and then give a gun or a calculator. I feel shame that I don't participate, but I would rather be in my house. What I like about living here is the comfort and economic things. I didn't come here to do volunteer work. We came to make money.

When I asked Madalena if she enjoyed leaving the compound at all, she replied: "Oh, I love it! We go to friends' houses. But to do as the Americans do, to go here and there buying crafts? I don't like that. I think they see it as exotic, to leave the compound and all. I don't go because it stinks." As she waved her hand in front of her face, she lowered her voice and leaned in, indicating that she was worried her Equatoguinean maid would hear her (as we were conducting the interview in a mix of Spanish and Portuguese). "The times that I have gone into town, I felt ill because of the heat and the smell of trash. I like crafts, but because of dust and trash I'll spend the whole day ill. And, because of social class distinctions in Brazil, this isn't exotic to me. Poverty doesn't choke me up. But what does choke me up is that the people here seem hostile. 'What are you doing here, whitey?' And they have the right to feel this way, because of the Spanish."

Again, we come to the affective ambivalences of these women's lives. They hoped that "more exchange" with Equatoguinean people would give meaning to their migrant existence, perhaps even afford more Guinean friends. But in practice, charity work outside the compounds produced a confused grappling with the contradictions of Equatorial Guinea's petro-project, which brought vast wealth to a few (including them) and dispossession, environmental degradation, and retrenched political repression to many. While these are not the terms in which the wives articulated their struggles, their embarrassment about pencil passing and invocations of the "white savior," and, clearly, Madalena's historical analysis of the continuities with colonialism, all show that they grappled in their own embodied ways with the contradictions of postcolonial racial capitalism. That philanthropy and volunteer work, in particular, generated so much confusion is revelatory. Enclaved life spatialized politics and interaction, so that philanthropy became, on the one hand, among the only accepted and permitted means of "reaching across the wall." But, on the other hand, it also became a series of ruptural experiences in which radical inequality, race, and colonial histories blipped into uncomfortable, embodied presence.

In my conversations and interviews with wives, and in questionnaires I distributed and they completed on their own, their yearning for more connection with "the outside"—Equatorial Guinea and Equatoguineans—was *not* a yearning to do more charity projects. Rather, their stated yearning was most often to learn Spanish, to make Guinean friends, and by extension, to be able to circulate more freely outside the compound. Women living in the Endurance compound were forbidden from driving company or private cars outside the enclave, and women associated with all three major companies were forbidden from taking public transportation, including taxis, which greatly limited their range of motion and meant that when they did leave the compounds, they often did so in large groups on company buses.[11] As one woman explained to me:

> We can't have cars—we have to be on buses as a spectacle. Everyone's staring. It looks weird. We go in as a pile of people, and it's embarrassing. You can't be a good ambassador for your country. . . . I don't know if a lot of the women are self-conscious, but I know I am. They're loud. . . . You're completely associated with whoever you're with, but you don't sometimes approve of the way they're acting. It's a tour group, and you're grouped in.

Women's movements were tightly controlled, and when they were able to move beyond compound walls, it was often as a conspicuous and, to many, embarrassing white group. Without language skills, or a physical presence beyond the walls that was able to transcend spectacle, the women's desires for more personal forms of connection remained thwarted. But note too, in this woman's account of her embarrassment at the spectacle of a bus full of white women, the tensions she experiences are tensions *between* the women on the bus. It is other women who embarrass her (*they're loud*), and it is differential privileges between women affiliated with different companies (*we can't have cars*) that frustrate her.

For migrant wives living in Equatorial Guinea, homosociality *within* the compounds, *within* the apparently homogenous whiteness of management wives, was deeply crosscut by multiple forms of difference, inequality, and prejudice.[12] These included, but were not limited to, nationality (Americans are loud, Brits are uptight); class and social capital (who had education, who was rural); differences between company policies and hierarchies within the companies themselves (who was married to whom, whose husband was whose husband's boss). In other words, on the whole, the women generally felt that there was *nowhere* comfortable to turn. Wedded domesticity in Equatorial Guinea was insufficient; reaching outside the compound via charity projects was fraught; and their relationships to one another were riven with tension and judgment. In this feeling of having nowhere to affectively rest, the compounds' luxurious infrastructure and the wildly inflated salaries paid to their husbands would often become, somewhat ironically, the *least problematic* psychic space. Remember the sentiments expressed by Madalena who, although she had originally volunteered as a nurse, renounced that start completely: "What I like about living here is the comfort and economic things. I didn't come here to do volunteer work. We came to make money." But this too was an ambivalent position.

For migrant wives in Equatorial Guinea, this ambivalence expressed itself as love of luxurious infrastructure and comfort on the one hand, and resentment and feeling caged by that same infrastructure on the other. They were thankful for the security and comfort the compound provided. Many explained that had there not been a compound, they would have been unwilling to come to Equatorial Guinea. Many wives also readily admitted that the comfort of the compounds bled into luxury; for example: "We're totally spoiled. Fabulous food, it's a life of leisure. We're taken care of with housekeepers and gardeners. It's a quality of life similar to the US, except that you socialize more because there's nothing else to do." For many women, this luxury extended to what they *were* able to do as they found ways to fill their time. One woman offered a particularly whimsical assortment of personal activities that kept her busy on the compound: "keeping fit, oil painting, printmaking, playing flute and small pipes, walking the dog, knitting and sewing, learning French, reading French novels, writing, different cultures, people watching, day dreaming, keeping in touch with friends and family, acupuncture, socializing." And yet, these feelings of security and comfort, luxury, and expansive free time were also bound up with guilt and the feeling of being judged, as one woman expressed:

> I love the greenery all around us. I love the fantastic houses and facilities. More than anything, I love the quiet. I feel protected. You feel like you're in the United States, and that's what the enclaves are for, so that the expatriates can feel like they're at home. But then when I talk to expats who live in Malabo, they say, "Of course you like living here! You live in Disney World!" They are jealous. It's as if we have to feel guilty for living well. You have to excuse yourself for living well here.

CHAPTER TWO

And another:

> When I moved in, the woman who lived in this house before me said, "This is
> the best prison you'll ever live in." It's almost embarrassingly nice. When you
> go into town and you see the abject poverty and then you come back here. . . .
> I mean I don't want to live in a dump, but it doesn't have to be over the top.

In our conversations and interviews, and in responses to survey questions,
the wives repeatedly described their experiences in the compounds using
metaphors of prisons, fishbowls, and microscopes. Interestingly, however,
they attributed their confinement and agitation *not* primarily to a threatening outside, but to the internal social divisions that I described above—
between Americans and Brits; between those with and without college educations; and between those whose husbands occupied different levels of the
corporate hierarchy. As Suzanne explained, "It's simple. Interacting with local people is rewarding. Interacting with other wives is frustrating." Or as
another wife put it perceptibly to me, "*Everyone* feels as if she is on the outside." Below I offer a sampling of these sentiments, culled from what could
have been an entire book about isolation and social distinction among
women living in the compounds:

> It can be a lonely place. Everyone's in their homes, and you don't want to
> disturb them. You feel eyes are watching you. When you go out, you feel
> others are watching you. When you're sitting by the pool, you feel people
> are watching you. "Oh, I saw you walking yesterday. I saw you playing ten-
> nis. Who were you playing with?" I'm guilty of this as well. I'll say, "I really
> would've liked to play tennis. Why didn't you call me?" A lot of ladies feel
> left out. I feel left out.

> You're under a microscope. It's an artificial life. If I lived back in Houston,
> I would not see the wives on a daily basis or on a frequent basis. Seeing the
> same people all the time is a little . . . I like 'em but that wouldn't be the way
> it would be if we were in the real world.

> There is a prominent difference in our compound between Brits and Ameri-
> cans. People go toward who they feel comfortable with, more so now than
> before because there are many Americans and fewer Brits. It gets very loud,
> so noisy. We Brits are less vocal. We're quieter as a nation; we don't express
> ourselves.

> There is a cultural difference. The rift is about the ones who have not had
> college education. For me to mix with those ladies, they're not sure of me.

People are paranoid, on the verge of hysteria. We're not one big happy family essentially. Human nature, health issues, menopause, all those things are within the equation.

Finally, regarding another common complaint about internal divisions in the compounds, a woman explained that "there are women whose husbands are managers and they feel more important. I have lived this in many countries." And, from the other side of that sentiment, a production manager's wife (the highest possible in-country position) talked to me about her peculiar position as "the pastor's wife":

I enjoy the infrastructure: water, electricity, telephone, internet, air conditioning. I never have to be lonely. But when you want privacy, it can be suffocating. There are women in the compound who like to know what everyone is doing. And there is some prestige attached to being the production manager's wife. People manipulate that relationship. So I don't like the lack of privacy, or people making assumptions, or having my behavior monitored. There's pressure to be the pastor's wife.

The wives also shared ways in which their migrant experiences in Equatorial Guinea and elsewhere had changed them. In particular, many articulated various moments of personal and intersubjective insight, whether about "culture" broadly conceived, race, or personal ignorance; for example: "The best thing that happens is the widening of my eyes, the re-seeing. It's the softening toward other people and other cultures." And another stated, "I was never 'American' until I got to Gabon. I was also never 'white.'" Still another woman explained her opinions on the "awful things" Chinese oil companies were doing to the local environment. "They're educated, don't get me wrong," she said. "But I don't think they care about the people and the environment like we [Americans] do." After that blunt judgment of an entire nation, she paused, and then chuckled, clearly at herself. "Maybe that's my arrogance coming out. Now you'll see how ignorant I was: When I moved to [my first overseas posting in] England, I thought I was moving to a third world country. Growing up in the US, I wasn't that interested in learning about the world." This reckoning with personal ignorance, and with the hermetic geographies of quotidian American exceptionalism, was also a common theme. Here is Cheryl, Donald's wife, for whom a posting in Equatorial Guinea came after long postings in Russia and Japan:

I told myself years ago I would never live in Africa. I would never come here because of the bugs and the diseases and all that. Never. EG would pop up every now and then in our previous years. I would think, "Oh forever more, no." I am never going there. When [my husband] came home I said *no*, I am not going there. Knowing that there was a Western-style compound, once I started thinking about it and had to make a real decision and think and choose, it was, OK, I can do that. And then it was, I'll have nothing to do. I'll be on a compound. But I got over that too. But then I thought I'll collect all my projects and come. I planned to keep myself happy, and I'll figure it out once I get there. I've come to Africa having never wanted to come here. Africa is malaria. Africa is bugs. Africa is those things that get in your clothes and eat into your skin, and National Geographic pictures, and things you read when you were a child. I just never wanted to go there. But I had the same reaction when he said Russia. Immediately, when he said Russia, I said no. I can't go there. We'll have to stand in bread lines. It's what you've learned previously that's not true in real life.

And finally, during negotiations with a woman to see if she would let me read a collection of detailed letters she had written home to family and friends in England, she acquiesced on one condition: "Provided that you understand that my views are not fixed and will change over time, and that I am striving to understand relationships and cultures here and how that reflects my own culture and beliefs." Whether or not this was a caveat strategically intended to please the anthropologist, it is important to me to take this provision seriously. In sharing their guilt, ambivalence, and frustrations with me, the women presented themselves as whole persons—neither corporate nor domestic automatons—feeling their way through the deep compromises and multiple hypocrisies that racial and gendered capitalism entails.

The women in their "golden cages" perched atop Equatorial Guinea's historic cacao plantations felt lonely, confined, and stuck in various forms of social tension and distance, not only with the Equatoguineans beyond the walls, but even more so between one another. Designed to partition, separate, and disentangle the oil industry from the place in which it happened to be operating, the enclaves also seemed to intensify the social partitions that divided these women one from the other by class, nationality, and education, leaving them resigned neither to humanitarian concerns nor lifelong friendships, but, as Madalena put it, to making money and getting out.

CONCLUSION

Solon T. Kimball was an anthropologist at Columbia Teacher's College. In 1955, the Arab American Oil Company (now Saudi ARAMCO) invited him to participate in a summer institute for teachers in Saudi Arabia, after which, unbidden by the company, he wrote a paper on his impressions entitled "Section of Anthropology: American Culture in Saudi Arabia" (Kimball 1956). Quoting from Kimball's paper, Vitalis (2007) retells his story of 1950s Saudi Arabia in interlaced infrastructural, emotional, and racial terms eerily similar to those I have explored in this chapter:

> Many of the men and women [Kimball] met appeared troubled. "Although . . . the phrase, 'we've never had it so good,' is used often, nevertheless there are deep currents of disquiet and frustration, and perhaps even more serious personal consequences." . . . "Even the lowliest" American worker had his or her sense of superiority reinforced through the spatial and racial organization of the camps. "The American position of pre-eminence is reflected in symbol and fact. It is also a position that imposes a high degree of cultural isolation . . . and contributes to an omnipresent sense of precariousness." When [Kimball] probed, even those who emphasized ARAMCO's contribution to Saudi development and the Americans' "sense of mission were prone to lapses of pessimism." Many if not most would say, "I'm here for the same reason as everyone else—the money." (254)

The migrant wives with whom I spent much of my time were sunk in this deep ambivalence, hopeful in one moment that for Equatoguineans, oil would be the salutary imposition of market rationality and benevolent development, yet defensive in the next moment: "Let's face it, we're not do-gooders. We're here to get oil out of the ground. I didn't come here to do volunteer work. We came to make money." Frustrated by the complexity of other desires, the spectacular profit to be had by collecting a 75 percent uplift salary, with all expenses paid, in the hardship post of Equatorial Guinea ironically became the least problematic justification for compound life. And, indeed, if one could characterize migrant management work in Equatorial Guinea by a unifying trope, it might be class mobility. One woman confided that she and her husband had $5 million cash in hand to buy an expansive ranch property in Texas; another worked diligently via Skype in her free time, monitoring the building of her dream farmhouse in the French countryside.

I have proposed in this chapter that the enclave is a framing device, a site of intentional disentanglement. At face value, this separation works toward two purposes. First, the enclaves provide residential comfort for migrant employees, shielding them from the discomforts of unpotable water, sporadic electricity, and typhoid and malaria that suffuse life outside their walls. Second, insofar as the enclaves enclose not only residential life, but also daily office life, they are a spatial representation of the extent to which the oil industry claims to ring-fence corporate practices away from an external business environment widely regarded by the industry as unpredictable, personalistic, and corrupt. Despite the apparently comprehensive partition between each enclave and its outside, the separation is partial, strategic, and performative, with many people doing a tremendous amount of work to maintain its boundaries, and others yearning for more meaningful contact across the apparent divide. Yet the *effect* of this work of separation is eerily successful: it allows companies—as individual employees living in Equatorial Guinea and as juridically disperse institutions with shareholders and Houston central offices—to inhabit a space of uneven disentanglement. They bemoan what goes on outside their gates as if they have nothing to do with it, when in fact their industry constitutes 98 percent of Equatorial Guinea's national economy with all the sticky entanglements that entails.

What is perhaps most striking about the compounds, however, is not their apparent partition from life outside the walls, but the ethnic, racial, class, gender, and kin partitions *within* them. As I have tried to show, these are not separate projects; they are intimately related. As I wrote in the book's introduction, the mobility of segregated domestic life within the compounds, and the mobility of the technical, legal, and infrastructural forms, like the rig through which oil moves to market, rely on and require one another for their licitness and their performativity. The industry's careful segregation of gendered and heteronormative married whiteness from "others," its cordoning-off, and its selective engagements via corporate social responsibility or philanthropy, sanctifies, and indeed *domesticates*, the power and sovereignty that US oil companies wield in Equatorial Guinea and elsewhere. Conversely, the felicity of the "technological zone" aids and abets the forms of segregation and racialized inequality present in both the compounds and in offshore life. The ability to appeal to standardized accounting practices the world over, or to enclaved corporate practices more generally that ostensibly separate hydrocarbon production from "local life," offer white supremacy what Cheryl Harris (1993) describes as "the legal legitimation of expectations of power and control that enshrine the status quo as a neutral baseline, while masking the maintenance of white privilege and domination" (1715).

And yet, as my work with the expatriate women in the compounds insistently showed, they live white supremacy ambivalently, redeemed only, and partially, through exorbitant pay. This is the simultaneity, or arguably the causal relationship, of transnational scapes of white supremacy to the isolation and emotional precariousness that Kimball's work also documented. The horrors of philanthropy—of passing the pencil—and the guilt of showering are moments of rupture. That rupture is partially bandaged by uplift salaries; but, in that moment of rupture, of shame and anguish and isolation for these white women, I'm reminded of Stefano Harney and Fred Moten (Harney and Moten 2013) addressing (on my reading) white people in white/black coalitions: "The problematic of coalition" they write, "is that coalition isn't something that emerges so that you can come help me, a maneuver that always gets traced back to your own interests. The coalition emerges out of your recognition that it's fucked up for you, in the same way that we've already recognized that it's fucked up for us. I don't need your help. I just need you to recognize that this shit is killing you, too, however much more softly, you stupid motherfucker, you know?" (140). (See also Fanon 1991; Césaire 1962; Nandy 1988; Stoler 2010.) The isolation, shame, and guilt produced in and around compound life are fleeting feelings and moments in which these women realized that it is fucked up for them too; in essence, that the ongoing work of white supremacy, colonialism, and resource extraction cannot be undone by philanthropy, but only by the realization and subsequent enactment of the fact that it's killing us white people, however much more softly, too.

Through ethnographic accounts of offshore rigs and onshore enclaves, chapters 1 and 2 have shown the centrality of racial capitalism (Robinson 1983) to the daily life of oil production. In chapters 3 and 4, the transnational histories and availability of racial differentiation, and imperial debris more broadly, remains a central theme in ethnographic attention to what is perhaps capitalism's central legal form—the contract.

CHAPTER THREE

THE *Contract*

The difficulty of writing about sexual and racial power today . . . is that it exists in
a context of formal equality, codified civil freedoms, and antidiscrimination legislation. People are thus encouraged to see any problems as a matter of discrete rem-
nants of older discrimination or the outcome of unfortunate, backward individual
attitudes. We tried to show how contract in the specific form of contracts about property in the person constitute relations of subordination, even when entry into the contracts is voluntary, and how the global racial contract underpins the stark disparities
of the contemporary world.
—Carole Pateman and Charles W. Mills, *Contract and Domination*

There are at least two kinds of "not-noticing" during fieldwork. There are
the things you don't notice because you rarely come across them, and there
are the things you don't notice because you come across them so frequently.
In the latter case, certain conversations, interactions, and experiences are
so common as to become a kind of background thrum. You know it's there
because it saturates your daily life, and, yet, you have to *not* notice, lest its
ubiquity leak into the foreground of your project. In my fieldwork, contracts
offer a good example of the first category in the not-noticing pair. Despite
the fact that the oil and gas industry is intimately structured by contracts,

they were not an obvious ethnographic object. I rarely came across the documents themselves, let alone representations, interpretations, or discussions of them. This is predictable enough in that contracts are confidential—especially at the state/company level—which can both preclude detailed study (Hardin 2001) and beckon it.

On the other hand, the background thrummed with a pair of discourses so ubiquitous that I had to stop noticing them for my own sanity: relentless "rule-of-law" and corruption talk paired with what I came to call the incantation: water, schools, electricity, healthcare—*waterschoolselectricityhealthcare*. In the case of rule-of-law and corruption talk, every day in the field was flooded with complaints, knowing glances, jokes, well-intentioned concerns, and hushed conversations about corruption and the lack of rule of law in Equatorial Guinea. (See Smith 2007 for this phenomenon in Nigeria.) Migrants who worked in the oil industry talked relentlessly about EG's "culture of corruption," the pains they took to avoid it, and the formal procedures they had in place to work around it, ensuring that their oil got to market "cleanly." Equatoguineans employed in the industry asked why money was deducted from their paychecks for healthcare if, when they went to Loéri Mba (the national clinic), it was always closed, or there was no subsidized medicine available. They asked why they were never indemnified after a contract termination, or even paid the overtime wages they understood to be their due.[1]

In the case of the incantation—*waterschoolselectricityhealthcare*—everyone, from my Equatoguinean friends and interlocutors, to my mother, to migrant oil workers, was constantly chanting some version of "This country is so rich, and yet clean running water, electricity, healthcare, and education are practically nonexistent! The suffering and poverty in the midst of such vast wealth is unconscionable." Rule-of-law talk was constantly paired with this incantation through the suggestion that it was the lawlessness and corruption of the state and/or the oil companies (depending on whom you talked to) that eventuated in Equatorial Guinea's ongoing lack of basic social services in the shadow of such vast wealth. Where, everyone wanted to know, was this elusive "rule of law"? Who or what determines the paths that oil money takes, and whether it ends in offshore bank accounts or in a national healthcare system? Who or what shapes the overlaid human processes and consequences of the oil industry in Equatorial Guinea? In this chapter, I begin to trace the ways in which the production sharing contract (PSC) is, and is not, the answer to these questions.

Back in Donald's office. Our conversations often centered on the specificities of doing business in Equatorial Guinea—the challenges and strategies; how it was similar to or different from the oil and gas business in Russia or Indonesia. During one of these discussions, I asked Donald how his previous experiences came to be useful in Equatorial Guinea: "Do you bring anything with you from place to place? How do you bring what you've learned into such radically different environments?" Donald responded that his "history of working internationally for many years," and the industry's history of working internationally, helped in every new place:

> Through work in remote, less-sophisticated countries, there's been a lot of learning. Before we start working in a country, we enter into a contract. Those contracts aren't simple one-page documents. They cover a lot of categories to shield us from the problems and inefficiencies. These contracts are set up to protect us from the vagaries of the environment. That's what gives us the comfort to come here.

For Donald, production sharing contracts are an itinerant, iterative form in which companies collate and codify the lessons they have learned across time and space. In Donald's estimation, these contracts, which often exceed one thousand pages in length, have the power to "protect" or "shield" companies from "local problems and inefficiencies," to give comfort to companies working in countries both geographically and politically remote from their Houston headquarters. Contracts, we might say, are legal enclaves. As with the making of offshore and enclaved spaces discussed in previous chapters, Donald suggests that contracts also aid in creating the separation or boundary between companies and wherever their extraction site might be today. This chapter follows Donald's provocations about the power of the contract form. Do contracts somehow protect, shield, or give comfort? If so, to whom and from what? How? And to what effect? As in other chapters, here I look at the work contracts do and how they do it. Rather than simply deconstructing contracts' obvious fictions, I show the empirical, ethnographically traceable effects these fictions have in the world. Like the offshore or the enclave, I explore PSCs both in all their ethnographic specificity *and* as a provocation to think about the contract as a more general form and process that facilitates diverse capitalist projects around the world.

After a brief introduction to the PSC and contracts more broadly in social theory, I combine a close reading of a sample production sharing contract

(República de Guinea Ecuatorial 2006b) from Equatorial Guinea's Ministry of Mines, Industry, and Energy (MMIE) with other fieldwork material to explore three interrelated themes. First, I look at the structural effect of making "the state" and "the company" into juridical individuals with unitary wills as parties to a contract—one of legal liberalism's animating fictions. The contract works, in part, to stabilize multiplicity, that is, to not only unify the radical heterogeneity of the corporate form in Equatorial Guinea (recall the dizzying figures 1.3 and 1.4), but also to convert a fractured and once-precarious state apparatus into a singular power. Indeed, for Obiang's repressive regime—once crippled by debt and threatened by an opposition coalition—production sharing contracts have been an unparalleled state-making project, the "oxygen balloon" for a sinking regime. Second, I address how, as juridical individuals in the contract form, the Equatoguinean state and the given company are ostensibly "equally free" to sign this legal instrument, yet both their incommensurability (one a state, one a company) and their inequality are consequential to the effects of contracts in the world. I am interested here in the conjuring power of legal liberalism, and the particular kinds of commensurations and equalizations it makes possible. How do "freedom" and "equality" operate in production sharing contracts and to what effects? Third, I turn from a discussion of ostensible equality to one way in which the contract claims to recognize difference, namely when it is used to "make up for" a host of qualities that investors deem Equatorial Guinea to lack (as signified by Donald's characterization of the country as "remote and less sophisticated"). While US-based transnational oil companies conveniently deem the Equatoguinean state sovereign enough to sign contracts, those same companies consider the state's substantive sovereignty for the purposes of capitalism—as daily protector of the rule of law or guarantor of property rights—to be inadequate. Using a discussion of rule of law on the ground in Equatorial Guinea, I show how companies attempt to use the contract to provide a proxy system of law in a state deemed unable to back up contractual obligations (the "protection" Donald was talking about). With an ethnographic focus on fiscal stability clauses and the productive confusions at the intersection of local law and transnational companies, I address issues of postcolonial sovereignty and power. Who rules in Contract Land?

(PRODUCTION SHARING) CONTRACTS,
CAPITALISM, AND ANTHROPOLOGY

Production sharing contracts are the primary contracts between the Equato-guinean state and major oil and gas companies. This contract form is also broadly used in oil and gas production sites across the global South. Nego-tiated and signed at the beginning of a company's relationship with a given sovereign state, each PSC details the production and dispersal parameters for billions of barrels of oil and billions of dollars in profit for companies and countries over the life of the concession, generally thirty to fifty years. The production sharing contract or production sharing agreement (PSC or PSA) was first used in the oil and gas industry in the 1960s, with the ancestral document considered to be a PSC signed in Indonesia in 1966. Before the PSC, license-concession contractual models were standard: a colonial extraction regime in which the state, in its administrative and authoritative capacity, granted private companies not only use, but also ownership, of the subsoil for extraction purposes.[2]

Formally, PSCs differ from concessions in three primary ways. First, they are contractual as opposed to concessional; the state becomes an "equal" commercial party to a legal agreement (at the same time as it is expected to maintain its authoritative and regulatory prerogatives). Second, PSCs explic-itly recognize that resource ownership rests with citizens, represented by the state, as opposed to with private extractive parties.[3] Third, the "sharing" in production sharing contract refers to the way in which risk and profit are distributed between state and company. In effect, states hire companies as contractors; the company brings all necessary investment, expertise, and technology to the project, and also bears all financial risk. If, after explora-tion, no oil is discovered, all "sunk" costs are the company's to bear. If the venture is successful, however, and production begins, the company recoups their initial investment and operating costs through "cost oil." Once they have recovered their initial investment, "profit oil" is split between the state and the company at rates determined in each contract (see Radon 2007 and Johnston 2007 in Humphreys et al. 2007).

Contracts are a material, legal, and symbolic presence everywhere at the formal center of capitalist relations, leading political economist Ronen Pa-lan (2006) to claim that "the distinguishing feature of a modern capitalist economy is that it rests on contracts" (85). The contract form—in which public and private, law and profit, are inextricably bound—easily belies neo-liberal fantasies of market independence from states, regulation, or judicial systems. Formally, the validity of any contract is backed by the power of a sovereign state, and hence capitalist society is, in theory, dependent on (rule of) law. "The contractual nature of market relationships," writes Palan, "reminds us that the capitalist market is essentially 'an endless chain of legal relations'" (85). Westphalian sovereignty then not only divides the world into geographic jurisdictions of authority and responsibility, but "also serves as the foundation of the national and international law of contract. . . . Contractual relationships, now spanning the entire globe, are rooted in a system of sovereign states" (Palan 2006, 86). In a world of deep postcolonial and colonial inequalities and their resultant interdependencies, one can anticipate the problematic fictions of such a system right away.

Some anthropologists may bristle at Palan's description of the global contractual economy, objecting that it reifies the contract form (not to mention sovereignty or capitalism's reliance on law), presupposing effects while neglecting the diverse social and political relations in which contracts are constantly negotiated, utilized, contested, and ignored. Indeed, many since Durkheim (1997) have drawn attention to the noncontractual universe of social obligation and cooperation present in every contract. I am sympathetic to these substantivist or neosubstantivist approaches and engage them throughout this chapter and book; but I hope to push them beyond an argument that what is framed as economic is, in fact, social (cf. Granovetter and Swedberg 2001). Today, the analytic task seems to be to move from "social explanation" to the enduring power of the form: how to account for the substantial effects of tools like contracts in the world (Latour 2005; Callon 1998; Çalışkan and Callon 2009). For instance, while the equality posited between states in the Westphalian system may be mythic—Equatorial Guinea and the United States are far from equal—that posited equality is deeply consequential. Contractual relationships like those established in PSCs depend on Westphalian sovereignty for their licitness, for their felicitous performance. At the same time, as this chapter will go on to detail, certain contractual parties are able to profit from the obvious inequalities that the Westphalian model ignores. In other words, normative models themselves—Westphalian ideas of sovereignty; neoclassical or legal liberal understandings of the contract form—are not "wrong" as much as they are partially constitutive of what anthropologists and others have often thought of as *the social*.[4] We know, contra Henry Sumner Maine (1861), that the contract does not intrinsically produce isolated, individuated *homines oeconomici*. What we still *need to know* is how to account for the effects of contracts that this chapter goes on to illustrate: How does the contract render blatant neocolonialism licit or legal? How does it frame multiplicities into legally recognized and politically consequential singularities, and contested political regimes into petro-powerhouses? Attention to the contract enables us to account for the power of formalizations without imagining that they have an essential nature (Bear et al. 2015). We can point to contracts' reliance on and manipulation of postcolonial inequality and sovereignty without imagining that these "social explanations" or "historical contexts" somehow undo their power. On the contrary, they are constitutive of it. Historical contexts of radical inequality and imperial debris (Stoler 2008) are the arbitrage opportunities in which inordinately corporate-friendly contracts are made.

Finally, contracts offer a mighty analytic threshold into the social life of not only capitalism, but also liberalism as a global form. Along with the rule of law and representation, *the contract* is part of the deep structure of both political liberalism (social contract theory, Hobbes, Rousseau) and, significantly for my purposes here, economic liberalism (Adam Smith). The contract forms I trace in this chapter and the next, *specifically the ways in which they organize inequality*, offer a productive glimpse into Mehta's (1997) empirical puzzle, wherein "something about the inclusionary pretensions of liberal theory and the exclusionary effects of liberal practices needs to be explained" (59). We begin to see this apparent contradiction come to life in Donald's description: Equatorial Guinea is "remote," "less sophisticated," full of "problems and inefficiencies." Where liberalism's theoretical vision is one of universality, these descriptors betray Donald's impression that Equatorial Guinea is *not yet liberal* (rule of law, sanctity of private property, representation); thus, the contract becomes liberalism's aggressive avatar, "covering categories to shield us from problems and inefficiencies." This is far from liberalism as theoretical universal, and is instead exemplary of the period of liberal history that Mehta (1997) traces since roughly the seventeenth century, which is "unmistakably marked by the systematic and sustained political exclusion of various groups and 'types' of people" (59). Via an anthropological reading of Locke, Mehta shows that rather than finding a universal conception of human beings as all equally endowed with rights and privileges, one sees a developmental teleology—a constant and aggressive differentiation between those whose endowments are always-already realized and those whose endowments must still be honed to be properly recognized as a liberal subject. Thus, liberalism indexes "a specific set of cultural norms [that] refer to a constellation of social practices riddled with a hierarchical and exclusionary density" (Mehta 1997, 70). As Wendy Brown (2009) explains, liberalism's promises have never been fully realized, but "have always been compromised by a variety of economic and social powers from white supremacy to capitalism. And liberal democracies in the First World have always required other peoples to pay . . . that is, there has always been a colonially and imperially inflected gap between what has been valued in the core and what has been required from the periphery" (51–52). Among other effects, contracts render licit, *liberal* even, that imperially inflected gap.

Oil contracts at the scale considered here—between a sovereign state and a given multinational company—offer a unique ethnographic intersection of political and economic liberalism. They become *both* a social contract and an economic contract. Following this line of thinking, I frame both this chapter and the one that follows on subcontracts with the work of Carole Pateman and Charles Mills (2007) on the contract and racialized/gendered domination. While Mills's work (1997, 2017) contemplates the social contract, broadly conceived as predicated not on inclusion and equality, but on European imperialism and racialized relationships of domination and exclusion, his work with Pateman (Pateman and Mills 2007) brings us closer to the material at hand here—actually existing commercial contracts. "Contract," Pateman and Mills argue, "is the major mechanism through which . . . unfree institutions [the modern state and structures of power] are perpetuated and presented as free institutions" (20). The "freedom" to sign contracts then is an illusory freedom, an always-already compromised freedom that is at the heart of liberalism.[5] Ethnographic attention to oil contracts between US-based transnational companies and the Equatoguinean state show liberalism's animating contradictions in action.

THE MAKING OF CONTRACTUAL PARTIES:
THE STATE EFFECT AND THE COMPANY EFFECT

Three parties sign each production sharing contract in Equatorial Guinea: the Republic of Equatorial Guinea, Equatorial Guinea's relevant National Company (GEPetrol in oil contracts, Sonagas in gas contracts), and the Company—the local subsidiary of the transnational oil and gas company in question (see the very bottom left of the conventional organogram—figure 1.2—in chapter 1). The making of "the state" and "the company" into juridical individuals with unitary wills is one of the productive fictions of the PSC form. Where much contemporary anthropology has sought to disassemble and disaggregate the fetishized surface of such fictions—showing the State (in this case), but also Capitalism, Development, Race, and Gender to be multiple, contested, and contingent—we have focused less on the practices by which those multiplicities, contestations, and contingencies are mustered into durable and consequential singularities. The production sharing contract is one such practice.

Timothy Mitchell (1991), in "The Limits of the State: Beyond Statist Approaches and Their Critics," suggests that we should approach the state as a structural effect, "that is to say, it should be examined not as an actual structure, but as the powerful, metaphysical effect of practices that make such structures appear to exist" (94). A single "state" signature on production sharing contracts is one such practice that lends obduracy and apparent unity to the state, one of the "modern techniques that make the state appear to be a separate entity that somehow stands outside society" (91). I want to extend Mitchell's analysis of the state as structural effect to the corporation as well. Like the state, and perhaps even more so, corporations are made through heterogeneous and geographically far-flung sociomaterial practices—shareholders, headquarters, stock prices, employees, production processes, transport, offshore accounts, raw material—that do not cohere within a clear external boundary. Yet, despite this multiplicity, we often imagine the corporation, like the state, to be singular and bounded. Stretching Mitchell's analysis, I suggest that corporate unity too be understood as "the powerful, metaphysical effect of practices" like the contract "that make such structures appear to exist."

On the first page of the "Model Production Sharing Contract," the first party to the contract is listed as:

THE REPUBLIC OF EQUATORIAL GUINEA (hereinafter referred to as the State), represented for the purposes of this Contract by the Ministry of Mines, Industry, and Energy, represented for the purposes of its execution by His Excellency Mister _____; the Minister. (República de Guinea Ecuatorial 2006b, 1)

Thereafter, the Republic of Equatorial Guinea is simply referred to as "the State." This contractual party fuses the citizens (who ostensibly "own" the resource), disparate legislators, officials, ministries, and the nation into a single figure, a juridical individual with a unitary will. This unification of those who sign the contract into "the State" is a form of nation-fetishism that legally masks the specific "parties" who, in fact, sign these contracts. In a country whose economy is 97 percent petroleum based, where, in other words, the differentiated and fractured state apparatus is wholly reliant on hydrocarbons, the contract form (who is privy to it and who signs it) gains inordinate power. To illustrate, a parliamentarian (technically, a member of "the State" signatory party) explains the secretive and sectarian history of oil legislation and contracting:

> In the early years, when American companies started production, the contract negotiations were done in a very private way (*de una forma muy reservada*), almost confidential. There were personal meetings set up by the American Ambassador, Chester Norris. It was as if they were dealing with private property, to such an extent that *no one knows what was in the first contracts.* What are the most important clauses? No one knew. No one even knew about the state's [percentage take] on each barrel of oil. And this has remained opaque until now. . . . In 1999, the president began to publicly denounce the contracts that his government had signed, [saying that] they were a disaster (*una ruina*) for the country. He said this publicly, that it was a shame. Since then, there has been one effort or another to improve [the state's] participation 3 or 5 percent. I don't know if today it reaches 35 percent. The petroleum business continues to be opaque.

According to this parliamentarian, the specific "parties" who actually negotiated and signed early contracts met in secret, in personal meetings set up by the American Ambassador. Although, by definition, the PSC (and EG's hydrocarbons legislation) designates all products of the subsoil as belonging to the citizens, "it was as if they were dealing with private property." When, years later, it became clear to the president—who was undoubtedly involved in the first contracts—that the oil companies had taken him for a ride, even he could distance himself from "the State" that had signed, claiming that early contracts had been "una ruina." Here, the contract facilitates contradictory claims. On the one hand, it unifies specific interests into a "state," which leaves even parliamentarians unclear as to what "the State" has signed. On the other hand, the contract allows room for those specific interests to renounce it as not of their making, to blame other parts of "the State" for signing ruinous contracts. The contract provides an authoritative frame within which the specific signatories can both locate and negate responsibility. But to what effect?

Recalling the days just before Chester Norris and the early oil contracts with American companies, an opposition politician waxed improbably nostalgic about what had been a time of economic and political crisis in Equatorial Guinea: "In the '90s," he explained, "the regime was drowning [and] the opposition was strong and growing. The country suffered a significant economic crisis with external debt among the highest in Africa. The country operated fundamentally on international cooperation, conditioned by the pressures and conditionality clauses [in the loans]. And that's when petroleum started. Petroleum was like a life jacket for the regime, an oxygen balloon to help it float." His nostalgia for a drowning regime is ironic; few pine for the days of deep debt and structural adjustment. But to the extent that the immiseration of the period destabilized a regime already in power for over twenty years by the early 1990s, for many it was a time of political foment and hope. And yet, despite active opposition movements and glimmers of conditionality-driven democratization, when "the State" signed the first contract with a US firm, the potential energy of the moment was quashed, and an eerily singular, and singularly powerful, "state" arose that endures to this day.

Coronil (1997) narrates a similar story of Venezuela at the turn of the twentieth century. In 1908, Juan Vicente Gómez had overthrown the ruling general and sought to restore order to the ensuing economic and political chaos partially through inviting foreign capital to invest. Oil companies responded, and with Gómez at the helm, "the State" signed a number of contracts with these oil companies. Coronil suggests that before the arrival of oil, "the state was so weak and precarious as a national institution that its stability and legitimacy were constantly at risk. Without a national army or an effective bureaucracy, in an indebted country that lacked a national road network or an effective system of communication, the state appeared as an unfulfilled project" (76). This could almost be a description of the Equatoguinean situation in the 1990s. Certainly, the state was as weak and precarious as it had ever been since Obiang took power in 1978. Like the Venezuela Coronil describes, Equatorial Guinea was highly indebted and lacked an effective bureaucracy or national road network. What legitimacy and strength the state had achieved emanated from a combination of fear and patrimonial sentiment, which many Equatoguineans, including the opposition politician quoted above, believed was eroding in unprecedented ways by the early 1990s. As with Venezuela, however, the opportunity to sign oil contracts provided unprecedented state-making effects in Equatorial Guinea, allowing a sinking regime to rise, as the opposition politician put it, as if attached to an oxygen balloon.

Tsing (2005) also writes about this structural effect of the contract in Indonesia, humorously referring to the contract as the COW, or Contract of Work:

cows have been magical tools of the national elite. Although merely paper and ink, they conjured a regular income for the Indonesian nation state. Their terms must be secure and attractive by international standards, or they will not draw capital. But if they meet these standards, they can conjure the funds that allow the nation-state to produce itself as what one might call a "miracle nation": a nation in which foreign funds support the authoritarian rule that keeps the funds safe. I have called this "franchise cronyism." *In exchange for supplying the money to support national leaders who can make the state secure, investors are offered the certainties of the contract,* which ensures titles to mineral deposits, fixes taxation rates, and permits export of profit. (69, emphasis added)

Tsing's franchise cronyism—foreign investment supporting authoritarian regimes rendered licit through contract—draws our attention to another aspect of the state/corporate effect of contracts, that of boundary making and separation. While the state and the company sign as separate parties, the contract in fact intercalates them in consequential ways, blurring any boundary between them. The parliamentarian and opposition politician's stories of private meetings between companies and state personnel, and oil money as an oxygen balloon for a sinking regime, illustrate both the permeability of state-society-corporate boundaries in practice and the political significance of maintaining those boundaries in theory (Mitchell 1991). Few US oil company shareholders could tell you where Equatorial Guinea is on a map, let alone which president they are effectively voting for there. The same opposition politician quoted above noted that at certain times, company representatives have been prohibited from talking to the opposition, not by law, but by the exigencies of the personal relationships on which contracts are based. "In a trip to the United States I tried to meet with company representatives but they escaped. They said, 'no, no, no, no!' An American company—which knows what liberty is, the usefulness of information—refused to speak with me. They realize that they accept these terms to preserve good relationships." In other words, where contractual parties appear as separate and individual, not only is each party far more multiple and contested on the ground, but they are also co-constitutive of one another in consequential ways. Here, US-based oil companies fund a repressive regime and refuse to speak with the political opposition. Liberty be damned.

"The Company" undergoes a similarly miraculous process of unification and separation in contract land.[6] Claiming a unified juridical identity as the third party in the contract, each version of "The Company" as it exists in Equatorial Guinea is, in fact, a subsidiary of a subsidiary of a subsidiary of

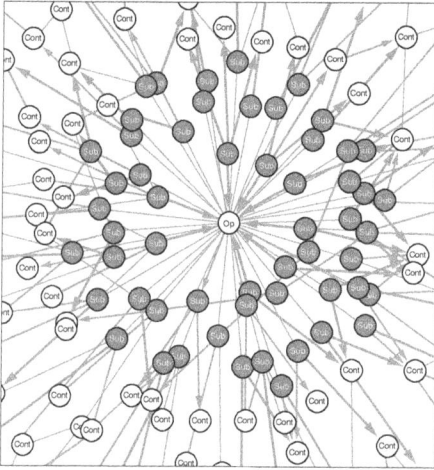

Figures 3.1 and 3.2. From corporate archipelago to corporate unity.

its parent company. Recall figure 1.2, diagramming the corporate geography of Endurance. The companies are located in one place, Equatorial Guinea, as a limited liability operating company; in another place, Houston, as corporate headquarters; and in still another place, Dubai, Cyprus, or the Cayman Islands, for financial purposes. Moreover, even The Company's most local contract-signing incarnation as an Equatorial Guinea–based subsidiary is itself a consortium—a group of companies who invest together to spread the costs and risk, and share profit. Like the state effect, in which disparate and indeed conflicting parties are unified in the contract form, so too with the corporate effect of the PSC, in which this archipelago of attenuated liability is felicitously unified, helping us trace the movement between figure 3.1 and figure 3.2.

As shown in chapter 1, where I diagrammed Endurance's corporate structure with Donald's help, the open, licit character of these state- and corporate-effect practices is important here. *It is the legality of these arrangements*—including dispersed corporate geographies unified as a responsible party in the contract form—that allows both the Equatoguinean contract signers and those who sign for oil companies to relinquish responsibility for local outcomes when necessary.[7] In the moments where any number of parties might be held accountable for failing to regulate money laundering or environmental and labor relations, let alone keeping a dictator in power, the disentangling authority of the contract form steps in. To illustrate, consider the story of Equatorial Guinea and Riggs Bank. In 1995, one year after the discovery of subsea hydrocarbon deposits off Bioko Island, Equatorial Guinea opened its first embassy accounts with Washington, DC–based Riggs Bank.[8] "Over the next eight years, the bank opened nearly fifty additional accounts and a dozen certificates of deposit not only for the government of EG but also for a host of EG senior government officials and their family members. By 2003, the EG account had become the bank's largest single relationship, with balances and outstanding loans that together approached $700 million" (Coleman and Levin 2004, 38). In 2004, the United States Senate Subcommittee on Money Laundering and Foreign Corruption released the results of their investigation into this relationship (as well as the bank's relationship with Augusto Pinochet). They concluded that everyone from the Riggs Bank employee directly responsible for the accounts, to his wife, to the Examiner in Charge from the Office of the Comptroller of Currency (OCC), to Federal Reserve regulators, to US oil company management, to President Obiang himself were guilty of anti-money-laundering infractions (Coleman and Levin 2004). Riggs shut down in the wake of the scandal, merging in 2005 with PNC Financial Services.

In the wake of the Riggs debacle, the Equatoguinean government *repeatedly* invoked the contract form in its own defense: to explain offshore bank accounts, direct payments from oil companies to government officials, and money in personal accounts that perhaps should have been in the national treasury. "All payments made by the companies to the State of Equatorial Guinea are *contractual, and therefore legal*, payments" (República de Guinea Ecuatorial 2005, 12; emphasis added here and below):

> It is important to point out that the funds which the state of Equatorial Guinea deposited in Riggs Bank were not in any way the product of money laundering or foreign corruption, but rather revenues earned by companies involved in the country's oil sector, *in fulfillment of their contractual obligations to the state.* (3)

> All the payments made by the businesses are *contractual payments, . . . to talk of hidden payments is to ignore the existence of contracts* which the companies have signed both with the State itself and with the Equatorial Guinean service companies. . . . *We confidently declare that we are perfectly able to demonstrate that all the payments made by the companies were contractual payments.* (15)[9]

In their headlong rush to use the contract form to demonstrate the licitness of the transactions and accumulations exposed in the Riggs case, the authors of the Riggs Report also divulged the kinds of contractual details that begin to flesh out the form's more quotidian entanglements in Equatorial Guinea's daily life. The Riggs Report cites a contractual clause in which the company is required to deposit a one-time payment of $50,000 into a Riggs account to "assist the state in the establishment of an Embassy in Washington D.C." (2005, 13). Thereafter, monthly payments to that account in the amount of $7,000 were to be made until the arrangement was canceled or the Embassy shut down. Also in this clause was the stipulation that the company will pay the state "up to $20,000 U.S. to fund the production of videos in order to assist the Ministry of Mines in marketing its mineral resources" (14).

The depth of the entanglements is striking; contract signatures not only buoy flagging regimes, but also pay to establish embassies abroad and produce videos to market national resources. These outcomes of the contract form easily belie separate, unified "parties." *And yet*, the licitness of contracts, the liberal and legal evidentiary regime in which they rest, allows the form to disentangle companies and states from these very relations, to move them from power to technique, from politics to procedure. The specificity of these entanglements also reveals the radical inequality between the two contractual parties.

TWO PARTIES EQUAL BEFORE THE LAW: COMMENSURATION, DIFFERENCE, AND INEQUALITY

Abstract assumptions about fair contracts, equal partnerships, and mutual profit have the familiar ring of liberal sentimentality—the desire to see equality in the face of obvious inequality. There is also, perhaps, the ring of faith in the conjuring power of liberalism itself, that is, the idea that two parties made equal before the law might then somehow also be more equal in substantive practice. And yet, in the contract form as a legal liberal instrument (similar to human rights as a liberal moral instrument), commensuration—all are equally free to sign contracts; all are equally entitled to human rights—does not equality make. To the contrary, these processes of commensuration (especially when they are as lucrative as a production sharing contract) often exacerbate and cement existing inequalities.[10] Again, this is contract as the aggressive avatar of liberalism, the contract as a form of licit violence where Equatoguineans have "not yet" developed their ostensibly universal liberal endowments.

As Donald intimated above, PSCs are a fundamental part of the modular bundle operating companies bring with them from site to site. An Equatoguinean judge who participated in contract negotiations confirmed this: "When they come, they already have contracts." Bringing precedents from previous contracts as they negotiate new ones, companies implicitly or explicitly equate Equatorial Guinea to Nigeria to Ecuador, at least insofar as the contract is meant to structure the same outcome—usable, profitable hydrocarbons—from site to site. And yet, what stands out about the contractual clauses on the establishment of embassies abroad and the bankrolling of promotional videos is their peculiarity, their idiosyncratic responsiveness to the needs of Equatorial Guinea at the specific moment the contract was signed. Within a standardized frame, these clauses index Equatorial Guinea's *difference* from other sites where PSCs might be in force, and index inequality (no American embassy; limited local media) as a certain kind of difference. Both agnostic toward and yet necessarily responsive to the differences among exporting sites, the making of licit and felicitous contracts requires different entanglements for different local conditions, "to produce functionally comparable results in disparate domains" (Ong and Collier 2005, 11). In the next section, I explore these dynamics of commensuration and differentiation, as well as the tensions between "equal parties" to the contract and the crystallization of their historical inequality. First, in thinking through what is "standard" in production sharing contracts, I look at how formal sameness produces difference when refracted across inequality. Second, I turn to the consequences of liberal sentimentality.

On Modularity:
Repetition and Difference in the Contract Form

Production sharing contracts around the world "give precise detail and legal specificity to the obligations of a state and the company or consortium of companies involved in an extractive project" (Maples and Rosenblum 2009, 15). While the content varies, all PSCs have a similar form or "boilerplate," with core sections covering definitions of key terms; the grant of formal legal title; rights, duties, and obligations; termination; confidentiality clauses; and dispute resolution, among other standard sections (Maples and Rosenblum 2009; Humphreys et al. 2007). In addition to this repetition of basic form, the sample Equatoguinean production sharing contract to which I had access was peppered with acronyms and proper nouns, including CIF, FOB, Platts, and LIBOR. Each references methods for global standardization in the hydrocarbon commodity chain, from shipping regulations to oil prices to interest rates. Many of the acronyms come from the International

Commercial Terms (Incoterms®) document, published by the International Chamber of Commerce (ICC) and closely corresponding to the UN Convention on Contracts for the International Sale of Goods. Self-described as the "rules at the core of world trade" and the "standard trade definitions that are most commonly used in international contracts," Incoterms® is a terminology standardization toolkit that finds its expression in English, considered the legally binding version (ICC 2010). In addition to Incoterms®, Equatorial Guinea's sample PSC mandates the use of the LIBOR rate, the interest rate offered in the London InterBank Market as published in the *Financial Times of London*, unless it fails to publish the rate for five consecutive days, in which case the *Wall Street Journal* is consulted (República de Guinea Ecuatorial 2006b, 5). For oil price negotiations, the contract specifies the use of Platts Marketwire, "the leading global provider of energy and metals information, and the world's foremost source of price assessments in the physical energy markets" (Platts 2010).

Incoterms®, LIBOR rate, and Platts Marketwire each seem at once trivial and natural. Of course there are standardized shipping terms, interest rates, and commodity prices, but are they even important? Anthropological work has long shown that no matter what Incoterms® has to say about shipping, for instance, individual people, ports, vessels, and histories will find their own ways of getting the job done (Çalişkan 2010, Dua 2019, Bishara 2017). These standardizations are, however, also power indices (Lampland and Star 2008). They index financial geographies—London and New York; specific tribunal locations for the settling of disputes (Riles 2011; Potts 2016, 2018); institutions such as the ICC, the UN, and NYMEX; the English language and the American dollar. And, needless to say, these apparently mundane rules are not always mundane, but can reveal themselves as cauldrons of sociality, inequality and collusion, as the LIBOR rate scandal did between 2012 and 2015 (McBride 2016).

In each contract, these standard sections and mundane proceduralisms commingle with far more consequential clauses that vary from extraction site to extraction site; for example, temporalities in production and payment processes; taxation guidelines, including tax deductions, exemptions, and reductions for the company; and the legal jurisdiction in which the contract is extant and where legal proceedings would take place should conflicts arise. In addition, PSCs include environmental guidelines and protection measures, and land use guidelines and rights to private and public land, including clauses on the displacement of local communities. In negotiating this more controversial content, the outcomes of which vary with the strength of local regulatory regimes, replication in the contract *form*—standard sections and terms—seems to offer a certain licitness. When, for instance, a Liberian newspaper published a report critical of Mittal Steel's contract with the government, the company responded that its agreement "mirrored essentially like it's done elsewhere" (Maples and Rosenblum 2009, 52). *"Essentially like it's done elsewhere"* draws our attention to the surface licitness and intended seduction of standardization.

Consider the phrases "good oil field practices" and "generally accepted practice of the international petroleum industry," for example, which each repeat on nearly every page of the 2006 sample PSC (República de Guinea Ecuatorial 2006b). The contract says that the Company will carry out operations in Equatorial Guinea according to "good oil field practices" (1); maximum production rate will be determined "in accordance with good oil field practices" (7); and perhaps most ominously, "The Contractor shall take all prudent and necessary steps in accordance with generally accepted practices of the international petroleum industry . . . to prevent pollution and protect the environment and living resources" (22). These quotes, and others like them throughout the contract, imply that there are standardized practices in the international petroleum industry enacted identically worldwide, and that these standard practices serve as guidelines or even regulations of some kind. As Reed (2009) clarifies, however, "Transnational oil corporations often declare compliance with global industry standards, but *there is no mandatory set of worldwide regulations,* and attempts at harmonization provoke a figurative race to the bottom" (66; emphasis added).

There are no mandatory global industry standards. Phrases like "good oil field practices" and "standard practices of the international petroleum industry" gesture, instead, to self-regulation and the prerogative of corporations to bring their own rules wherever they go. In practice, these phrases do not denote replication in oil fields from Norway to Equatorial Guinea. Instead, they render licit *dramatic* international differences, predicated on the specificities of each extraction site. Environmental regulations; labor regulations; accounting practices; and capacities and incentives of states to monitor, supervise, scrutinize, or audit oil field practices vary radically from state to state. "Standard oil field practices" vary accordingly. Ethnographically, then, "good oil field practices" means that there are countries in which the petroleum industry can get away with almost anything—the more underdeveloped and newer to the extraction process, the better—and countries where they can get away with less. During my fieldwork, Equatorial Guinea, perhaps needless to say, was definitively in the former category. Consider a longer version of Eugenio's story, which I introduced in chapter 1.

Originally from Equatorial Guinea, Eugenio had gone to school abroad to study petrochemical engineering and returned home to work in the industry. He was employed by a chemical processing company, which I'll call CPCO. This company had the monopoly on chemical processing in offshore production in Equatorial Guinea. Globally, they are considered the top company in this line of oil service work. The Major Corporation subcontracted CPCO to test the water off their platforms and to clean petroleum by-products from the water used in the drilling process, so that "clean" water could be returned to the ocean. During his petrochemical master's program, Eugenio learned that according to what one might call "standard oil field practices" for CPCO's work, the more petroleum extracted from beneath a given platform, the harder it is to clean the surrounding water. Hence, the amount of oil extracted is to be conditional on surrounding water quality, tested daily for pH levels. Once the water reaches a certain level of contamination, extraction must be slowed or stopped altogether. In Equatorial Guinea, however, where CPCO is not regulated by any outside body, there was no enforced conditionality between pollution levels and extraction rates. According to interviews with men working on offshore platforms, including Eugenio, workers were told sporadically to prepare for environmental assessment, but someone from within the company always carried out these assessments, and the results were always positive. As Eugenio starkly put it: "In the history of oil exploitation in Equatorial Guinea, there has not been a single environmental inspection by competent authorities. At the very least, I can confirm this fact during the three years I worked with CPCO." Equatorial Guinea's lack of regulatory oversight—its difference from Mexico or Norway or Malaysia—allows pollution off its platforms to go completely unchecked. In these moments, standardizations become more ominous because of their radically different execution across space.

Liberal Sentimentality, Equality, and the Making of Contractual Difference

The laws of contract and international commercial relations generally suppose two corporate entities doing business with each other, both seeking profits and answering to shareholders. This makes sense unless one of the parties is not a corporate entity, but rather a government answerable to citizens (itself, of course, a liberal supposition). Even as they conduct business, governments have duties, obligations, and interests that go well beyond profit maximization (Maples and Rosenblum 2009, 11).

After peaking at $18.5 billion in 2008, Equatorial Guinea's gross domestic product (GDP) hovered at $10.18 billion in 2016. By way of comparison, Gambia—a country of a similar size, although without oil—had a GDP of $964.6 million in the same year. The Major Corporation's revenue for the same year was over $200 billion, placing it among the world's largest economic entities, ahead of Belgium and Norway, just smaller than Sweden (and significantly smaller than Walmart). Major's 83,000 worldwide employees exceed the entire population of Equatorial Guinea's capital city. The Major Corporation and others at work in Equatorial Guinea have at their disposal vast teams of experts in everything from chemical and mechanical engineering to contract law, deepwater geology to cutting-edge financial investment. They also have global telecommunications and information technology networks, and expansive public relations and corporate social responsibility departments. During my main fieldwork in Equatorial Guinea, the country had one university that was without electricity, let alone computers. These inequalities come to life in the contract form (Radon 2007; Zalik 2009).

Consider another version of oil's origin story in Equatorial Guinea. Just after Obiang came into power, Spanish oil company Hispanoil (now Repsol) began to explore. According to government documents, the World Bank came on a technical assistance mission to the oil sector at the same time, introducing the PSC model to the country in 1981 and offering the oil companies "very wide and advantageous margins, so as to make worthwhile the exploration" (República de Guinea Ecuatorial 2005, 12). While the tales circulating about the outcome of Hispanoil's exploration varied, a particularly popular version was that Hispanoil indeed discovered deposits with commercial potential; but then they stipulated as a condition of their production that Equatorial Guinea's Minister of Finance and Minister of Mines be replaced with Spaniards. When the local government rejected this offer, Hispanoil wrote reports claiming to have found nothing. One year later, after drilling two successful exploratory wells, wildcat American company Walter International put the first gas and condensate field into production, and with it the first PSC, with an appallingly low 3–5 percent profit share to the state. As Mauricio, the Equatoguinean man who had worked for the Major Corporation and then moved over to the government, explained: "We didn't have another option. Take it or leave it. You'll die of hunger with food in the ground." Soon, larger US companies came in, buying out Walter International and signing up "to the excellent special advantages offered by the Model Contract suggested by the World Bank . . . in order to attract foreign capital investment in the country" (República de Guinea Ecuatorial 2005, 12).

Reminiscent of Tsing's cows described earlier in this chapter, Mauricio continued:

> In order to attract foreign investment, we had to really make things easy. We didn't have a history of oil. We wanted exploration. Very few people were willing to take the risk. It takes a lot of investment to find oil, and if you don't find it, it's down the drain. All countries around us had oil. What is the incentive for a company to jump from Nigeria? *Contract terms.* You have to really lay down at a certain level terms that attract investment. Equatorial Guinea did that. Once the oil was found, slowly the government had to build up to the level that is appropriate for the industry. . . . It's just equalizing the playing field for an area now considered a producing country. Even so, Equatorial Guinea continues to provide a better business environment than our counterparts. . . . Hannah, have you been to Nigeria?

More attractive contractual terms provide the incentive for companies to explore. And indeed, government documents claim that of all the PSCs in West Africa, Equatorial Guinea's conditions remained the most favorable to the companies well into the 2000s. As of 2005, a document claimed that the "rate of profit tax levied in the sub-region is of the order of 50%, while in Equatorial Guinea it is 25%. The State share in the sub-region is over 50%, while in Equatorial Guinea it is still below 20% in the fields currently under exploitation" (República de Guinea Ecuatorial 2005, 13). The same report goes on to claim that Equatorial Guinea offers "*the most flexible fiscal environment in the world*" (República de Guinea Ecuatorial 2005, 16, emphasis added). Viewed from another perspective, Mauricio wryly noted of this flexibility that "one day they will add to the contracts that the foreign business owners are permitted to hit Africans. They have freedom to do anything. They can transfer all the money they make without any problem. These things have to change."

In the daily life of contracts then, from the earliest moments of their negotiation, long and ongoing histories of inequality are never "written out." Contracts are first negotiated according to, and will later capitalize on and further institutionalize, those inequalities. I have repeatedly drawn the analogy to Maurer (2008) and Hudson's (2017) work on the processes by which small island economies in the Caribbean become tax havens, ostensibly via their participation in a market of equals. More accurately, Maurer suggests, their particular insertion into this market relies on histories of imperialism, colonialism, and slavery, the aftermath of which makes them more attractive repositories for fleeing capital (see also Hudson 2017a; Rodney 1972). I quote Maurer (2008) here, but replace "offshore finance" with oil, and "small island economies" with African economies:

> [Oil] has become the lifeblood of many [African] economies, where legacies of slavery and colonialism are not imagined to remain in the distant past but rather always in evidence in the conduct of international affairs and everyday encounters in tourist or financial settings. Here . . . reputation is about hierarchies of rank, not about equivalence or a market of equals. . . . Coming to grips with [oil exploitation in Africa] may mean coming to re-appreciate, analytically and politically, the politics [and histories] of those hierarchies of rank. (172)

That Equatorial Guinea continues to provide "a better option than our counterparts," as stated by the Equatoguinean official quoted above, suggests just such a hierarchy of rank. Weak regulation, tax laws, and low state percentage takes of profit become something to promote: "the most flexible fiscal environment in the world." *This is imperial debris proxying for market terms, becoming the shape the market takes.* Indeed, at the end of 2012, Equatorial Guinea's US-based lobbyists purchased a full-page ad in the *New York Times* that read: "The new commercial hub of central Africa is a paradise for investors."

The discovery of oil and gas opened new opportunities for Equatorial Guinea and turned it into a "pillar of stability in the region," as the text in figure 3.3. states. ("Pillar of stability" here refers, somewhat ironically, to the fact that Obiang is still in power, gesturing to the state-making effects of the contract form.) Imperial debris (Stoler 2008) and hierarchies of rank suffuse the contract form and its effects in the world. And yet, Equatorial Guinea's early 3–5 percent share of the profits from their resources is couched in the procedures, standards, and aspirations of legal liberalism—fair contracts "the same everywhere" and mutual profit. Imperial debris here is constitutive of legal liberalism, not a deviation from it.

In the next section, I revisit a particularly tenacious metric in these hierarchies of rank: corruption and rule of law. At the intersection of the "local logic" by which Equatorial Guinea's judicial system works, or doesn't, and transnational corporations with billions of dollars at stake, I focus on the power of "corruption" to organize consequential contractual outcomes. How do hierarchies *make* the contract form, and to what effect?

Figure 3.3. Full-page advertisement in the *New York Times*: "A New Paradise for Investors."

WHO RULES CONTRACT LAND?
LOCAL LAW, TRANSNATIONAL COMPANIES,
AND STABILIZATION CLAUSES

Ferguson (2006) has written that the spectacular wealth pouring into Equatorial Guinea, Nigeria, Sudan, or Angola contradicts the neoliberal development nostrum that corruption diminishes foreign direct investment (FDI). In this section, I explore how the contradiction between neoliberal development nostrums and actual FDI is not without consequences in Equatorial Guinea. What happens when spectacular amounts of FDI come into a place widely understood *by the people doing the investing* to be without the rule of law? What instruments and processes "protect" (to use Donald's formulation) foreign investment in places whose states companies deem incapable of or untrustworthy in the implementation of that protection? The sovereignty of the contract form—given moral and conceptual space by ideas of absence and lack—emerges as one answer. "Since most developing nations do not yet have established practices and stability in the rule of law and its application, oil companies seek to create a stable working environment through contractual means. . . . The less reliable the legal system, the more issues need to be addressed in the [PSC], as this contract effectively becomes a self-contained law" (Radon in Humphreys et al. 2007, 95, 100).

To reframe an argument about corruption and "the African state" that I laid out in the introduction, consider Transparency International, the World Bank, Freedom House, and the Economist Intelligence Unit, which widely rank Equatorial Guinea among the most corrupt and kleptocratic dictatorships in the world. In 2010, Equatorial Guinea occupied the 168th spot among 178 countries in Transparency International's influential Corruption Perceptions Index. In 2011, Freedom House put the country in its "worst of the worst" category, along with North Korea and Turkmenistan; in 2014, Equatorial Guinea did not even receive a ranking from Transparency International, as the organization deemed the country "too opaque to rank" (EG Justice 2014).[11] Anthropologically, this becomes interesting because "corruption" and "weak rule of law" are characterizations based on historically and geographically specific ideologies of power and governance (Elyachar 2005; Gupta 1995; Grovegui 1996; Mbembe 2001; Roitman 2005; Smith 2007). Thus, "corruption" or "weak rule of law" are not explanations in themselves, but characterizations to be explained. First, they are characterizations unified by *lack*: Equatorial Guinea *lacks* rule of law, transparency, and good governance. Consequently, as Mbembe (2001) explains, they tell us not what Equatorial Guinea *is*, but what it is not. "The upshot is that while we now feel we know nearly everything that African states, societies, and economies are not, we still know absolutely nothing about what they actually are" (12). One classical anthropological move here, and the one Mbembe encourages, is to produce work on what African states actually are—not what they lack, but what they have and how they work, and by what logics, philosophies, and histories. I agree that this work is crucial. Indeed, I do some of it in the next section on local law. However, at the same time it is also crucial to note the effects of these fungible categories—corruption, lack of rule of law—in circulation. While we anthropologists are busy showing the oversimplifications and commensuration-conjuring power of the World Bank or Transparency International, their metrics are having effects in the world. Notably, the daily making and remaking of Equatorial Guinea as a place that lacks rule of law provides the expansive PSCs with their conditions of possibility. These metrics create market conditions. The lacks and absences posed by this relentless thrum clear the way for inordinately powerful contracts that one would not find in Mexico, Malaysia, or Angola. First, then, I turn to an ethnographic account of rule of law on the ground in Equatorial Guinea, before shifting to the precise contractual tools—fiscal stability or stabilization clauses—through which oil companies seek to create "stable working environments" in Equatorial Guinea and beyond.

Lithium Batteries and Legal Ambiguity

In 2008, a major oil services firm was expecting a shipment of lithium batteries into Malabo's port. Instead of a notification from the port authority that the batteries had arrived, the company instead received a letter from a highly ranked customs official accusing them of bringing missiles into Equatorial Guinea. ("The tubes *do* look like projectiles," a manager of the firm's local subsidiary offered.) The letter notified the company that their property had been confiscated and was being held in a bunker in Fishtown, one of Malabo's peripheral neighborhoods. Explaining that there was a $500,000 fine attached to the import of this material, the letter concluded with a request from the customs official to set up a meeting in order to clear up the problem. Two company representatives (including the manager with whom I was in contact) attended a meeting with the official, during which he accused them of "wanting to poison our people, our country," and went on to explain that if they wanted their property back for less than $500,000, he had an environmental company that could pack the missiles/batteries and export them on the company's behalf. Following the meeting, the company representatives called their regional office in Douala, Cameroon, and the company's legal team dispatched a lawyer to Equatorial Guinea to look for a law relevant to the import of lithium batteries. The only law they found was vague, and the lawyer was unable to determine if the battery import was indeed in violation of the law or not. The situation ended when the oil services company agreed to pay $500,000 to get their lithium batteries out of the Fishtown bunker.

That local law exists in theory is widely acknowledged, and (as was the case with this oil services firm) foreign companies generally contract-out the practice of engaging it to local subsidiaries of international consulting and law firms, including Miranda & Associates, Ernst & Young, Pricewater-houseCoopers, and Deloitte, among others. As this story illustrates, however, in daily practice there is a tremendous amount of ambiguity and muddling through among both migrant company management and local officials about Equatoguinean law—Does it exist? Does anyone follow it? Do they have to follow it? Where the answers to these questions are unclear—*Is* there a law relevant to the import of lithium batteries in Equatorial Guinea? Where did the $500,000 figure come from?—companies often find themselves in legally murky dealings with Equatoguinean officials. When circumstances demand a decision more quickly than the timeframe within which legality might be established, Equatoguineans and foreign workers often act by mobilizing moral or "cultural" reasoning. Confronted daily with competing and partially overlapping regimes of "compliance," migrant oil personnel and the Equatoguineans who encounter them in various ways stitch together multiple legal codes, moral judgments, and ideas about "the other" into what amounts to self-fashioned frameworks for action, only "compliant" in the most superficial sense. The Smith Corporation's country manager described this improvisational decision-making process to me as "the rules of the game":

> It starts with the law in this country. What is the law? [This question is] associated with land title, the ownership of assets including oil and properties, the judicial system and dispute resolution processes. In West Africa there is the illusion of [resolution via] connections to power and intimidation rather than the merits of the case. The FCPA [Foreign Corrupt Practices Act] processes, that helps us here. It helps us reinforce that although local officials may want a special deal or money under the table, there's compelling legal constraints on the company and individuals that make me not do that. For us it's very helpful. Criminal, civil, environmental law—we have to understand those rules and regulations. But even if the law exists, it's hard to see supporting regulations. Oil and gas companies operate according to international law, or their own internal "what makes sense."

This manager describes simultaneous sources and scales of authority that unevenly inform his work: Equatoguinean law, on the one hand, and the perceived lack of enforcement on the other; "West African illusions" about personal connections and intimidation in dispute resolution; the Foreign Corrupt Practices Act;[12] international law; and what he called oil and gas companies' internal "what makes sense." Note that while these "rules of the game" start with Equatoguinean law, they end on a sense-making terrain *internal* to oil and gas companies. Quite literally in one sentence this man moves from the most "standard" and "global" to the most internal and par-ticularistic: "Oil and gas companies operate according to international law, or their own internal 'what makes sense.'" As with the "good oil field prac-tices" discussed earlier, this manager is intimately aware of competing reg-ulatory landscapes, many of which are at the end of their tether in Equa-torial Guinea, where there is little official oversight of any kind, whether national or international. Companies most often reconcile these unevenly overlaid sources of authority through *internal* processes, made licit by rhe-torical recourse to *external* authority (like international law or the FCPA), when necessary.

Certainly, oil and gas companies' internal ideas of "what makes sense" are just as "illusory" as what this manager called "West African" sense-making via connections to power and intimidation. Both are context-specific, reli-ant on localized moral cosmologies, hierarchies of authority, and notions of value. What often makes sense to oil industry actors is to find a way to keep extraction, production, and export running smoothly, in an effort to meet targets and keep profit, and ultimately stock price, up. What often makes sense to powerful Equatoguineans is figuring out how to insert themselves and their interests into these processes, so that all of this obvious profit doesn't pass them by. While equally context-specific and intelligible, these two forms of sense-making are not equal. The internal "what makes sense" to oil and gas companies is contractually buttressed not only by sweeping international legislation like the Foreign Corrupt Practices Act, or "inter-national law" more generally, but also by the bureaucratic micropractices of audit, record keeping, and signature, which, as we saw in chapter 1, have taken new form in the post-Enron era. In Equatorial Guinea's post-Enron moment, the internal "what makes sense" by which oil and gas company personnel seemed to abide was made licit by these mundane microprac-tices. Proliferating signatures, exhaustive documentation, and budgets transparent to audit seemed to be the practices that rendered compliant what was, in fact, messy and compromised daily decision making that led, for instance, to paying a $500,000 fine. The messy and compromised daily decision making of Equatoguinean officials, however, does not have an evi-dentiary regime for lithium batteries that is comparably licit beyond the country's borders.

The combination of mundane bureaucratic procedure, powerful fram-ing devices like the FCPA and Sarbanes-Oxley, and orientalizing ideas about "African" ways of doing things offers oil company personnel a tremendous amount of interpretive space in quotidian decision-making practices about whether or not to pay for those lithium batteries. "Corruption," in other words, is hardly avoided in daily practice, and yet the ensuing entanglements are compliant within extant post-Enron regimes of proof. The profitable perception that local law cannot be followed produces two consequences that seem contradictory, but actually reinforce one other: first, a tremendous amount of freedom to negotiate the licit/illicit on the ground, while *at the same time* making way for a powerful contract that gives companies inordinate power over local law. Before turning back to the contract itself, I dwell for a moment longer on local law, this time from the perspective of someone far more intimate with it.

Tarzan and the Archive Cemetery

Concrete walls and wooden louvers welcomed me into the familiar, cool dimness of a colonial-era government building, in this case, Malabo's aging Supreme Court. Angled just so, the louvers let in enough light to see without electricity, but not so much as to allow the sun to heat the long hallways and small offices. My footsteps echoed up the stairs as if I were alone in the building, and indeed it felt empty inside, an experience I had come to expect of most state buildings excepting the bustling police station. I had come to meet with a judge with whom I'd had passing, circumstantial conversations over many months (over dinner with mutual friends, for example), but had not yet spoken to alone or in an interview setting. As I wandered down the long hallway looking for his name on various doors, the first and only person I passed was an old man walking in the other direction, yellowing files in hand. I found the door with the judge's name ajar, poked my head in, and a secretary explained that he was not in, instructing me to take a seat back down the hall. The old man and I sat together in the foyer, looking out at the city between the louvers, me silently practicing what I was going to say, until the boisterous judge arrived. He welcomed me and gestured that I should follow him, and I ran to keep up with him down the hall as he shouted greetings to people I didn't see. I began in my usual way, reminding him that I was an American anthropology student interested in the oil industry. As often happened, this statement immediately yielded far more about "anthropology" as a traveling signifier than about the industry itself. After an interesting, if unsolicited, lecture on "black African" concepts of kinship ("in Fang there is no word for uncle") and group versus individual rights that were to him at the heart of the problems with postcolonial African law, I asked him about the implications for oil and gas investment in particular. "Ha!" he laughed. "Someone has to create a finishing school for Spaniards and Europeans!" He continued:

> The investors who come to Equatorial Guinea think that they are going to encounter Tarzan. Tarzan lives in the forest, where there is no law. So, because they don't have the proper information, investors go directly to the treetops. Then they get into problems and call me to try to help them find a solution. If I try to help them, they will talk about me as if I'm doing a side business. [The same foreign investors who get into trouble and ask me for help] are the first people to write internationally that there is no independence of the judiciary. Banana republic, this is the mentality that they have. These investors install themselves here illegally, without respecting a single law, and when things go badly, they say, "judicial insecurity."

The judge's critical assessment of FDI in practice offers an on-the-ground account of the circuitous logic and processes by which ideas about lawlessness circulate productively and reproductively in Equatorial Guinea's hydrocarbon industry. Foreign investors come in with a priori assumptions of lawlessness, which they then act on by installing their investments without regard for extant law. If there are legal consequences to their actions (they are kicked out of the country, or the state garnishes revenue), only then do they look to this judge for help. If he chooses to help them on their terms—a consultancy arrangement in which they need his expertise in local law—they often turn around to their communities of origin and use his intervention on their behalf as proof of judicial corruption and dysfunction. The judge's critique is useful to show how facile oppositions—us : them :: rule of law : corruption—fail to account for the co-constitutive and reproductive mobilizations of these categories in practice. Like the expatriate managers above who resort to self-regulation and murky practices in the name of "rule of law," the judge explains how foreign investors both circulate and reproduce their own ideas of judicial insecurity. After this beginning, however, the judge continued in a different direction. "But neither does reality here belie what the investors think," he said. "Local law is a real problem in Equatorial Guinea."

At the time of independence from Spain, the constitution of 1968 stipulated that Spanish law would apply until Equatorial Guinea began to create its own laws.[13] But the coup attempt the following year led Macías to annul the constitution. By 1973, Spanish law no longer existed in Equatorial Guinea. According to the judge, *from 1973 to 1979 there was no law*. Spanish law had been annulled. Macías issued decrees but never legislated, producing what the judge called "a legal vacuum" that lasted until 1980. After Obiang's 1979 coup, he reinstated Spanish law and began to legislate regularly, but the judge lamented: "Now there is confusion. People don't know what law is applicable here. When I go into court, we don't have archives or documentation. We have only been trying to get these together since 2004. This is essentially since yesterday. The principal problem is that no one knew what law was extant in Equatorial Guinea. Then along comes oil. What law applies?"

The implications of Equatorial Guinea's historic relationship to codified information changed with the coming of oil. Codified law, for instance, became newly important or useful. How information is or is not recorded, archived, or documented began to matter in novel ways. The new hydrocarbon industry became a situation in which, as the judge put it, "It's easier to find petroleum than it is to find information." He explained:

> The archives are a cemetery of records. So what happens? Bad lawyers introduce a different recourse; sometimes they invent laws. How far do they go? They'll say, "Law number such-and-such," and this law doesn't exist. Information is a real problem. In one case, I had to go look for a law in the archives. But I couldn't go into the archives in my tie. I came back in boots and a T-shirt with my nose covered, and I still got sick afterward. As I went through, I found laws that I didn't have. The archives are a general problem in the whole administration.

Postcolonial legal disarray, coupled with incipient projects to archive law and establish precedents, overlap uncomfortably with a transnational industry heavily and powerfully reliant on specific kinds of information and methods of codification and archive. The consequential intersection of Equatorial Guinea's postcolonial legal history with the transnational oil industry calls to mind Stoler's (2008) "political life of imperial debris, the longevity of structures of dominance, and the uneven pace with which people can extricate themselves from the colonial order of things" (193). Certain forms of imperial debris—here, Equatorial Guinea's legal system—add immeasurably to the longevity and strength of structures of dominance. That dominance finds expression through specific moments and mediators—here, contract in general, and below, fiscal stability clauses more specifically. If, to paraphrase the judge, *along comes oil and no one knows which law applies*, companies are only too happy to bring their own law in the contract form.

Fiscal Stability Clauses

I often thought that Donald would grow tired of me, or at the very least, offended that I wanted to discuss what I understood to be some of the more controversial aspects of the oil and gas industry abroad. To my surprise, quite the opposite occurred as our research relationship stretched over fourteen months. Despite my perception that my inquiries sought answers about increasingly controversial subject matter, Donald seemed ever more supportive as I posed questions requesting greater detail and specificity. Karen Ho (2009) makes a similar observation about her work on Wall Street. She was concerned that bank employees would not want to talk about downsizing, mass layoffs, or the expendability of employees, topics she understood as politically sensitive and potentially damaging to the image of the companies with which she worked. On the contrary, she found people talked about these strategies openly, and even those who had been negatively affected by them often narrated their personal experiences of layoffs and unemployment in terms of efficiency and shareholder value. In talking to Donald about contracts, corporate power, and fiscal stability clauses, I had much the same experience. Where I was concerned that my interest in fiscal stability clauses, in particular, would somehow expose a little-discussed and ignominious corporate practice, Donald spoke openly about the contractual tools. Like the licitness of disperse corporate geographies for tax "planning," here contractual clauses that override local law are a surprisingly unproblematic technique for those who use them.

Hotly debated in the late 1990s in international investment circles, and reanimated in recent years by the Trans-Pacific Partnership (see Stiglitz 2014), stabilization and fiscal stability clauses are related to the practice of "regulatory takings," or the right of corporations to sue states for regulations that affect their profit margins. In Equatorial Guinea and elsewhere, these contractual clauses attempt to guarantee two types of "stability" on the companies' behalf. First, they stipulate that the legal and fiscal regimes in place in the supply site *at the time the contract is signed*—environmental law, labor law, tax codes—will not change over the life of the contract. Second, these clauses stipulate that if those fiscal and legal regimes *do* change, and if those changes reduce companies' profit margins, the state is contractually obligated to indemnify the corporation. "If, for example, a new environmental law—even if it is of general applicability to all companies and is adopted to bring the country into compliance with international treaty obligations—would increase the cost of oil development or operations, then the oil companies would automatically be exempt from complying with such a law. Or, a government would have to compensate a company for the cost of compliance" (Radon in Humphreys et al. 2007, 96). While public policy in a variety of countries can trump the practice of regulatory takings—one cannot find these clauses in contracts with Canada or Norway—this is certainly not the case in Equatorial Guinea. With PSCs in particular, stabilization, fiscal stability, or indemnity clauses are arguably the most egregious contractual methods companies use to profit from inequality. "Through stability clauses oil companies limit the normal prerogatives of any legislature and government, such as their right to enact and issue protective environmental, labor, and other regulatory laws. These clauses are immune even to judicial challenge by the host country's domestic courts. . . . Stability clauses are too often contractual colonialism, the modern world's legal answer to a discredited system" (Radon in Humphreys et al. 2007, 95–96). Stability clauses are the ultimate "shield" to which Donald referred, through which companies claim to protect or disentangle themselves from local state lawlessness, arbitrariness, or avarice. It is through fiscal stability and stabilization clauses that companies claim to import or guarantee a "rule-of-law-based" environment.

Continuing to think with the judge's comments above, if stability clauses stipulate that the legal and fiscal regimes in place *at the time the contract is signed* will not change over the life of the contract, imagine Equatorial Guinea's legal and fiscal situation in the early 1990s, when the state signed the first PSCs. As a technocrat at the Ministry of Mines explained to me: "There was no legislation that dealt with the hydrocarbon industry in Equatorial Guinea. We didn't know anything about it. You could not draft an oil contract based on local law because there were no relevant local laws." Given this historical situation, what exactly is "stabilized" in the contract form? Most basically, oil companies and their profits become central to legislative and regulatory processes that are newly urgent, given the massive influx of cash, construction, pollutants, and humans that an oil boom always entails, stabilizing a peculiar and perverse relationship in which state actors have to haggle with oil companies to change laws.

What do fiscal stability clauses look like in Equatorial Guinea, and what are their effects? How do they work on the ground? Donald explained:

Our fiscal stability clauses [say that] the laws that govern this contract are those that are currently in place. The tax will not be more than 35 percent. If there's a change in law that requires the company to pay more, the government will step in and pay that. The clauses are mainly fiscal: taxes, import duties, the ability to control who you contract with and how you do the contracting. Here [in Equatorial Guinea] and elsewhere, the government will ask you to do business with specific companies. But our contract says we can contract with whoever we want as long as it's legal and ethical.

As he continued to elaborate the content of these clauses, Donald explained acts by the Equatoguinean state, including new laws or regulations, higher taxes, or requests to do business with specific contractors, as "encroachment" and "step by step nationalization." In response to these acts, the company could put up the "shield" of the stability clauses as a temporary solution while they figured out what to do. For example, in 2008 the Ministry of Health and Social Welfare issued a decree stating that all oil companies needed to start paying social security tax for migrant employees, equivalent to 26 percent of their salaries. Before this decree, the companies paid social security tax for their Equatoguinean employees and a 35 percent income tax on their own salaries. While companies with newer contracts had specific clauses shielding them from paying social security taxes on migrant employees, Donald's company did not, and they had to find what he called "general language" in the contract to help them avoid paying this new tax. "That's the fiscal stability part," he explained. Because this was a fiscal issue, the company was able to counter with the power of their stabilization clauses. In other circumstances, however, this was not the case. For instance, GEPetrol (the national oil company) formed a local insurance subsidiary in 2008, and companies were then required to obtain insurance through them. As Donald explained it, Equatorial Guinea "had learned from their neighbors." In Angola, oil and gas companies are required to buy insurance from local insurance companies. Because Angolan insurance companies are often without the collateral to underwrite the massive insurance costs required in the industry, they contract through international underwriters, effectively doubling the cost of insurance for transnational oil and gas companies. As with the tax issue above, Donald explained that companies with newer contracts in Equatorial Guinea have clauses that specify that they can get insurance from whomever they want. "They have shields. But ours doesn't have that specific language. Then the question is, is this a fiscal issue? It's not taxes. So then you try to work with ministers and officials, to tell them it doesn't make sense to get the same thing at twice the cost. For local business development we're willing to pay a little more, but we're not willing to pay twice the amount. This is encroachment."

Broadly speaking, stability clauses are at the center of what becomes a protracted, constantly negotiated relationship between companies and their representatives, and state representatives (generally from the Ministry of Mines) and their consultants. "Contracts," as one informant put it, "are like marriages. Just because there are difficulties, or even if one party breaks their vows, the marriage is not necessarily over. The relationship is too important and you're too much involved in one another's lives to end it just like that." Indeed, contracts worth billions of dollars annually, based on enormous outlays of capital and technology and subsea hydrocarbon that cannot simply be relocated, will not be annulled (to continue the analogy) just because the government wants to change an import duty or insurance policy. Instead, the party representatives will meet and negotiate, albeit on the radically unequal ground described above. Contracts yoke both sides into a protracted relationship of negotiation that seems to have its own teleology, in which Equatoguinean contract signers fight to regain the sovereignty and potential profit they signed away in the exploitative early contracts.

CONCLUSION

Contracts-as-marriage returns us to Durkheim ([1893] 1997), who famously declared that "everything in the contract is not contractual" (211). He continued, "Contracts give rise to obligations which have not been contracted for [and] 'make obligatory not only what there is expressed in them, but also all consequences which equity, usage, or the law imputes from the nature of the obligation'" (212). And the parties to a contract "are not only in contact for the short time during which things pass from one hand to another; but more extensive relations necessarily result from them, in the course of which it is important that their solidarity not be troubled" (217). With the luxury of more than a century of interceding social theory, this chapter's too-simple response to Durkheim is, *of course*. Of course, everything in the contract is not contractual, and a production sharing contract whose duration is the life of the concession, sometimes nearly fifty years, generates as much overflow as it does felicitous framing of risk and responsibility (Callon 1998).

I want to conclude, however, with a rejoinder to Durkheim and to neosubstantivist approaches more generally: that everything in the contract is not contractual is *not* a radical insight of critical theory. On the contrary, this was the assumption from which oil companies started. Working in central Africa or Indonesia or Kazakhstan is complex, overdetermined by intimate and complicated ties among corporations and local people and institutions and environments. That was Donald's point. Distance from local complexity, replicability, and standardization are *goals* of Donald's work. More accurately, they are *desires, aspirations.* He and his company started from the assumption of Equatorial Guinea's complexity and unknowability, and the contract became an admittedly imperfect effort through which to *aspire* to capitalism in its own image, to specify and standardize profit distribution or taxation rates, to peg interest rates to LIBOR or oil prices to Platts Marketwire. The one-thousand-page production sharing contracts, in other words, exist *precisely because* oil companies understand the wide-ranging social and political entanglements of their work, and they are constantly trying, mostly in vain, to control that overflow, to contain it, to anticipate it, and to write contractual clauses broad enough to account for its unaccountability. I insist on this point because I think there is a danger in offering a "social explanation" without (a) understanding that as shared terrain and (b) understanding that the work of the contract may be social through and through, but this insight does not by itself destabilize its formal effects in the world—for accumulation, for sovereignty, for the retrenchment of inequality, for the endurance of apparent singularities like *the corporation* or *the state.*

The banal provocation of this chapter, then, is that contracts work. They are one of the procedural cornerstones of this project we know as global capitalism. The work of contracts—their ongoing negotiation, their documentation of the desire and aspiration to approximate qualities thought to inhere in capitalism itself, as a stage for legal liberalism—is productive. The contract is a charismatic and felicitous form, and we fail to account for it at our analytic peril. We see, for instance, how the felicity of the contract form can yoke it to other charismatic processes like development or state-making. Moreover, contracts are good to think with because they show how capitalism is dependent on and constituted by its "externalities"—here, a repressive political regime and corporations eager to take advantage of the patchy legal landscape they find there. Finally, contracts help to draw our attention to capitalism's licit life; that is, the contract as ubiquitous practice that is legally sanctioned, widely replicated, and ordinary, at the same time as it makes markets out of inequality, domination, and imperial debris.

We stay with contracts in the following chapter, shifting to the subcontracts that intimately structure the lives of rig workers and oil service workers. Continuing to think with Pateman and Mills (2007) on questions of the contract and racialized domination as constitutive of contemporary capitalism, chapter 4 also examines subcontracting as a practice in and against teleologies of late capitalism.

CHAPTER FOUR

THE *Subcontract*

By the time I came to know Eduardo in 2008, he had been a Voxa employee for three years. Voxa is an Equatoguinean-owned subcontracting firm whose business it is to provide local laborers to international oil and oil services companies, a labor-brokering niche widely referred to in the global oil industry and beyond as a "body shop."[1] Foreign firms with operations in Equatorial Guinea pay Voxa per worker hired; Voxa takes a cut of each payment, and then pays their employees what remains. Among other contractual stipulations, the body shop guarantees the cleanliness, sobriety, and punctuality of the workers they provide.

Eduardo's first job with Voxa was on an offshore rig in drilling and lab analysis. After one "hitch," or *marea*, the contract between the company that owned the rig, SeaTrekker, and the operating company renting the rig, the Major Corporation, expired, and the rig moved out of Equatoguinean waters. After a mere two weeks on the job, Eduardo was out of work. His fellow subcontracted workers on the rig—Filipino, Scottish, and Venezuelan men, among others—stayed with the rig as it moved on to Angola or Nigeria or Ghana, because their subcontracts were with Laurel, an international oil services body shop whose contracts remain extant irrespective of national borders. Guineans, including Eduardo, were left behind, as their subcontracts were a fleeting product of "local content" requirements written into production sharing contracts (PSCs). Local content requirements are contractual clauses that require international firms to employ a certain number of local workers, clauses that lose their power once the relevant contract (in this case, a rig rental) moves beyond national borders.

After Eduardo's first job sailed, Voxa sent him to work for the Breffield Corporation in the construction of TurboGas, a gas-to-electricity power plant on the Endurance compound. Eduardo worked there for four months, up to the completion of the TurboGas construction project and the end of Breffield's contract with Endurance. From there, Voxa sent Eduardo to Hume Tools, another oil services firm, this time to work as a security guard for four months, after which that contract with Voxa expired as well. After the Hume job, Voxa sent Eduardo and other employees home, because there was not any available work. When I met him, Eduardo had again been contracted by Voxa and had been working for two months as a security guard outside a gated ShaeferCorp residence, another global oil services company with small private compounds in Equatorial Guinea. On the job a mere two months when we met, not only had Eduardo already been switched from one SchaeferCorp house to another, but he had also been told that he now no longer even worked for Voxa, whose contracts had been taken over by another local body shop. "This month," Eduardo explained with a wry grin, "we are with Silvano."

It is difficult to overstate the extent to which subcontracts and subcontracted workers saturate every level of the hydrocarbon commodification process (Maples and Rosenblum 2009). Once a PSC is signed, the proliferation of subcontractors in the given concession stretches from tool companies to catering companies; rig companies to road builders; transoceanic tankers to sea water sprayers; audit, tax consulting, and financial advisory services, to shipping and supply chain management companies. Subcontracts literally *people* the oil industry in its production sites in a way that PSCs do not. The actual commodity chain of a barrel of oil or 6,000 cubic feet of natural gas, which the general public understands to be a Shell or British Petroleum (BP) product, is, in fact, produced by hundreds of companies and thousands of employees, only a tiny fraction of whom are directly contracted by the operating companies or their subsidiaries. Consider again the FIPCO 330 from chapter 1—employing 115 men from twenty different nations, only four of whom worked directly for the Smith Corporation, and only twenty-five of whom worked directly for SeaTrekker, the rig company. The remaining eighty-six men were hired from fifteen different subcontracting companies, bringing the total number of companies at work on *one* rig to seventeen.

Ethnographic attention to subcontracting highlights the processes that bring workers to rigs and determine how long they stay there, how much they are paid, and even the comfort of their living quarters while at sea. Insofar as these conditions vary along predictable lines of racialized global inequality, subcontracting arrangements, like the PSCs that provide their conditions of possibility, still have us dwelling with imperial debris and hierarchies of rank, but now as lived in individual lives. As I explore in this chapter, for Equatoguinean rig workers like Eduardo, hierarchies of rank manifest in low-level industry jobs hoped for, gained, and lost; in children's school fees paid or not; in profound disgust and disappointment at the complicity of absolute rule and corporate freedom in Equatorial Guinea. For Filipino rig workers, hierarchies of rank manifest in their perceived "obedience" and "docility" as workers (Parreñas 2015; McKay 2007, 2014; Appel 2018b); their ostensible fungibility with US workers (one-tenth the value, Standard & Poor's tells us); their long rotation hitches; and their historical relationship with African American navy soldiers and US imperialism. For North American and Western European managers, hierarchies of rank are made manifest in uplift salaries; new vacation homes that await them upon their return from Equatorial Guinea; and radical upward class mobility. Given these distinctions, this chapter focuses on subcontracts as another ethnographic threshold through which to understand racialized inequalities as *constitutive* of capitalist markets, rather than merely exacerbated by them, or the "context" in which they operate.

Transnational labor markets of the kind I explore in this chapter—in which the relative value of workers is calculated by ratings agencies like Moody's or Standard & Poor's—are an exceptional example of Tsing's (2009) assertion that "no firm has to personally invent patriarchy, colonialism, war, racism, or imprisonment, yet each of these is privileged in supply chain labor mobilization" (151). Following Karen Ho (2016) and the work of the Generating Capitalism Group (Bear et al. 2015), and in dialogue with the Black Radical Tradition, my argument here and throughout the book is that global capitalism is made of, in, and through inequality (including racial fraternities and exclusions) and imperial debris.[2] This means, for instance, that the relationship between corporate freedom and absolute rule in Equatorial Guinea is not incidental to, but *constitutive of*, the daily life of hydrocarbon capitalism. The contract form, and the subcontract in particular, comes to play a meaningful part here. Just as the Equatoguinean government mobilized production sharing contracts as proof of licit financial transactions in response to the Riggs Bank debacle, so too with subcontracts, through which companies mobilize the contract's "neutral" and "impersonal" qualities to legitimate exploitative and racist labor practices.

My research revealed two particularly egregious forms of inequality and discrimination in and around US oil firms in Equatorial Guinea, and I attempt to chronicle both in this chapter. The first was the racialized "skills hierarchy" that structured life (and pay) for mobile, transnational workers in the industry, which we saw a glimpse of in chapters 1 and 2. This first form of discrimination was characterized by the mobility of strikingly unconcealed Jim Crow segregation among workers—in living facilities both offshore and on, in pay, and in rotation schedules. I specify Jim Crow here, following Vitalis (2007; see also Butler 2015), because these arrangements moved with and were reproduced by US companies. The second form of egregious inequality was the particular ways in which Equatoguinean workers were (and were not) included in the industry. In all but exceptional cases, when Equatoguinean workers were hired at all, it was for the most transitory and ill-paid positions, as Eduardo's story chronicles. Worse still, these positions were structured by *sub*-subcontracting arrangements, in which locally owned subcontracting companies like Voxa often forced would-be employees to pay for the mere chance of a job. Local workers in these arrangements widely critiqued the Equatoguinean state for keeping their wages from foreign companies down and their labor grievances unmet. This second form of discrimination, then, involved the apparent complicity of foreign firms and the Equatoguinean government in keeping Equatoguineans' relationship to the industry precarious and unequal. While these two forms of discrimination are linked through the figure of the subcontract, and hence I always imagined writing about them together in this chapter, there are other ways in which I kept them analytically distinct. Initially I understood the first form to be about histories of race peculiar to the US, made mobile in the transnational oil and gas industry, whereas I understood the second to be about histories of colonialism and its aftermath in forms of global inequality, where multinational companies and rapacious postcolonial statecraft meet.

Drawing on Jemima Pierre's (2013) work, this chapter moves away from that analytic distinction. My impulse to keep these forms of discrimination apart was based on what Pierre describes as a problematic disciplinary split, wherein "African diaspora studies generally concerns itself with articulations of race and Blackness but not directly with Africa, [and] African studies generally concerns itself with Africa but not directly with race and

Blackness" (xii). But if, as Pierre demands, we start from the *longue durée* of European empire making, we see that "conquest, the commerce in Africans, slavery . . . and the colonization of the Western hemisphere, the African continent, and Asia are . . . an interlocking set of practices that have cemented the commonality of our modern experience." "What is significant here" she writes, "is the racial dimension of this international system of power and the attendant global White supremacy through which it is enacted and experienced" (3). In sum, we can and should understand the mobility of Jim Crow and the particular kinds of exploitation visited on black Equatoguinean workers as historically related in "the nervous system of the liberal diaspora" (Povinelli 2006, 225). This is a global system in which white supremacy—again, "a political, economic, and cultural system in which whites overwhelmingly control power and material resources, conscious and unconscious ideas of white superiority are widespread, and relations of white dominance and nonwhite subordination are daily reenacted across a broad array of institutions and social settings" (Ansley 1989, 1024)—remains the norm. Tellingly, in Pierre's invocation of "global white supremacy" she cites Charles Mills, whose own work has chronicled the relationship between contract and racial domination (1997, 2003; with Pateman 2007). Having discussed the work of Pateman and Mills (2007) in chapter 3, it is their specific attention to contracts about property in the person that is directly relevant here—a contractual form that "constitute[s] relations of subordination, even when entry into the contract is voluntary" (3). Subcontracting becomes another entry point into their argument that "the global racial contract underpins the stark disparities of the contemporary world" (3). Contracts and subcontracts offer a stunning empirical site for understanding how race and other forms of postcolonial inequality are constitutive of both capitalism and the liberal political and legal theory from which capitalism draws much of its moral and historical justification.

In the sections that follow, after defining subcontracting, I work through the familiar teleologies of late capitalism by which both my migrant industry interlocutors and critical theory (albeit, in slightly different ways) often make sense of subcontracting. I note the ways in which these teleologies miss empirical questions of inequality and instead rely on ideas of capitalism in its own image—efficiency, shareholder value, progress. After showing how subcontracting arrangements long antecede (and seed) late capitalism, both in the oil industry and in Equatorial Guinea, I turn ethnographic attention first to Laurel Incorporated, a transnational body shop, and then to local body shops and the relationship between corporate freedom and absolute rule.

ON SUBCONTRACTING

A subcontract is a legal agreement between a party to an original contract (in this case, the subsidiary-many-times-over that signs PSCs as "the company") and a third party, which is contracted to provide all or a specific part of the obligations (work, materials) specified in the original contract. Standard interpretations of the business incentives for subcontracting include cost reduction (sweatshop labor being a classic example here, such as H&M subcontracting manufacturing to Cambodian children) or risk mitigation. Risk mitigation brings us back to the discussion in chapter 1 of the thinning of liability across an extensive web of contracts. Think of the finger pointing among BP, Halliburton, and TransOcean in the wake of the Deepwater Horizon conflagration, in which each company was able to claim that their specialized fragment of the production process was not causal to the explosion and then point the finger at someone else's fragment. This dispersed liability ("risk mitigation") is an intended effect of subcontracting arrangements.

The wider anthropological and critical theory literatures, and indeed my management informants, often contextualize subcontracting within the now-familiar epochal story of "late capitalism," "post-Fordism," "flexible accumulation," or supply chain capitalism (Harvey 1990; Jameson 1992; Comaroff and Comaroff 2001; Thrift 2005; Ong 1999; see Tsing 2000a, 2009 and Bear et al. 2015 for critiques). From both celebratory and critical perspectives, subcontracting here is narrated as part of a suite of (late) capitalist practices—just-in-time production, global supply chains, outsourcing—that emerged at a specific time. The *Oxford English Dictionary*, for example, dates the term "outsourcing" to 1981. When narrated teleologically, we are asked to understand these practices both as relatively new and as inevitable. But discussions with migrant oil company management in Equatorial Guinea and the country's own labor history reveal different temporalities and possibilities. What drops out of these epochal histories of late capitalism as management and even critical theory tell them? And even as ethnography allows us to see those elisions, what effects have they had, qua elisions, in the world? What does it mean, for instance, that Wall Street narratives of shareholder value aren't narrated alongside histories of colonial cacao plantations in central Africa (Hudson 2017a; McKittrick 2013; Robinson 1983, Williams 1944)?

Interview material from migrant managers and historical records from Equatorial Guinea allow us to work through these questions. While some of these narratives repeat predictable and already-known histories of late capitalism, they also contain moments of rupture. I want to start here, with predictability and rupture together interrupting subcontracting's epochal history, so that ethnographic surprises later in the chapter—for instance, why Filipino men make up one in every three workers at sea—can be seen not as anecdotal flourishes on an already-known teleological story, but as central to the ongoing project of global capitalism.

Predictably, my management informants narrated the subcontracted labor regime in which workers (although not them) are employed on a rotating, as-needed basis as an ever-intensifying arrangement guided by efficiency, industry-wide standardization, and the maximization of shareholder value. "Subcontracting is the way the business has evolved," one manager put it:

> It has to do with Wall Street and profitability. [It is] not cost-effective for one company to do everything. We don't need everything all the time. We don't need all these drillers when we are finished drilling. They're paid a good wage while they're here, and then when they finish they go home. Same thing for the Indians [doing] the construction. They follow [Breffield] around and do plants. They don't stay in the countries. They have engineers and civil teams who go in and do what they do over a two- to four-year period and then they go on to the next project somewhere else.

Or another:

> If you look at the oil industry almost thirty years ago when I started, employees of the companies performed most of the work. . . . Endurance produces the same volume of products as we did when I was first hired, but our workforce is probably 20 percent. If I'm in West Texas out in the sagebrush, you've got a field, and you've got wells every half mile or so. You build roads, power lines; you drill the wells; [there are] people out collecting oil. In the olden days, twenty-five or thirty years ago, company employees performed all that work. They dug dirt and got trucks for roads; you dug holes and strung your power lines. Over time, a lot of that has been outsourced to service companies, to subcontractors. It's more efficient. You don't have to maintain a huge inventory. You have a separate subcontractor that has an inventory for roads, another for tools, etc. [This setup] reduces the workforce so you don't *always* have to have road builders. The workforce decreased and service companies

and subcontractors increased. But we've gotten to the point of outsourcing accounting, IT, and it's maybe not the best situation. I personally think that we've taken it too far.

Wall Street, cost-effectiveness, the segmented character of the production process, discrete forms of technology and expertise required at different moments, and minimal inventories and workforces all come together in these managers' descriptions. Workers on rigs or in road-building companies move around the world, following industry-wide demand. "We don't need all those drillers when we are finished drilling," but the assumption is that someone else, somewhere, does. Whether the need is plant builders in Ecuador, rigs off the coast of Angola, or shipping in the North Sea, operating companies enter into subcontracting arrangements with oil services firms who deploy mobile personnel, expertise, technology, and infrastructure (Barry 2006). As discussed in chapter 1, these arrangements intensified in the industry with the advent of offshore production. Even in famously nationalist oil industries like those of Norway and Mexico, you find in offshore settings "a radical contracting-out of the production process" (Woolfson et al. 1996, 322).

If only a handful of companies float rigs around the world or provide deepwater drilling support, then those companies have technology and personnel in or near all major extraction sites around the world, organized by repeating contractual terms with many of the same major operating companies. Technologies, people, and contractual terms circulate, and the intended outcome is to "maximize standardization and repeatability in design, procurement, and construction, to introduce fit-for-purpose functionality into codes, specification, contracting and procurement documentation" (Woolfson et al. 1996, 311–312). Although this arrangement is flexible, varied, and reliant on unique entanglements at each site, its attraction for operating companies is in the repetition, the reliability, and the boilerplate functionality of the same people and technology moving around the world, most of whom they never contract with nor contact directly. When things go wrong, as they often do, *part of the reliability* here is both the licitness of the subcontracting form and the thinning of liability that comes with it; with so many companies involved, responsibility (or risk, depending on your perspective) is radically decentralized. Notice too that the empirics of this circulation or repetition do *not* create replication, or even deep standardization, on the ground in supply sites. On the contrary, they enable companies to import experts, technologies, and processes while having *minimal* engagement with local systems. The ability of companies and technologies to "operate worldwide" through extensive subcontracting systems takes geographic or political differences among extraction sites as a given, to be anticipated and managed through global subcontractors, whose processes are legalized, standardized, and often already in place. In this way, companies can appear ever-distant from local political, environmental, or labor concerns. So far, so familiar.

The migrant managers' accounts of this shift toward subcontracting in the industry foreground efficiency, just-in-time production, and an emergent relationship between Wall Street and American corporate profitability in the 1980s—often referred to as the shareholder value revolution—in which downsizing and mass layoffs first became commonplace (Fligstein and Shin 2007; Ho 2009; Khurana 2007). Both managers quoted above seem to buttress this periodization, implying a distinct epochality in the industry's use of subcontracting. The first manager comments that the business has "evolved" this way, relating that evolution to Wall Street. The second tells an evocative story of West Texas in which each company did everything for itself until a march toward specialization and outsourcing, which "may have gone too far." However, leaving the analysis here would lead us to miss other temporalities, other rhythms. First, the history of subcontracting in the transnational oil industry long predates the shareholder value revolution. Second, the teleology of "late capitalism" allowed industry personnel to explain as "inevitable" exploitative labor practices that, in fact, long predated any specific shift, and then to tautologically justify those practices in the name of inevitability. For instance, the workers that the first manager refers to as "these drillers," or later "the Indians," gestures to the fact that the segmented character of the production process takes preexisting forms of postcolonial inequality *as means of production* for "efficiency" or "shareholder value." "The Indians" contracted around the world for plant construction are "engineers and civil teams"—often formally educated and highly skilled, yet miserably paid and housed en masse because they are "Third Country Nationals" or TCNs. Third and finally, the epochality of late capitalism effaces labor histories in Equatorial Guinea and on the African continent more broadly, where various forms of subcontracting and labor mobility have been common at least since the colonial era, and have long linked colonial extraction with capitalist accumulation. I explore these interfering temporalities and rhythms below.

ON EPOCHALITY AND SUBCONTRACTING'S
MULTIPLE HISTORIES

The international contracting-out of the oil production process started in the industry in the 1920s and 1930s, anticipating by half a century the global commodity chain organization that came to typify other transnational industries in the 1980s or 1990s. Bechtel, for example, "had been building pipelines for Chevron since the 1920s, and by the 1930s had spun off one of the world's first full services firms for the industry" (Vitalis 2007, 67). Bowker (1994) writes of Schlumberger's work in the USSR, Venezuela, and Burma in the 1920s and 1930s. Vitalis (2007) also notes that "most of the first geologists, drilling crews, and camp bosses who worked in Saudi Arabia . . . learned their trade as wildcatters and contract employees elsewhere, notably, in the South American fields of Colombia, . . . and Venezuela" (54–55). If production in Saudi Arabia started in earnest in the 1950s, then these contract employees got their starts in Colombia and Venezuela in the 1930s and 1940s. From at least the 1930s onward, then, major oil services firms, as well as geologists, drilling crews, and camp bosses, sold their labor and expertise through subcontract to operating companies around the world.

In this history of subcontracting that long predates 1980s Wall Street, we can see the relationship between subcontracting and increasingly specialized forms of expertise—Bechtel's in the construction of oil infrastructures and Schlumberger's in geologic exploration and survey. But we cannot allow explanations of the technology or expertise-intensive character of hydrocarbon extraction to naturalize the forms of labor discrimination that have come to typify the industry. The industry uses the "natural" properties of the commodity—its geologic depth and pressure, its inaccessibility—and the specialized skillsets those qualities require to justify not only discriminatory labor hierarchies, but also the attendant practices of providing certain benefits (housing, insurance) to certain workers (often the white, directly contracted ones) and not to others (often the nonwhite, subcontracted ones). Subcontracting, in other words, has always been a contested terrain where inequality, expertise, and power meet. Consider a comment dating from the 1950s from Abdallah Tariki—the first Saudi to obtain an advanced degree in oil geology in the US, who then served as Director General of the Saudi Oil Ministry—on ARAMCO's use of subcontracted labor:

> ARAMCO found that applying the Law of Work and Workers to all the workers they needed in their operations would cost them a lot, so they introduced

the concept of contractors and vehicle owners. . . . And the mission of those new contractors was to collect willing workers, and the company would bring materials and engineers to train the contractor and his workers to do the work required of them. In this way the company was not responsible for arranging accommodation or health insurance or the care of the workers and their families. (Quoted in Vitalis 2007, 135)

In other words, the discriminatory subcontracting practices that I will explore at length in this chapter were already solidly in place in the industry by the 1950s. Yet, there was undoubtedly an intensification of subcontracting practices from the early 1990s, a timeframe to which both of my management interlocutors above refer. With oil prices at record lows, and with production costs escalating as easily accessible onshore supplies both began to shrink and were met with increasing resistance, companies were desperate to stay profitable and to find the surplus value to invest in new offshore extraction technologies. And yet, even considering this intensification of subcontracting, technological expertise, and global procurement chains over the last twenty years, labor and supply chain processes today still vary radically among supply sites, *belying ideas of inevitability and industry-wide standardization.* Consider Donald's description of his work in Russia in the early 2000s:

When I was in Russia, there was quite a bit of pressure to bring in global processes, global procurement, and after very long discussions we ended up doing everything 100 percent locally. [Everything was] available, half the cost, quality was adequate, and you avoided import logistical issues. That's just not the case [in Equatorial Guinea]. It's a small industry, relatively new, doesn't have manufacturing capability. In Russia [they] manufacture their own rigs, drill bits, everything. Ireland and Equatorial Guinea are similar because they have small industries and most of the service contractors are foreign companies that come in and set up local branches. [But it's] more cost-effective to bring things in locally.

In just a few brief sentences, Donald confirms that massive subcontracting is *not* inevitable; it is *not* neatly epochal; and it is *not* necessarily cost-effective or efficient. On the contrary, if the country can produce what the industry needs, it can be cheaper for the industry to source everything locally. Specific industrial and political histories in each place—in Russia, communist industrialization and competitive Cold War technological development—come to play a formative role in the ways the US-based oil and gas industry can, or cannot, operate in a given extraction site.

The oil industry is not alone in having a much longer history of subcontracting than epochal claims about late capitalism might suggest. Subcontracting has long genealogies in the extractive industries of southern Africa, in particular, where investors preferred migrant mine labor brokered through recruiting bureaus to avoid local entanglements (Van Onselen 1986; Moodie and Ndatshe 1994). Equatorial Guinea's colonial cacao economy also was sustained by a labor system that looks much like the rotating, as-needed labor of today's itinerant oil industry workers (Martino 2012; 2017; 2018b). Starting in 1906, the Spanish administration passed a series of labor regulations intended to control hiring processes, primarily through the creation of a mediator organization known as the Colonial Conservatorship. Like the body shops that administrate Eduardo's labor and the labor of those on the FIPCO (discussed at length below), the Conservatorship was to act as an administrative intermediary between employers and local and foreign workers (Campos Serrano and Micó Abogo 2006). The Conservatorship signed agreements for massive labor imports from Liberia in 1914 that were canceled five years later. After the 1930s, labor recruitment turned to Nigeria and Nigerians, whose contracts were to last two years "with the aim of not generating a new group of deep-rooted population" (Campos Serrano and Micó Abogo 2006, 31; Ejituwu 1995). As we will see below, this rationale is eerily similar to that of the oil industry, in which rotators are expected to engage exclusively with their work, regardless of where it is taking place. The implications of temporary, imported workers for labor organizing and workers' rights are clear: those without roots or rights where they work are less likely to organize.

With the "efficiency" that subcontracting promises now situated within historical time, and in relation to colonial and imperial relationships in particular, I turn to Laurel Incorporated and the daily life of the foreign subcontractor market in Equatorial Guinea today.

LAUREL INCORPORATED

Laurel Incorporated is a Scottish company with main offices in Houston, which, for tax and litigation purposes, is registered and operates out of Cyprus.[3] From their office in Equatorial Guinea, Laurel Incorporated matched transnational laboring bodies with labor needs of the local oil and gas industry. Paolo, a finance manager for Laurel, originally from Ecuador, explained, "Here we provide manpower, which is called in the worst kind of way 'body shops.' I don't feel proud of that." He continued:

> We recruit, look for people. We bring the person here via air transport. We secure visas, Letters of Invitation, Equatorial Guinea [work] papers. We pay employee expenses, give them cash advances which are deducted from their salary. . . . People work for the client, and we simply keep track of the work days, travel days, pay them, [and] process payroll. They're covered with insurance against all sorts of fatality or work-related accident. . . . But the employees are not covered when they get home. They only have health insurance as long as they are working in EG. When they get home, it's their responsibility.

Reminiscent of the Colonial Conservatorship that brought Liberians and Sierra Leoneans into Equatorial Guinea's cacao economy, Laurel recruits and administers labor from around the world for operating companies and oil services companies via subcontract. Once workers are hired, Laurel coordinates their logistics, from plane tickets to paperwork, and provides intermittent health insurance. Companies pay Laurel for workers' costs plus an additional 18 percent; after taking their cut, Laurel then pays the employees. Those who find work through Laurel can be fired without notice, and they are likewise "free" to quit at any time. Reflecting on how often people quit and how often they were fired, Paolo commented that "you play with people like chess pieces. The lie is that you work for stability, benefits, social security, but here nothing. They just pay you what you earn that day. The business doesn't take anything from your check, you don't take anything [from them], and they pay you a little higher rate, but not enough to make up the difference." The temporalities of this payment setup are worth reiterating: the men are paid and insured *only when they are working*. Because they work on rotating schedules, this means that they are paid and insured for two months while on the rig, for example, and then go home for two months without pay or insurance.

Taken together, people as chess pieces, intermittent pay and insurance with no guarantee of work stability, and vaguely connected webs of operators and subcontractors mean an attenuation or distancing between the worker and the employer. Indeed, this effect of distancing seems to be the key intention of what Paolo referred to as "the contract instrument." As he explained:

It's way easier to have control and expect results with the contract instrument in the middle. It's a client relation. [It's] much better to have contractors. You can demand results and track performance. In a contract, there's a structure; everything turns out to be converted into money. If you have a bad performance, I will pay you less. [This structure] incentivizes the company to give better service and the operators to get better service. To get things done [it is] better to give instructions, send out regulations, give orders. It is better to give to the subcontractor *the figure of contractual clauses in the middle*. Either this works out or it's over, and it will involve law and lawyers that no one wants. (Emphasis added)

Paolo suggests that having "the contract instrument in the middle," by which he means a subcontract, specifically, transforms the employer/employee relationship into a client relationship. He says that this contract instrument makes it easier for the company to have control and expect results by changing their relationship with labor from employer : employee to client : service provider. With the subcontract instrument in the middle, the operating company moves from employer (with all the attached rights and duties) to *client*, with all the attached privileges and entitlements. The employee becomes a service provider, upon whom it is incumbent to provide good service or risk diminished pay or termination. Note, however, that Paolo's understanding of the contract instrument is almost anthropological, insofar as he suggests *not* that these qualities of distancing and attenuation inhere in the contract form, but rather that contractual clauses become *figures*—*symbols* that invoke the threat of law and lawyers, expendability, the monetization of everything—and it is that symbolism, then, that guides the actions of the subcontractors.

Among the approximately one thousand workers Laurel Incorporated managed during my time in Equatorial Guinea, company documents show that roughly 40 percent were Filipino, 20 percent were British, 15 percent were Indian and Pakistani, 10 percent were from the US, and another 10 percent were South American (Paolo explained: "The majority of them are Venezuelan, because they have oil skills, they're good labor, and they're cheap; and also because Chavez fired 18,000 employees three years ago who tried to unionize.") The remaining 5 percent were a mix of Lebanese, Italian, and French. Nationality is central to this form of labor organization. As each worker is brokered through a body shop, his pay and rotation schedule are calculated according to his nation of origin. In Laurel's case (and as described in chapter 1), American and British laborers work a "28/28"—twenty-eight days on in Equatorial Guinea, and twenty-eight days off at home—considered the best schedule. Filipino workers have the least desirable schedule: eleven weeks on and three weeks off (an "11/3"), recently switched from a 14/3. South Americans work 8/3s.

When I asked Paolo why schedules varied by nationality, he replied: "They say that it's the market. . . . Companies take advantage of inequality in the economies of the world. Some people say that it is discrimination and it is, up to a certain point. But it is also working with the rules of the economy. . . . You bring in ten Filipinos for one American guy. Same human being working the same ten hours, with equal or better knowledge, and your business is running."

In paying and scheduling employees differently according to nationality, the industry is responding to a specific kind of difference—global inequality between nations—rendered profitable. In assembling a mobile transnational workforce, companies take difference into consideration, work with it, and profit from it, while ensuring they are absolved from responsibility for promoting or reproducing it. "You are paid according to passport," explained Paolo. Even he, as a finance manager, was paid as an Ecuadorian national. "I am an administrative manager very high up in the company, but I'm paid as a 'third country national,' even though I don't have a house in Ecuador, and I want to live in Australia." Despite his management position, his tertiary degree in finance, and the fact that he had no home in Ecuador and was in the process of migrating to Australia as his permanent residence, Paolo was paid according to his passport, guaranteeing a lower wage. Conversely, many US passport holders who worked as subcontractors in the industry actually lived in the Philippines or Central America. As US oilmen, they had met their wives while on previous assignments in Venezuela or Indonesia, and now rotated to those sites between hitches. But their passport guaranteed that they were still paid the US wage, details that belie any industry attempt to explain wage and scheduling inequalities through skills hierarchies.

The preponderance of Filipino labor in Equatorial Guinea's oil and gas industry is particularly illustrative of how histories of colonialism and racialized imaginaries constitute global markets. One out of three workers at sea today is Filipino (McKay 2014). In explaining this phenomenon to me, a migrant country manager offered his own naturalization:

> Worldwide in shipping there's a lot of Filipinos. Why? They're English-speaking. [They have a] willingness to work, good attitudes. [They are] good workers, friendly. If you go to Saudi, Kuwait, UAE, they have millions. That whole society is built on the back of imported labor, from Pakistan, Bangla-

desh, India. That's what these countries export. For our LNG facilities they went to the source of inexpensive but English-speaking, highly educated workforce. Over the years it's become a tradition [with Filipino labor], and it's almost generational. Their grandparents, aunts, and uncles were all involved in this sort of industry. In the early years it was exploitation: low wages, poor living conditions, but you found people willing to do it. What we find now is that the wages for these third-world people are creeping up worldwide.

In this man's explanation, Filipinos are English-speaking, highly educated, willing to work, and friendly. Moreover, it is "traditional" in their extended families to work at sea. How did this "tradition" of maritime work and this widespread characterization of Filipinos as docile workers come to be? These naturalized and racialized explanations of the preponderance of Filipinos in a US-dominated industry in Equatorial Guinea have a specific history. The US Navy was resegregated after World War I, and Filipinos took the place of African American seamen, receiving English-language nautical training in American colonial institutions in the Philippines. At the time, three years in the Navy qualified Filipino men for American citizenship, and over 100,000 per year applied for menial steward jobs. With their demographically dominant but hierarchically subordinate position in global shipping well established by the time the Philippines gained independence from the US in 1946, the newly independent state began to *market* its population as good subordinates, from men in shipping to women in nursing and domestic work (Parreñas 2008). The state marketed its citizens as possessing "inherent" traits of docility and loyalty, as the Philippine Seafarer Promotion Council claimed in its motto, "in loyal service to God, Country, and Company" (McKay 2007, 2014). The global labor market, in other words, is *made* through the colonial relationship of the Philippines to the shipping and military industries (McKay 2007), or Chavez's firing of unionized workers. Supply and demand are *made* by Jim Crow segregation of the US Navy, which created a labor demand that was ultimately met through the coercive promise of citizenship-for-exploitation to Filipino colonial subjects.

In the oil industry broadly speaking, there are unmistakable continuities between subcontracted labor organized, scheduled, and paid by passport, and earlier versions of the racial wage, both of which are intended to keep costs and worker organizing down, and profit up. And yet, I want to think beyond functionalist explanations of deterring unionization or even maximizing profit, toward this convergence that Paolo articulates of discrimination and what he calls the "rules of the economy"—of fungible Filipinos and global oil prices. To do so, I start with an expanded version of Paolo's explanation of the practice of hiring by nationality:

> Part of the business here is that manpower is provided by nationality. [Companies including ours] take advantage of inequality in the economies of the world. Some people say that it is discrimination, and it is, up to a certain point. But it is also working with the rules of economy. You have first, second, and third world. Of course, if you're a US citizen going to South America, you have more spending power. Your money is more solid than local money. A barrel of oil is sold in dollars, also quoted in pounds. If you have the possibility to bring people from other parts of the world and hire them by paying them what would be considered an acceptable wage for their position back in their home country, you really at the end don't have a problem: you're satisfying their needs—having a good wage—and the company is making millions of dollars. Endurance pays our company so much money in commission fees, but they are making all of the money because they are not paying people benefit plans, as they would pay an Endurance employee. Whatever money they spend, they get five times more money because of having different nationalities. You bring in ten Filipinos for one American guy. Same human being working the same ten hours, with equal or better knowledge, and your business is running. All the machinery is running. You're pumping oil and gas, and you're selling it abroad. Oil price is standardized all around the world, so there is the gap. People are expendable.

Paolo notes that "some people say it is discrimination, and it is, up to a *certain point.*" For Paolo, the certain point past which these hiring processes are not discriminatory is to be found in "the rules of the economy." Vitalis (2007) also notes that, when pressed on labor discrimination, the firms he studied insisted that "markets, not hierarchy, dictated that some workers received their pay in dollars, others in riyals" (23). The fungibility of ten Filipinos for one American off the shores of Equatorial Guinea, we are told, is a matter of markets, not hierarchy. But markets are *made* of hierarchies.

In Paolo's description, "the rules of the economy" are about the simultaneity of inequality and fungibility—ten Filipinos for one American; the spending power of US dollars in a South American economy; barrels of oil sold in dollars and quoted in pounds, hard currency denominations that index the geographies of power in which the oil industry operates. *Oil price is standardized around the world, so there is the gap. People are expendable.* Where labor value varies radically across the furiously maintained borders of nations, genders, and races, the price of oil—while unstable over *time*—is largely stable across *space*. The LIBOR rate or Platts Marketwire, discussed in chapter 3, determines the contractually sanctioned interest rates or price per barrel for oil, and yet there is no analogous procedure for labor. Or perhaps more accurately, there *is* an analogously licit and "market-based" procedure for determining labor value, with a radically different outcome.

Laurel uses a ratings system devised by Moody's and Standard & Poor's (S&P) to determine wages by nationality, from Americans and Brits at the top to "TCNs," or Third Country Nationals, at the bottom. On the one hand, then, an idea of *the market* absolves charges of discrimination or racism, while on the other hand, the very methods, tools, and metrics that make "the market" rely on and reproduce already existing categories of inequality, as in S&P's use of first, second, and third worlds. Following Cho (2008), Ho (2016), and others, this pushes us past both the social embeddedness theory of markets (in which "the social" can offer an explanation for that which was previously assumed to be "economic") and the social studies of finance theories of markets (in which devices, expertise, and economics in particular make markets). The point is neither that the global labor market is responsive to postcolonial inequality, nor that devices and theories imagined to describe markets in fact create them, but rather that the market is *made* by inequalities. Accreted histories of racialized inequality—including African Americans being kicked out of the US Navy and replaced by Filipino colonial subjects, or Paolo being underpaid for his skillset as a financial manager because of his passport—"proxy," in Karen Ho's terms, for rational, neutral market behavior, or "the rules of the economy." The market is not taking advantage of these circumstances; it is constituted by them. The contract form, here as subcontract, is a legalizing frame that offers stability and licitness to that constitutive process.

ON LOCAL BODY SHOPS

As rigs off the coast of Equatorial Guinea today leave for the Congo or Ghana tomorrow, the Filipinos, Brits, Indians, and others contracted by Laurel stay with the rig, rotating home as if, in many ways, nothing has changed. But the Equatoguineans, Nigerians, Gabonese, or Angoleños do not often find themselves moving around with the rig once it has left their waters, as their employment positions are most often the result of "local content" contract requirements, an additional subcontract removed from any type of job security. As the first significant production company in Equatorial Guinea, the Major Corporation brought subcontracting norms and practices with it in the 1990s. Shortly thereafter, Equatorial Guinea's Ministry of Mines set up an Agency for the Promotion of Employment (APEGESA) based on the body shop model. Filtering Equatoguineans who wanted to work in the industry according to family relationships and political affinities, APEGESA and other early employment agencies routinely took 40–70 percent of workers' pay, often charging Equatoguineans interested in working in the industry a fee for the privilege of accessing their services, before they even had a job. A local lawyer explained that "companies of this type proliferated, all of them [related to] the president. The first belonged to the Minister of Mines. The ministers were the ones who started them all. Catering companies, service companies, all were in the hands of children and nephews of the president and his ministers."

Tight control of local industry employment and the profits to be gained therefrom remained typical throughout the 1990s and early 2000s. However, as the industry boomed and foreign companies came in greater numbers, the hiring processes and even the local body shops themselves slipped beyond such centralized control. One Equatoguinean interlocutor who left the industry to go back to cacao farming described this change through his own experience: "I worked for Major in 1998, but I had problems because I wasn't in the party.[4] I am apolitical; I am professional. Joining the party to get a job isn't convenient for me. At the beginning jobs were more controlled in this way, but now there are more jobs, more companies, more demand than supply for qualified local workers. And the government is richer now, so there is less need to have their people in the industry."

Certainly, part of what motivated this change was the government's increasingly substantive and lucrative ties to the industry, rendering whatever profit or intelligence could be had from control of local workers insignificant by comparison. In addition, some operating companies began to organize direct recruiting, bypassing local body shops and offering "aptitude tests" and competitions, often recruiting Equatorial Guinea's "best and brightest" for jobs that turned out to be little more than menial labor. Finally, the Law on Employment Promotion established that all Equatoguineans who wanted work in the industry should register at the Ministry of Labor, where they would be given access to free employment-promotion services. Despite the creation of this law, however, the lawyer quoted above noted that "the private [subcontracting] firms still dominate, making money from the money of workers. They rob the workers. I have the records of groups of people trying to reclaim the money that was robbed; not only salaries, but also other unknown taxes."

Indeed, local body shops are hyperexploitative to varying degrees, and securing a job still almost always requires either money to the body shop up front or a personal connection of some kind. Consider the stories of Sara and Gloria, two domestic workers who were employed through a local subcontracting firm to clean houses on the small compound of an oil services company. They were in the compounds every day, where they cleaned windows, swept the sidewalks, mopped, and, as Gloria put it, "looked for dirt." The constantly running air conditioners brought dust into the houses in an uninterrupted stream, which they persistently fought, in between washing and ironing the residents' clothes, occasionally going to the market, and even cooking for the compound's inhabitants if they were asked. Sara got the position through her sister, who had the cleaning job before her; her sister left toward the end of her pregnancy, passing the job on to Sara. In explaining this story to me, Sara insisted that had her sister not been able to pass her the job, had she been "on the outside," as she worded it, she would've had to pay the equivalent of US$200 to $400 to access the job. "Supervisors negotiate on the side," Sara explained. "There is one supervisor for the whole agency and paying her obliges her to give you work. Every month you give her money until you get your job." When I asked if these payments were explicit body shop policy or mutually understood if unofficial norms, she replied: "This whole industry is about unofficial norms. Our bosses [at the oil services company] know. Each one is filling his pocket. There are official norms but nobody fulfills them. It is a norm to have the right to rest, to take medical leave, but they cut your pay, and your salary is left at nothing. So you are here, sick, working."

Gloria had a similar story to Sara's. She got the job through her brother-in-law, who worked for the local body shop. Her husband told his brother that Gloria was looking for work, and they hired her when a spot opened. She too insisted that it was only because of her family relationship that she didn't have to pay, and indeed that her brother-in-law accepted money from others as a matter of routine, most of which he gave to the same supervisor Sara spoke of (both claiming that this particular supervisor had a considerable amount of power). Even when subcontracting arrangements ended, which they frequently and unpredictably did as we saw in Eduardo's case, as Gloria explained, "you have to give her something, bribe her so that she will call you [if there's more work]. If you make an agreement with her that you are going to give her money and at the end of the month you don't pay, she won't call you."

Sara and Gloria each had a job that many Equatoguinean women would covet, due to its more or less steady paycheck and the occasional perk of surreptitiously doing your family's laundry in the washing machines at work, as opposed to by hand in the river or using the spigot behind your house. While they each recognized the benefits of their jobs, the women also felt disgusted and exploited by the subcontracting setup. Gloria explained that she was paid 200.000 CFA every month (roughly US$400), but that was after the subcontracting company took the first 50 percent of her original salary, paid to the body shop by the oil services firm for which she worked. Once that half had been taken, the government took half of what remained to pay for "the highways, the sidewalks, social security, and taxes." The money that she has left, she says, "gets you as far as your family situation permits." As the mother of two children, her monthly bills included 40.000 CFA for rent, 11.000 for electricity, 3.000 for a landline, and 2.000 for her mobile phone. Considering these bills plus the cost of food, Sara exclaimed, "To reach the end of the month, witchcraft has to be done! The children need transport to and from school, school fees, shoes, and you yourself have to buy clothes and braid your hair. With this salary you can work your whole life and have nothing." "Imagine!" she demanded, musing about where the body shop's 50 percent of her salary went before she even saw it. "I don't know how many of us work for Voxa. They don't only have SchaeferCorp, they have [contracts with] Hume Tools, Regal Energy, Expor, EGLNG. [They have] people who work in offices, as drivers, as logistics. We would like to know how much Voxa earns off our work. We want to know!"

The intricacies related to salaries between contractors and subcontractors—who made how much off of whose back—was a widely contentious issue for nearly all my Guinean interlocutors who worked in the industry, from those like Sara and Gloria who worked as house cleaners, to Eduardo and others who worked as security guards or semi-skilled labor on rigs, to Roberto and Rogelio who had earned advanced degrees abroad and come home to work in petroleum engineering and accounting, respectively. While Roberto and Rogelio had access to direct contractual relationships with operating companies, Rogelio explained that he had many highly educated Guinean friends who refused to come back, despite aggressive recruiting by oil companies. "They will always be strapped to a certain income," he explained. "They will always be Guineano." Equatoguineans with the initial capital needed to start their own businesses preferred that option to working for "Guinean" salaries in an American industry. "If you can offer more qualifications or more skills, it's demoralizing to work for them."

Subcontracted workers most often explained frustrations similar to Gloria's, that the body shop was making an undisclosed amount of money off of their labor, seemingly without having to lift a finger, and the government was removing "taxes" from their checks, but not providing services in exchange. Ramón, however, had a more specific story to tell. A temporary position as an administrative assistant in payroll for the Major Corporation gave him access to certain empirics of the subcontracting setup:

> The objective is that the parent company doesn't want to be responsible for all the labor costs; for example, transport, food, the problem of accidents. Anything of this sort, the company gives a bill to the [subcontracting] agency. According to law, contracting agencies cannot take more than 30 percent of their employees' salaries. So if Major pays one million CFA, the agency cannot take more than 300.000. But what happens is the exact opposite. What happens is that the agency pays me 300.000, and they can't justify it. There's no demand for them to justify it. Both the parent company and the agency have the idea that we know nothing, that we don't have the education to investigate this. But I know because I have worked in the parent company's administration. I have handled bills, and I know lots of things. There have been moments where I've seen an expatriate who does the same work as a Guinean, and he makes 10 million CFA each month. The same Guinean makes 500.000. The law talks about salary parity. If the expatriate does the same work, the Guinean cannot receive less than 50 percent of what the expat receives. If the expat makes 1 million, the Guinean should not make less than 500.000. But in practice it is incomparable.

In addition to noting how the subcontracting relationship allows "parent companies" or operating companies to disseminate liability for worker reproduction, Ramón points to the ways in which local body shops easily evade extant-but-unknown and unenforced law (see chapter 3). Despite the labor law which decrees that body shops cannot take more than 30 percent from each salary, leaving 70 percent (pre-tax) for the employee, Ramón insists that in practice some local body shops reverse the percentages, taking 70 percent of workers' salaries. Moreover, Ramón points to another law in which "salary parity" entitles Guineans to no less than 50 percent of a migrant's salary for comparable work. Already inequitable on its own terms, Ramón insists that this law is rarely followed either, and that salaries for equal work are "incomparable." I asked if he could give me a specific example of this situation, and he described a Guinean friend who had been trained in Auto-CAD, technical software for computer-assisted, three-dimensional design and modeling used in the industry. Ramón explained that Major previously contracted an American engineering firm to do this specialized work, but once his friend had mastered the program, he took over the design and planning for the tubes. "The American company charged a fortune," Ramón explained, "about £20,000 per month. [The Guinean man's] salary doesn't even reach two million CFA [the equivalent of £2,600 pounds or roughly 1/10th] per month."

Despite these radical salary inequalities for comparable work, £2,600 per month is an *exceptionally* high salary in Equatorial Guinea. Even the salaries of the semi-skilled, subcontracted rig laborers with whom I spoke, which averaged 700.000 CFA (US$1,400) per month, were exceedingly high relative to the paucity of other options outside civil service, where official pay was notoriously low. Security guard salaries were often a small fraction of that, on par with Sara and Gloria's wages at roughly 200.000 CFA, or US$400, per month. However, because of the relative size and reliability of these salaries compared with other options, Guinean's complaints to their migrant bosses of unequal pay for equal work, or unexpected and unremunerated contract termination, were most often met with familiar refrains: "*Locals who work for us make a lot more than they would have otherwise.*" Or "*Why do you complain? You have it so good compared to most people here. There are so many others who would jump at the chance to take your job.*" "If you complain," said Antonio, "they call you 'troublemaker' or 'problematic.' *If you want the job take it. If you don't want the job, you're free to quit.* That's the philosophy. [I do this job] just to maintain my family. The company knows that you don't have another option."

Paolo's reasoning above, that paying a Filipino worker one-tenth of an American worker's salary abides by "the laws of the economy," aligns with migrant managers' refrains of "*Why do you complain? You have it so much better than most.*" Inequality is tautologically justified by the inequality that preceded it; "ethical variability" (Petryna 2005) is naturalized as both market variability and human variability, which sees different standards of living for different people as natural, or always-already there (Benson 2008). We might also return here to the double meaning of freedom discussed in chapter 3. The "freedom" to quit, like the "freedom" to contract, is an illusory freedom. Without access to capital or the basic necessities of life, the choice is between being exploited and being hungry. In Pateman and Mills's (2007) words again, "Contract is the major mechanism through which these unfree institutions [the modern state and structures of power] are . . . presented as free institutions" (20).

CHAPTER FOUR

The realities of a $1,400 per month salary (let alone $400) in Equatorial Guinea are somewhat grim, especially when one imagines that this is Equatorial Guinea's aspiring urban "middle class." With their regular paychecks (if ephemeral subcontracts), Sara, Gloria, Ramón, and Eduardo all aspire to own a home, put their children through school, support wide extended families both in the city and in rural areas, access regular medical care, and more. The rig worker's words from chapter 1 come back to haunt this question of salary: "We are working like Americans but being paid like Africans."

Consider Antonio, who was sent to Canada by an operating company for training. While there, Antonio and a cohort of others were given a salary of $1,800 per month to pay for room and board, clothing, and other personal needs. He said:

> If you translate 900.000 CFA to dollars [$1,800/month], an American will say, "Why do you complain?" But I have five children, and I have to maintain them. It's like saying, "Which do you prefer? Abandon your family to study? Or abandon your studies for your family?" I ate bread with sugar in Canada to send $200 or $300 for my children. I was saying to my wife, "Bear it, it will get better." When my child was sick, I said, "Don't tell me; let me keep studying. I can't handle it and I don't have the money." When you talk to the company about your situation, they say that you signed something that said you agreed to this amount.

Antonio's last sentence captures the bitterness of subcontracting: "*When you talk to the company* [to ask for help], *they say that you signed something that said you agreed to this amount.*" The contract comes to be the figure in the middle, as Paolo invoked it, through which legal codes facilitate "the isolation and partitioning of responsibility" (Benson 2008, 209). In Benson's formulation, the legitimizing force of the law, here in the figure of the subcontract, "is experienced as a double negative" (209). Salaries for intensive work on which people cannot support their families are not *not* regulated; rather, they are sanctioned by the subcontract. They are licit. That the bosses' unsympathetic responses are, in fact, protected by the contract form "compounds the sense that depravity is sanctioned, even deserved" (209). Personal appeals for help from subordinate to superordinate—the conventional way of accessing resources for many people in Equatorial Guinea and elsewhere (Ferguson 2013)—and their potential for moments of mutual responsibility and understanding are short-circuited by the contract form.

The legalization of disregard for Equatoguinean industry employees' cost of living is brought into sharper relief when put next to migrant manager salaries and costs of living in Equatorial Guinea. Gloria listed the petro-inflated prices of rent, electricity, phone, and food, not to mention payments to the body shop manager, all of which came out of her monthly salary, leaving her in need of "witchcraft" at the end of the month to make ends meet. Migrant managers, on the other hand, pay none of these expenses for their luxurious facilities in Equatorial Guinea. Their gated compounds, electricity, satellite-based phone service, SUVs, and food costs are *all paid by their respective companies*. None of those costs come out of their salaries. As Gloria put it:

> The money the company spends *weekly* for food in this compound is equal to three and a half months of our salary. And they spend nothing of their own salary! They don't even bank it here! The company pays for their house and their electricity and their car and their gas, even the water they drink! Credit on their cell phones! We make so little, and we have to pay for everything. The expatriates are making a killing here. And us? I have been working for six years, and I have nothing. They work here for six years and become a multimillionaire in their country.

Indeed, for agreeing to work in a "hardship post," migrant manager salaries were often raised by as much as 75 percent, with cost of living nearly 100 percent subsidized. Banking all of these exorbitant salaries abroad, nearly all of my migrant management interlocutors found in Equatorial Guinea a ticket to radical class mobility. As I wrote in chapter 2, migrant management couples shared with me their plans for investing in a retirement ranch in Texas, a romantic farmhouse in France, and a surf retreat in Costa Rica. Recall one US woman who described shopping for a retirement home with her husband with "$5 million cash in hand."

    This juxtaposition of "uplift" salaries (a deeply ironic term, given its history in African American struggle) for white American and Western European management with the sub-subcontracting salaries and conditions for black Equatoguinean workers brings us back to Pierre's (2013) assertion with which I opened the chapter. Where we have the option to see the practices of Laurel Incorporated as the mobility of Jim Crow white supremacy, and those of local body shops as rapacious, postcolonial corporate- and state-craft, we can and should understand them together in "the racial dimension of this international system of power and the attendant global White supremacy through which it is enacted and experienced" (3). These stories of Eduardo and Paolo, Voxa and Laurel Incorporated, Sara and Gloria, Filipinos and the US Navy, and uplift salaries for white people are linked in what Pierre (2013) refers to as the longue durée of European empire making. In the last section of this chapter, I turn to the imbrication of the Equatoguinean state and the multinational oil companies in the licit organizing not only of exploitation, but also of abdication of responsibility.

ON RESPONSIBILITY AND THE OBSTACLE
OF ABSOLUTE RULE

Frustrated that the "social security tax" removed from his paycheck every month didn't seem to be showing up in any INCESO (social security) account, José sought to trace the question of responsibility through his contractual relationships. In the maze of operating companies, contracts, subcontractors, and body shops, the possibilities and avenues for redress—already tenuous in Equatorial Guinea—get further muddled. José traced the circuitous problem from the subcontracting agency, to the social security administration, to the contractor and around again:

> The agency [body shop] is taking advantage of the workers. For example, INCESO takes money from your check, but you go to INCESO and none of the money they've taken is there. But they say the problem is with Voxa. So right now I want to see the director. Every day they tell me, "come tomorrow." They begin to give you the runaround. At the end, you get tired and you leave it [*lo dejas*]. Concretely, we don't know how much SeaTrekker pays us. We've gone to them to ask, but they say they can't reveal it, that it's a secret of the contract. Now this I don't understand. Each of us would like to be contracted directly. I don't know the agency system, but it clearly doesn't work, the pay, everything. We should complain to Voxa, but we don't go because it's an agency of SeaTrekker, so we go directly to them.

The figure of the contract emerges again here, as José tries to determine how much SeaTrekker pays Voxa. "It's a secret of the contract," he is told. Also secret in this web is who, ultimately, is responsible if José's social security taxes do not appear in his account, tied also to his ability to access subsidized medical care and medication. José's frustrated routes trace contractual relationships designed to abdicate responsibility for him: from INCESO that sends him to Voxa, from Voxa that sends him to SeaTrekker, from Sea-Trekker that tells him what he wants to know is secret.

While workers held production companies and subcontractors primarily responsible for their secretive and convoluted salary situations, they also (with fear in their voices and looking anxiously around my apartment to see if someone, somewhere had installed a bugging device) frequently cited the government's role in allowing the companies to get away with it. As Ramón put it, "You go along with it because you don't have anywhere to talk about it." He continued:

> We have a saying here: No one can come to cut plantains behind your house if they weren't permitted by someone in the family. This means that everything needs the permission of the owner. The company knows the living conditions here, but if they encounter the father of the family—those with power here—they will tell them, "*You don't have to pay our people.*"

Ramón's parable is a thinly veiled allusion to the president and those in power. If *they* allow companies to pay incomparable salaries, or body shops to take up to 70 percent of workers' salaries, where can I seek redress? Or, as Gloria explained:

> The government and the expatriates share responsibility. Both of them lack humanity. Even though the expatriates are making so much money, you ask for a raise and they tell us we already make a lot. We have appealed to them so many times to see if they can help us resolve our issues, but nothing comes of it. They know our situation, and they do nothing. You realize that they are in cahoots with those in the regime to cheat the poor people.

Of labor organizing in the face of industry exploitation in Saudi Arabia in the 1950s, Vitalis (2007) writes: "Conditions for the nascent Saudi labor movement and the relative handful of officials who sought to move Saudi Arabia in a more inclusive and redistributive direction were, to understate the obstacle of absolute rule, inauspicious, and the firm there had a freer hand to deflect, ignore, and counter demands for fairness and human capital development" (24). In Equatorial Guinea, as in Saudi Arabia, there is an intimate relationship between absolute rule, labor organizing, and corporate freedom to contract labor on the companies' own terms. The president and his regime use their absolute rule to sign production sharing contracts that enable near-absolute freedom for the oil companies, producing the treaty-like capacities of PSCs and the subcontracts they proliferate. It is fair to say that all branches of the government under Obiang's absolute rule have failed industry workers in collusion with US oil and gas companies: from government-related body shops that pay exploitative wages; to the failure to enforce already inequitable salary parity laws or laws dictating what percentage of wages can be garnished by body shops; to taxes on already low subcontracted salaries that are said to go to social security or "training funds," when no medicine is available at the public hospital and training funds seem to go directly to the ruling party; to the absolute failure of the judicial system to address workers' complaints against subcontractors.

Many workers even claimed that the government had mandated a legal *maximum* salary for Guineans working in the industry of 900.000 CFA per month, prohibiting companies or subcontractors from paying more than this, regardless of training or skill levels. After hearing this from a handful of interlocutors, I finally asked Ramón why the government would create a legal maximum. He responded:

> Here there's so much money, but citizens don't have it. Salaries in the public sector are low, too low. So the only way you can have money is to engage directly or indirectly in corrupt practices. If the salaries are too high in the [oil] companies, this will create discord because most people in the administration don't have the opportunity to take money either. Why does the government step on/put pressure on [*pisar*] oil companies so that they don't pay us? Simply so as not to create discord, inequality between the administration and the companies. But it's something that's not written.

I was unable to verify from either the operating companies or government personnel if there was, indeed, a de jure or de facto salary cap. But the mere *plausibility* of this explanation, and the fact that workers I came to know widely understood it to be true, attests to citizens' relationships to and understandings of their government—its willingness to depress oil industry salaries in an effort not only to retain the attraction and power of government work, but also to slow the growth of a middle-class independent from the ruling party. Given the handsome profits to be had by oil companies from subcontracted (uninsured, precarious) labor, we see again here that the relationship between corporate freedom and absolute rule in Equatorial Guinea is not incidental to, but constitutive of, the daily life of hydrocarbon capitalism. Where companies would like to frame "government corruption" or the "inability to enforce regulations" as conditions external to their operations (as "local" or as state versus company), on the contrary, they are at the very core of these companies' daily affordances in Equatorial Guinea.

Because conversations about government complicity (or critique of any kind) were effectively illegal in Equatorial Guinea, in practice not to be had outside of one's most intimate circles, I came to know Guinean industry workers only through other workers. One would come for an interview only because someone he knew had vouched for me, and then he would come back and bring a friend. The next day the friend would bring another friend, and soon I had an extended network of industry workers I came to know over fourteen months. Routinely, however, they came in groups of three or four to my apartment, often two people I knew already initiating another friend into the odd ritual of talking to an anthropologist. At one point, a conversation in which Ramón, José, and Antonio were narrating endless labor exploitation nightmares and bemoaning government complicity seemed to trail off into discomfort, all of us looking at our hands, not knowing what to say to ourselves or to one another about the ugliness of what we were talking about, so rarely heard out loud. In our earliest meetings, I had been clear with workers that I could offer no guaranteed help beyond listening to them, recording their stories, and thinking with them about our unevenly shared situation. In that quiet, I was revisiting my pitiful caveats, sickened that the only thing I could think to do was to offer more juice when Antonio said, "Hannah, I know you can't help. But can you tell me, *is this normal? Does this happen in other places?*" I said that I wasn't sure, that I thought it probably did, and that labor exploitation in various forms was common everywhere, including where I come from. But then I continued that in other places, *people often feel they have recourse: they go to the media; they file a lawsuit; maybe they belong to unions or other solidarity organizations; maybe the church can help. They are not so afraid of telling their stories.* Elsewhere, I thought but did not say out loud, the convergence of corporate and governmental absolute rule did not seem so hermetic. The workers continued to look at their hands, obviously unhelped and perhaps sick of hearing, of knowing, that "elsewhere" things were different.

The relevant "elsewheres" for Equatorial Guinea, particularly in relation to the industry, were often Nigeria and Angola, regularly invoked in comparative frames by locals and migrants alike. Angola was widely recognized for its strong national oil company (Sonangol) and enforcement of training and nationalization plans for workers. Even Nigeria, considered a "model failure" in certain respects, was considered *better than here* for the substantial role that the Nigerian government and workers played in the industry. Rogelio, who had worked in accounting for multiple firms, explained that "in Nigeria and Angola, the driving force comes from the state [who says], 'We own this industry.' Whereas here the idea is, 'We own this resource.' Owning the industry requires the orientation of putting things and people in place to make it work." One operating company sent Rogelio to Texas for six months of training, and while he was there he met Angolans who had been sent by the same company for *five years* of training. The discrepancy, he explained, was not related to the company, but was between the governments. The Angolan government has a nationalization plan in place through which migrants can work in Angola; however, they can't stay for more than two or three years, after which a local has to replace them. A migrant logistics manager for the Smith Corporation had just come from years in Angola, where he worked closely on their nationalization plan. In Equatorial Guinea, he said, "there's no requirement to assure that service companies have something in place for locals. In Angola, there was the requirement to show a plan to advance the local population. Here, those opportunities have passed us by."

FUTURES

I like the work. Everything that we're doing is what I wanted. I love working in the field and not in the office. I love to learn, and I always want to learn more. [But] they're impeding me from learning what I want to learn. I feel frustrated. They tell us that we're learning really fast, as if that's a bad thing. I would like to be a mechanical engineer, but now they're saying that they won't offer that training. Now you are *stuck* at the level you're at. This is completely frustrating, the most painful. You start a job with your objectives perfectly clear, but then they tell you it's impossible, and you've lost eight years of your life believing something. This is the most painful. They promised us. . . . They just don't want educated local people. There is no explanation. . . . Major came and they told us if you want to be an engineer, you can do it, according to your abilities. But then, poof, it's all gone.
—Antonio

Frustrated by exploitative and racist subcontracting arrangements, low salaries, and a high cost of living; expatriates "making a killing" in a fully subsidized lifestyle; and no viable options for redress, many subcontracted Equatoguinean workers expressed a final, lost hope in the broken promises of training, education, or marketable skills that their time in the industry might give them. If the industry couldn't remunerate them fairly, perhaps at least when oil dried up, they would be fluent in English or have skills in AutoCAD, Excel, or engineering. But in the absence of government enforcement, early company promises—"*if you want to be an engineer, you can do it, according to your abilities*"—have disappeared. "Once we were in a meeting and the [American boss] said to us, 'I've contracted you to be capable of two things: if they call you on the radio, you need to understand [what they're saying], open and close the valves, and that's it.' The boss said that to us plainly. From there, well, to now . . . When I started in 1999, the idea [was to educate us], to give us the competence levels that expats had." One company agreed to finance the construction and organization of a technical training center, but then pulled out or asked to renegotiate when it became clear that the government expected them not only to finance it, but to build it, staff it, and run it themselves.

Perhaps that technical training center will be built one day. Perhaps, as many governments have before it, Equatorial Guinea's government will come to see "the most flexible fiscal environment in the world" moniker not as an asset, but as a euphemism for exploitative contractual terms, rampant environmental degradation, labor abuses, and corporate sovereignty at the expense of national interests. Perhaps, as many governments have before it, Equatorial Guinea's will change hands. Or perhaps the training center will end up where others have, in Equatorial Guinea's relevant elsewheres. "As if to mock the sad fact that [Nigeria's first oil town] is now a sort of fossil, rotting detritus cast off by the oil industry, a gaudy plaque dating from a presidential visit in 2001 sits next to Well No. 1. It is a foundation stone for Oloibiri Oil and Gas Research Institute. . . . But the ground has not been broken, and never will" (Kashi and Watts 2008, 37).

Futures always seem at once urgent and fated in the oil industry. The workers I spoke with knew that their opportunities to make money or receive education from the industry were finite and dwindling rapidly with each passing year. While many had effectively given up their own dreams of achievement in the industry, what remained in the more wistful moments was always the future without oil (Limbert 2010)—for their children, their homes, maybe even a life in agriculture. As one worker expressed:

> When [petroleum production] ends it will be worse. Because of this I am building [a house]. Even if you have worked, if you are without a house you have nothing. That's why I'm always building. You have to build so that your children have something. There is also information that other countries have suffered. I would like to work in agriculture, but right now agriculture can't feed my family. . . . But if agriculture is ever industrialized in Guinea, that's where I'm going, even with little money, that's where I'm going.

The persistence of the management refrain that those with ephemeral and underpaid subcontracts shouldn't complain brutally overlooks this future. It is a future in which the only sure thing is that oil will be gone. To accept these working conditions despite the fact that they can't prepare you for that future, despite the fact that they don't contribute to the vision of the country you have for your children, is to foment bitterness of our contemporary world's most violent kind:

> My fight is that my child has a good life. It doesn't matter to me if I have work or not. That's what I said to an American who asked me, "Why do you care what's going on? You have work!" I said back to him, "You didn't make America what it was. Your ancestors did. I want that for my country." This type of hate makes terrorists.

CONCLUSION

From the fungibility of Filipinos to future Equatoguinean terrorists, the subcontract is a frighteningly productive form. One could certainly argue simply that both the companies and the government want to make money, and workers—as always—are left by the wayside. Certainly that story would be true. But I have tried to attend in this chapter, and in the previous one, to the ways in which contracts offer licitness, legality, and legitimacy to those outcomes. Both production sharing contracts and subcontracts offer ethnographic entry points into imperial debris and hierarchies of rank. Again, in the words of Pateman and Mills (2007), we see that "the global racial contract underpins the stark disparities of the contemporary world" (3). More specifically, I have argued that both PSCs and subcontracts, and the long life of imperial debris in which they are forged, are productive ethnographic thresholds through which to think about racialized inequality as *constitutive of* capitalist markets, rather than merely exacerbated by them or the "context" in which they operate. If we consider subcontracts only within the teleological stories of late capitalism through which they are so often narrated, we miss not only the multiple rhythms and temporalities of global petro-capitalism—what happens in Russia is radically different than what happens in Equatorial Guinea—but also the way that global inequality *proxies* for ideas of capitalism in its own image: efficiency, shareholder value, and "the rules of the economy."

And it is to these "rules of the economy" that I now turn.

CHAPTER FIVE

THE *Economy*

BLACKOUTS AND SKYSCRAPERS

The first months of 2008 were dark in Malabo. The capital city went for days without electricity, stretching at one point to two weeks. Those who could afford it used private generators in the days *sin luz* (literally, without light) to keep businesses running, keep homes cool, or allow electric light, music, or television. The city filled with the clattering roar of generator motors fighting their flimsy steel containers, along with the stench of diesel exhaust. My neighbors—a Lebanese-owned restaurant and nightclub complex—had a powerful generator, the noise and fumes from which sometimes filled my small apartment so completely that staying inside became unbearable. Unable to sleep on one such generator-filled night, I opened my door to look for air, and to share water and complaints with Moussa, the Senegalese watchman who spent every night on the sidewalk outside the Lebanese complex. We chatted about the blackout. He said that Senegal provides electricity for many of its neighbors—for Guinea Bissau and as far away as the Ivory Coast. We laughed and said that Senegal should consider providing electricity to Equatorial Guinea as well. But, for all of Senegal's apparent success in the realm of electricity provision, Moussa spent every night sleeping on cardboard laid over broken concrete on Malabo's sidewalk, inhaling generator fumes, covered head to toe in clothing and plastic sheeting to fend off the malarial mosquitoes in the eighty-plus-degree heat. Even without electricity, and sleeping on the sidewalk, he seemed to think that Equatorial Guinea provided better prospects than his native Senegal.

In the first decade of the new millennium, Equatorial Guinea was among the world's fastest-growing economies. It is now the wealthiest country per capita in Africa (ahead of the Seychelles and Mauritius), and in 2013 saw more investment as a percentage of gross domestic product (GDP) than any other country in the world. It is eminently reasonable to assume, as Moussa did, that even sleeping on the sidewalk where the streets are paved with gold might get you a little closer to it.

On my block in the old colonial center of Malabo, what one might call economic life in a petro-boom was vibrant, if quotidian. In the Lebanese complex on the corner, the large restaurant with a menu priced for foreigners and the wealthy most often seemed empty, but the nightclub was popular with a young local crowd on weekends, and their corner soft-serve machine with strawberry or vanilla for the equivalent of fifty cents did steady business throughout the week. On the opposite corner was a Senegalese-owned restaurant popular with locals for the afternoon meal, outside of which a Beninois cobbler sat on the sidewalk, expertly fixing shoes for next to nothing. Behind him, a small door led into a darkened room, where another young man from Benin sat at a desk adding credit to people's mobile phones, changing money, and apparently selling some of the jewelry and home décor items displayed around him, although I never witnessed those transactions. Down the block, a local man who owned a small grocery store sold dry and canned goods, sodas and beer, candy, pastries, and basic cleaning supplies. The steps in front of his business served as a gathering place for the neighborhood's older men, who would stand talking in groups of two and three as afternoon cooled to twilight. Next door, Maria—an Annabonés woman—sold bananas, avocados, coca, soursop, tomatoes, and atanga according to the season from a low wooden table set up on the sidewalk in front of her house. A handful of neighborhood children would congregate and play on this corner, sometimes selling for Maria, and other times encouraging passersby to purchase a handful of peanuts from Maria's table so that they might eat them.

There was a lawyer's office on the other side of my apartment, and the lawyer therein had a reputation for his willingness to take on clients and cases opposed to the regime. A kind, old, and often drunk Fang man served as watchman for this office, although the post was clearly a ceremonial one. He and I would often exchange pleasantries in Fang, after which he would switch to Spanish and promise to bring me plantains from his farm outside the city. In the apartment above mine lived a local businessman and his impossibly elegant wife, recently returned to Equatorial Guinea from lives in London and Nigeria. She decorated their spotless apartment with imported IKEA furniture, while he worked to create a local merchants' association and run a computer store. Down the block in the other direction, there was a shoe store that sold cheap imports from Spain at incredibly high prices, although they were open to bargaining, and across the street was a small hair salon where I would get my hair washed when I had missed the twenty minutes of running water at 6:00 a.m. or was too lazy to dump buckets of cold water over my head.

Around the corner, the Ministry of Finance and Budgets was also routinely without electricity, despite the multiple generators (presumably broken) visible in its courtyards. With the street-facing door to the Ministry's archives constantly ajar, precariously stacked files and papers escaped into the street, caught by harmattan winds or soaked in the rainy season's mud. Further down the block sat Martinez Hermanos, a Spanish-owned, Indian-staffed supermarket that started as an import-export business in the colonial 1940s. In both its Bata and Malabo locations, Martinez had that hyper-air-conditioned, too brightly lit supermarket feel that might be found almost anywhere. But the display cases full of cellular phones selling for the equivalent of US$1,000, in contrast with the frozen imported meat that had clearly been under deep freeze for months, perhaps distinguished it as being in Equatorial Guinea. While Martinez's clientele were mostly wealthy, the city's majority shopped in smaller stores scattered throughout the city's neighborhoods and in the open-air markets of Los Ángeles and Semu. Even those who frequented the supermarket for imported packaged goods most often bought fruits, vegetables, meat, and fish at these other markets, where the staples came not from Spain, but Cameroon. Smaller French-owned shops always carried delicious, fresh-baked goods, but nearly everyone subsisted on government-subsidized bread—small, nearly hollow baguettes that sold for five cents each. Lines in front of these bakeries—run by Lebanese merchants who were widely rumored to have a monopoly on the import of oil and flour—were routinely fifteen to twenty people deep. Patrons would buy anywhere from one to thirty loaves at a time, while harried Equatoguinean employees would count out change with lightning speed. Outside of the supermarket and other establishments frequented by the wealthy, male youth between the ages of about ten and twenty would offer to carry your groceries, wash your car windows, or help with any other task either of you might dream up. Just behind Martinez Hermanos, you could look over a low wall down toward Malabo's port, into which large wooden boats overflowing with Cameroonian produce would motor on a daily basis to unload their goods.

While prices on staples like tomatoes and plantains fluctuated with border politics—Is the border closed with Cameroon? For how long?—and prices at high-end restaurants and stores were staggeringly high atop petro-inflation, day-to-day currency transactions were relatively stable. As a member of the Central African Economic and Monetary Community (CEMAC) region, Equatorial Guinea (along with Cameroon, the Central African Republic, Chad, the Republic of Congo, and Gabon) uses the Central African franc (CFA), itself pegged to the Euro and issued by the Bank of Central African States (BEAC) in Cameroon. While this arrangement provided certain stabilities, based on cross-border and postcolonial relationships that tie Equatorial Guinea to wider regional and transnational networks, other aspects of the daily life of money contrast with CFA convertibility. For instance, despite petro-inflation that made some aspects of daily life in Malabo outrageously expensive, in 2010 Equatorial Guinea remained an all-cash economy. It was not uncommon to see nattily dressed men and women walking out of the bank wheeling what any onlooker would know to be suitcases of cash. Credit or ATM cards of any kind were useless within national borders, from the fanciest hotel to the supermarkets. While one had to carry cash at all times, petty crime and theft were rare. It was safe to walk the city's streets day or night, despite the shared awareness that many people in the city carried large amounts of cash at all times.[1]

But none of this was why Moussa was in Equatorial Guinea. Moussa was there for the future. On my block and far beyond in the greater Malabo area, stretching unevenly outside the city into Luba, Riaba, and Moka, and into the continental region in Bata, Niefang, Evinayong, and Mongomo, the Equatoguinean landscape was plastered with elaborate signs. These detailed and colorful billboards depicted the large buildings that would soon stand in this or that patch of forest: the new BEAC Regional Bank, the new headquarters of the national gas company, a new refinery, or a new series of mansions and apartment buildings. In front of two matching luxury chalets newly built by the First Lady on the airport road, the sign read: *For Rent to Businesses or Individuals. Equatorial Guinea, Growing Day by Day.* Chinese construction workers lingered smoking by the unfinished elaborate fence going up around the mansions. Thousands of Chinese workers erected the new skyscrapers of

Malabo II seemingly overnight, and thousands more fanned out through the country paving roads and building dams. Arab Contractors—Egypt's largest parastatal construction firm—also had workers throughout the country, building a stadium here, a governmental palace or ministry there. Prominent signs for the efforts of major oil companies also dotted the landscape touting community development projects: *Building the Future of Equatorial Guinea Today.*

In a country the size of Delaware with roughly 700,000 inhabitants, new infrastructure saturated daily life. It is difficult to overstate the pace, the feel, the proliferating sites of physical and infrastructural change in Equatorial Guinea, and in Malabo in particular, during my fieldwork. For Malabo residents, the experience of these projects was visceral, sensory (Larkin 2013; Mrázek 2002; Mba 2011). We watched skyscrapers grow overnight and new roads unspool beneath them; heard the endless thrum of jackhammers, bulldozers, and trucks too big for old colonial roads; smelled the air full of cement dust, felt it on our skin, and tasted it in our mouths. Where in 2006 there was a huge billboard depicting a department store with a glass-front façade, with a Ferrari and its stylish white owner in the foreground (figure 5.1), by 2008 the blue and yellow Lebanese-owned Ventage store appeared, selling home appliances by Whirlpool, Black and Decker, and General Electric.

On the Sipopo road, where massive Ceiba trees climbed through dense equatorial forest and whales migrated not twenty meters offshore, signs sprouted for a new hospital. Sipopo le Golf—a luxury hotel and golf complex—grew there in 2011 to host the African Union Summit. Land was cleared as fast as you could blink, and signs declaring government ownership blossomed in the newly exposed red earth. Expropriation was rampant and all but incontestable, except for those most intimately connected with the regime, and even for them the process was protracted, cumbersome, and most often futile. In an almost farcical move, the president went to great lengths on television to explain how *even he* had been expropriated by this unstoppable future, as thousands of acres of his private land became "state property," rented to the large US oil companies for enormous sums of money. Huge tracts of new housing were springing up—small white homes with blue roofs as far as the eye could see. The homes were said to be subsidized by the state and intended for those who had been dispossessed, but were widely known to be for sale by members of government to the highest bidder. The development was called La Buena Esperanza, or Good Hope (Mba 2011; Appel 2018a).

The discovery of commercially viable hydrocarbon deposits brought in

Figure 5.1. Signs of things to come.

Figure 5.2. La Buena Esperanza.

its train not only innumerable infrastructure and construction companies, not only oil production companies large and small from the United States and Western Europe, Brazil, China, South Africa, Russia, and Malaysia, but also countless oil services companies (as chapter 4 describes) offering seismic studies and exploratory drilling; plant construction and rig rental; well heads, casing, and completion services; transport and shipping, submarines, and fireboats; catering and accommodation; and the list goes on. With new companies coming in at a dizzying pace, and all of them wanting to know how to invest, and how to navigate de jure and de facto laws and procedures, state functionaries were well placed to serve as gatekeepers and *socios* (associates) for transnational business ventures. Equatoguinean government appointees and others who worked with or for the state would moonlight as business and investment consultants. The petro-boom also allowed Equatorial Guinea to all but pay off outstanding debt to international financial institutions, which held nothing of the power or influence they once wielded. While World Bank or International Monetary Fund (IMF) delegations passed through Equatorial Guinea repeatedly during my time there, they were always on "technical support" missions and were rarely able to broker actual loan deals with any meaningful strings attached. Even USAID, attempting to operate a social fund in the country, was under an unprecedented arrangement in which the Equatoguinean government *paid* for their services, essentially using USAID as a development subcontractor (an agreement that had officially failed as of 2011).

In short, daily economic life in Equatorial Guinea was, on the one hand, quotidian—from blackouts to Moussa's migration, from sidewalk vendors to overpriced shoes from Spain. On the other hand, daily economic life was extreme—from the felling of swaths of equatorial forest overnight, to the tragicomic act of the president dispossessing himself, to the marginalization of international financial institutions. As anthropologists, we rightly approach these empirics through all their embodied frictions, refusing to flatten them into decontextualized analytics of growth or decline, statistics, demography, or other tools of social engineering. *The* economy is perhaps chief among the flattening analytics available to capture daily economic life as I've described it here (Mitchell 2002; Young 2014, 2017; Appel 2017). Rather than refuse this simplification, however, this chapter starts from the teeming scene in Malabo to trace the making of this unified object called "the national economy" in all its authoritative and fetishized surfaces, and that apparent singularity's role in the licit life of capitalism.

SERIALITY, INEQUALITY,
AND ECONOMIC EXPERTISE

> What is a national economy and who controls it?
> —Jean Comaroff, with David Kyuman Kim,
> "Anthropology, Theology, Critical Pedagogy"

In Equatorial Guinea as elsewhere, the economy is a privileged object, perhaps *the* privileged object, in official discourse. State actors and multilateral institutions articulate futures in its terms—development, diversification, growth. In 1983, after his third trip to Equatorial Guinea, United Nations Special Rapporteur and Costa Rican law professor Fernando Volio Jimenez reported that "one official after another all the way up to Obiang himself" justified the limitations on the press (there was no press) and on political participation (political parties were banned) "as being necessary for the focusing of attention on economic issues" (Fegley 1989, 220). Over twenty years later, during my time in Equatorial Guinea, the persistent unreliability of electricity, education, potable water, and healthcare, not to mention the continued draconian limits on press and political organizing of any kind, were similarly justified by the need to focus first on economic development. Here, the economy is both the object of the future and the justification of the present's constant deferral.

Where the first four chapters of this book refused the timeworn oil-as-money approach, this chapter and the one that follows return to this question. But, having now looked at the offshore, corporate and residential enclaves, contracts and subcontracts, we can return to the question of oil as money with a much more capacious sense of the processes and projects that long precede oil's specie transubstantiation. In the expanded understanding of oil's conversion processes that I've explored thus far, transnational corporations come under direct ethnographic scrutiny. In contrast, in the oil-as-money literature (both its anthropological and resource curse variations), the industry recedes. As I argued in the introduction, in the oil-as-money literature, the industry itself is all but invisible, merely a revenue-producing machine, a black box with predictable effects (Appel 2012c). And, once the industry has disappeared from view, the well-documented pathologies of oil-exporting places then appear to reside only in state mismanagement of oil money—here, *the pathological African state*—rather than at many different points within the carbon network, from racialized labor and domestic intimacies to contracts and infrastructures. In resource curse theory in particular, state mismanagement of oil money is made visible in something called *the national economy*, which can then seem like something "out there," somehow separate from local experiences of state violence and corporate power. (Recall from the introduction my haunting experience with the HR manager who passed me an article on the resource curse, lamenting state graft but overlooking the agency of his firm entirely.) In this chapter, in my effort to understand the power of the national economy form, I follow the work of resource curse theory in the world via an ethnographic account of Equatorial Guinea's "first" and "second" National Economic Conferences, held in 1997 and 2007, respectively. The chapter weaves in and out of conference proceedings to examine a constitutive tension in the idea of "national economy," where *national* signifies a naively Westphalian framework within which "economies" are still conceived, and *economy* signifies an imagined space of private accumulation from which the state must recede. I explore each of these terms in turn. First, with regard to *economy*, I look at resource curse theory, economics "in the wild" (Callon 2007), and the embodied fantasy of an idealized private sector. In the second half of the chapter, I turn to *national* as a signifier in this pair, focusing ethnographically on the aftermath of the Riggs transnational banking scandal and national budget documents, to think through the relationship between public office and private gain in the wake of oil.

Taking something called a national economy as self-evident—something "out there" that must be made to grow, and on which the future of any and every given place depends—asks that we overlook the translocal histories and political processes of its creation (Mitchell 2002, 246). As has been the case throughout this book, part of my project in this chapter is to narrate those histories and processes as they exist in Equatorial Guinea, showing them to be integral to the making of an object ostensibly a priori to them ontologically. But more pressingly, I want to show how this ability to refer to the national economy as an objective measuring tool or space of intervention, in fact, creates much of the object's power. By insisting on "the economy's" power at the same time as we attend to the processes of its making, we begin to see the national economy as part of the project in which Equatoguinean petro-capitalism is made *licit*.

In its bluntest representation, a given economy is simply a statistical aggregate, and one widely acknowledged as distressingly approximate and incomplete at best. At the same time, *the economy* is arguably the most privileged epistemological and political object of our unevenly shared modernity. Part of this privilege comes from the seriality of national economies, their comparability from one place to another *as if the same*. This seriality and comparability efface histories of empire and the radical inequalities produced therein (Speich 2011; Young 2017; Mitchell 2002; Appel 2017). As Equatorial Guinea's national economy was remade with the arrival of major US oil and gas firms, these histories of domination were the terrain on which its attendant fallacies of liberal equality and the spectacular accumulation of petro-capitalism were built.

Having returned to oil-as-money, we will see in these last two chapters how the mundane bureaucracies of capitalism—accounting, statistics, metrics like GDP, national budget documentation—are constituted by postcolonial, racial, and deeply fraught local politics which form the ground for economic expertise. While I go into an extensive historical and theoretical account of "the economy" elsewhere (Appel 2017), this chapter focuses on how the national economy form can render the radically unequal postcolonial order licit and apparently subject to scientific management. Because the particular science in question is economics, this chapter returns to the spaces of "as if." As I asked in the introduction, how do oil and gas emerge *as if* untouched by the messy frictions of cultural production that we see from enclaves to subcontracts? "The" economy—its reliability as a serial transnational form and apparent separability from corporate and state power—is one of the most powerful "as ifs" routinely available to the project of global capitalism.

NATIONAL ECONOMIC CONFERENCE

In 1997, Equatorial Guinea held its first National Economic Conference. There had been national-level conferences before, but the 1997 conference was distinguished by the fact that it was conceived, documented, and publicized by its state organizers as Equatorial Guinea's *first* National Economic Conference. Oil had been discovered three years earlier by an independent US oil company, and small amounts of money from exploratory contracts had just begun to circulate back into Equatorial Guinea. This was a dramatic turn of events for a microstate characterized by unprecedented economic collapse in the late 1970s, creeping into a crippling debt burden by the early 1990s. In December 2007, two months into my fieldwork and ten years after the first National Economic Conference, Equatorial Guinea, now flush with oil wealth, held its second National Economic Conference. Here again, despite conferences on other themes in the interim decade, the "second" designation was an official part of the conference's full title: National Economic Conference II. National Plan for Economic Development. Agenda for the Diversification of Sources of Growth. Equatorial Guinea toward Horizon 2020. How might we understand the idea, implied in those "first" and "second" descriptors, that a national economy, or at least a conference on it, was something new in the wake of oil? Seemingly ahistorical or amnesiac, the idea that a 1997 conference could be called the "first" regarding something called a "national economy" comes into sharper relief by going back briefly into Equatorial Guinea's colonial and immediate postcolonial history, some of which I narrated in the book's Introduction.

Recall, for instance, that as Macías progressively lost control in the first years of his rule, "after 1970 there was not one reliable economic figure, government statistic or census report to be found in the country" (Fegley 1989, 72). After eight years of *la triste memoria* under Macías, in which being identified as an intellectual was reason enough to be put to death, and statistics were officially illegal, Obiang's coup brought foreign observers back into the country for the first time. Dr. Alejandro Artucio, an Uruguayan lawyer invited to Equatorial Guinea to witness Macías's trial by the new regime, offered a firsthand account of the economic situation he found in 1979:

> It can fairly be said that the economic, social and cultural situation I found in Malabo was bordering on a national disaster. A few data will suffice to explain. The economy was to all intents and purposes paralyzed. The basic services—electric power, transport, the post, banking, communications, etc.—were virtually at a standstill. It was the practice for all citizens, including salaried government officials, to go into the forest in search of fruit and other food for themselves and their families. Macías, as Head of State, took the national treasure to his palace of Nzeng Ayong, [and] . . . he administered the funds of the state from his house. He had not ventured into the capital for five years. . . . Commerce is practically at a standstill in Malabo. There are only a few small shops . . . still open but their shelves are empty. There is not a single restaurant in the whole town. There has been no electric power since 1978. For four years or so, there has been no written press in the country because, according to the government, there has been no paper. (Artucio 1979, 14–15)

As Fegley summarizes, "Nowhere at any time had an economy collapsed in the sense that Equatorial Guinea's had by 1978" (1989, 155). A separate sphere called a national economy—recognizable and circulatable beyond national boundaries—is predicated first on the act of money changing hands, and second, on certain forms of representation of those acts in the aggregate: record keeping, accounting, statistic and demographic data—and paper.

Although foreign aid poured into Equatorial Guinea with the advent of Obiang's rule, and the country joined the CEMAC region in 1988, those renewed ties had little effect on an already-fraught relationship to certain forms of knowledge production dating back to the colonial era. Even in 2009, the IMF wrote of the country that "data on the national accounts, balance of payments, and inflation all have significant limitations and make it difficult to get an accurate representation of the performance of the non-oil sector and domestic price developments" (16). In addition to noting lacunae in the specific forms of knowledge production required to *make* a national economy in the normative sense, this report also draws our attention to an interesting detail: the IMF had the hardest time finding data for the *non*-oil sector. Although they don't specify why, one can infer that oil industry accounting makes up for the data gap, and that Equatorial Guinea's public administration is obliged to keep certain kinds of records in the oil and gas sector that are not obligatory outside those boundaries.

From erratic colonial administration, to the Franco regime's complicity in official silences, to the national treasure in Macías's Nzeng Ayong home, to the Ministry of Finance and Budget's records soaked in the street, Equatorial Guinea has had a long, fraught relationship with the kinds of enumeration and recording practices constitutive of the national economy form. And yet, with oil, that form was interpellated in new ways. Contracts with major multinational firms, required the making of a series of documents—budgets, reports, economic forecasts—and procedures for accounting, accountability, and audit, among others. An acceptably documented national economy became both newly demanded and newly possible, hailed by the visibility of a single, and singularly profitable, global commodity. The economy, in other words, did not predate the commodity in any simple way, as a separate sphere ready to be populated. Rather, the commodity's circulation *made* the economy both possible and necessary in new ways.[2]

UTOPIA AND DYSTOPIA I:
THE RESOURCE CURSE AND SALT INTO GOLD

The future remains a stranger to most anthropological models of culture. . . . Economics has become the science of the future, and when human beings are seen as having a future, the keywords such as wants, needs, expectations, calculations, have become hardwired into the discourse of economics. In a word, the cultural actor is a person of and from the past, and the economic actor a person of the future.
—Arjun Appadurai, *The Future as Cultural Fact*

In the year 2020, Equatoguinean society will be dominated by the middle class, with work and regular income available to all citizens, who will own their homes, whose children will be educated, and who will be capable of meeting their health needs.
—República de Guinea Ecuatorial, "Plan Nacional de Desarollo Económico y Social"

The 2007 Horizon 2020 Conference was held in Bata, Equatorial Guinea's second city. Unlike claustrophobic Malabo, where you could barely tell that the ocean was just over there unless you peered over a wall, Bata felt expansive, with wide roads, fewer people, and room to breathe. Although daily activities of governance took place in Malabo, the public administration moved, working periodically from a second set of ministerial and administrative buildings and palaces in Bata.[3] Thus, for those living in Malabo, if one needed an official to sign a document or issue a permit, it was not uncommon to hear that "the government is in Bata," suggesting that unless you wanted to fly or take an irregular boat to the continent, that need wouldn't be filled any time soon. During preparations for the National Economic Conference, radio announcers and empty streets reminded Malabeños daily that the government was, indeed, in Bata.

At the conference, the president and the prime ministers, ministers of government, oil company representatives and their local liaisons, national and international business owners hoping to invest, the World Bank, the IMF, USAID, Washington lobbyists, United Nations Development Programme personnel, European Community and Spanish Cooperation delegates, diplomats from throughout the region, local business men and women, and who knows who else were all brought together in the name of the future: Horizon

2020. After an initial day of registration, opening ceremonies, and closed-door meetings, Day 2 included four concurrent, day-long presentation sessions by conference participants on infrastructure, social sector, public sector, and private sector. I followed the decidedly largest portion of the crowd into the private sector session, held in an opulent conference hall with seating for perhaps five hundred people. The hall's front wall was painted floor to ceiling with a romantic mural of our host city. Leafy cacao and ceiba trees framed the foreground, inviting the eye in toward an idyllic rendering of the newly constructed port just behind. The harbor glinted clean at the edge of shimmering blue water, plied equally by oil tankers and *cayucos*—dugout canoes in common use throughout the country's coastal regions for fishing, transport, and recreation. With Bioko Island and Malabo visible in the background, the entire painting glowed a warm pink-orange, lit by the flaring oil platforms painted pastorally at the mural's bottom left. With the mural as a backdrop, men in suits began to fill a series of chairs at a long table behind large white computer monitors. Oversized plastic flowers and plants lined the floor in front of the table. This artificial hedge was interrupted only in front of the speaker's podium, inexplicably embellished with a three-foot-tall plastic martini glass placed on the floor in front of it, with two shorter versions on either side. As we took our seats, there looked to be about two hundred or so people in the room, about 80 percent of whom seemed to be Equatoguinean, mostly men, talking with one another in Spanish or Fang. The first two rows of the auditorium were filled with white men, and young Equatoguinean women in conference uniforms scurried around the front of the auditorium distributing water bottles and seeing to the needs of the dignitaries in the audience.[4]

The printed materials we were given as conference attendees were startling. As we entered, we were each handed a navy blue plastic carrying case, modeled after a briefcase, containing more aggregated information about Equatorial Guinea than had arguably ever been released into the public. Each plastic briefcase included three main booklets of about seventy-five pages each. Printed in vibrant color with extensive graphs and charts, the first booklet offered an economic and social diagnosis of the country; the second, strategies for improving that diagnosis over the next thirteen years; and the third, a poverty profile. Each briefcase also contained five smaller booklets of demographic and statistical analyses, including one documenting the results of the country's third general census in 2002. Within the context of Equatorial Guinea's long and fraught history around these forms of knowledge production, and the judge's quip—"it is easier to find petroleum than information"—audience members marveled quietly at the material. *Where did it come from? Who wrote it? Who produced the statistics? Is there any reason to think that any of it is accurate?* The materials quickly became a hot commodity after the conference and were essentially impossible to get a hold of had one not been among the lucky few in attendance. Even at the conference itself, they ran out early.

What is it about oil, if anything, that produces such unprecedented displays? Certainly, there is a performance for the international community; and as Callon (1998) would have it, and Butler (1993) and Mol (2002) too, there is performance going on here. But what exactly is "in formation" or "being formed"? As I will suggest below, weaving through the themes of the resource curse, an idealized private sector, and the pairing of utopic futures with dystopic presents, there is something about this new, separated space of the national economy that seems "safe" for the ruling regime, for transnational corporations, and for civil society participants. When something called the "national economy" is laid out in colorful graphs and charts as an entity out there in the future to be diversified, it allows a space to open up between the world and its representation, between the people in that room and the reform agendas set out in those unprecedented booklets. "The idea of 'the economy' provided a mode of seeing and a way of organizing the world that could diagnose a country's fundamental condition, frame the terms of its public debate, picture its collective growth or decline, and propose remedies for improvement, all in terms of what seemed a legible series of measurements, goals, and comparisons" (Mitchell 2002, 272). Rather than the anthropological task being to collapse that distance, I want to acknowledge its power by exploring the conditions of its making and effects. How is the distance between absolute state/corporate power and complicity, on the one hand, and these colorful graphs and charts, on the other, *made*, and what does it, in turn, make in the world?

An interlocutor who had been involved in conference preparations, including the drafting of these documents, explained that the Ministry of Planning, officially in charge of the event, had subcontracted the research and publication work to a Senegalese consulting company specializing in economic development. As a result of this subcontract, early meetings, research, and drafting work were conducted in French. Although French is spoken widely in Equatorial Guinea, given both the country's position in a francophone region and the number of people who lived in exile in Gabon, many Equatoguineans don't speak French. My interlocutor recounted that those early meetings were far from participatory and full of grumbling by

Spanish speakers about their inability to participate meaningfully. Meanwhile, the leadership in the Ministry of Planning was desperately underprepared to assume their share of conference preparation. They didn't have office space to hold meetings or accommodate workers; they didn't have paper, photocopy machines, or ink. Hence, much of their contribution to the conference literature was eventually forged in the offices of an US development subcontractor, in town for other purposes. The documents were then printed and duplicated in Spain before making their way back just in time for the conference. The transnational, subcontracted, and linguistically and processually messy process of making of these documents redoubles the theme of separation at issue here. From Senegalese and American development experts came a series of unprecedented documents filled with predictions about the resource curse and statistics of unknown origin. Processes of economic theory-making and the production and circulation of oft-repressed forms of information turned these documents into what my informant called "dream papers," a widely shared sentiment at the conference and after that the materials' contents were more daydreams spun atop economic theory than routes through petro-capitalist complicity between foreign companies and the Equatoguinean state.

The dream papers came to life through the long hours of the private sector session, as the Minister of Mines, representatives from the state oil and gas companies, and a German agro-businessman seemingly in the process of brokering a large deal with the government, one after another, narrated a future at once utopic and surreal:

> In the energy sector, the government will build two conditioning plants, one in Bata and one in Malabo, to make gas available for local use and diminish ongoing reliance on foreign processing of their products. They will also construct a modular refinery in the country to bring down prices. By 2020, state-of-the-art gas processing facilities in Equatorial Guinea will monetize the gas that is currently burned off in the petroleum production process not only here, but also in Cameroon and Nigeria.

> The potential for the fishing industry is colossal for industrial, artisanal, and farmed fish in the ocean and rivers, with an estimated local capacity of 65,000 tons per year, the equivalent of $100,000,000 US. The government will build two industrial centers and ice factories in Malabo and Bata to service this industry. They will educate oceanographers and boat engineers. By 2020 there will be a fleet of industrial fishing ships, and an industry producing value-added products for export including salted, dried, and smoked fish,

canned and packaged products, and modern fish farms. By 2020 Equatorial Guinea will be the commercial center in the region for sea products. Annobón will be the center of the industry, where women will be transformed into commercial fishers.

Equatorial Guinea has the richest soil in the world for tropical cultivants. By 2020 local agriculture will supply one hundred percent of local nutrition needs and there will be more for export. The forests of 2020 will be conserved. There will be a restriction on timber exports, shifting to add-value products to preserve the ecosystem. Tourism too is an important world industry that generates employment and eradicates poverty. Equatorial Guinea will be a destination for luxury and business tourism, offering endemic-species safaris in the Bioko Caldera, sumptuous lodges on remote islands, and innumerable opportunities to invest while you recreate.

Small businesses and the entrepreneurial sector are the priority in this process. The private sector must drive the economy. It should be the right of small and medium size businesses to avoid being in an uncompetitive economy. This will require microfinance, small business loans, export infrastructure, easy and transparent access to credit for small businesses.

But the "pure gold," the German businessman suggested as the session concluded, *is in salt.* "Equatorial Guinea has excellent conditions to produce salt. The production of salt here, alongside caustic soda and chlorine, is the first step in modern industrialization, and then into the real chemical industry: plastic and other petroleum products that turn salt into pure gold." As he hurried to get through his last improbable sentences, having been told repeatedly and loudly to end his presentation, the Minister of Mines finally stood to conclude the session: "By 2020, Equatorial Guinea will be a successful African model of the transition from a petroleum economy to a diversified economy. Equatorial Guinea will be the first country in the global South to have avoided the resource curse."

The salt-into-gold future presented to the audience of this conference was premised to a surprising degree on resource curse theory. Even more specifically, the erratic utopias of economic diversification were premised on an anticipation of "Dutch Disease" (Ebrahim-Zadeh 2003; Sachs and Warner 2001), an economic theory that takes its name from the Netherlands, where, after oil and gas discoveries in the North Sea in the 1970s, revenue from the non-oil manufacturing sector started to plummet. The "disease" now refers to the pattern in resource-rich countries of exchange rate appreciation; a decrease in domestic manufacturing, agriculture, and nonresource exports; and an increased reliance on imports. Economic diversification away from oil is widely prescribed as the "treatment" for this malady, thus the 2007 conference title—"National Plan for Economic Development and Agenda for the Diversification of Sources of Growth."[5]

That the Dutch Disease was presented as a looming fate for Equatorial Guinea by both conference presenters and the printed materials was incongruous in multiple ways. In the first instance, the country had *long* been without a productive agricultural sector, let alone other industries for either domestic consumption or export. Far from a disease contracted in relation to oil, the rise and fall of a cocoa industry had everything to do with colonial and postcolonial relationships (not only between Spain and Equatorial Guinea, but also with Nigerian and Liberian labor and *Fernandino* landowners), contemporary political struggles over imports from Cameroon, and internal private property issues. Moreover, with once oil-rich Gabon just across Equatorial Guinea's eastern border, and with the offshore platforms of Nigeria visible on clear nights from the shores of Malabo, Equatoguineans did not need a World Bank or International Monetary Fund expert to help them understand the perils of becoming an oil exporter, or to realize that these perils had everything to do with the toxic cocktail of local politics and our world's most powerful corporations. And yet, a narrative that could lay these outcomes at the foot of a resource, and not at the feet of power, was welcome by many in attendance at the conference. The resource curse provided a safe, causal, and authoritative narrative that replaced translocal histories of power and ownership, offering a modular explanation for social and economic ills that moved responsibility, in both time and space, away from grounded histories and toward futures already seen elsewhere, ostensibly inherent in qualities of the resource. The conference became about avoiding a potentiality "out there" as opposed to reorganizing power distributions "in here." The national economy becomes another offshore, another enclave, another stage for the performance of liberalism—market rationality and diversification—erected on the terrain of postcolonial inequality and capitalist violence.

The work that resource curse theory in particular was doing at this conference was astounding, *specifically* in relation to time: a future in which the resource curse "might" befall us; a deferred future in which we can be everything that we are not now; a future framed by economic theory in which we don't have to think about the historically and politically laden past or present. And yet, each of the conference's utopic visions was burdened by layers of dystopic presents and pasts. These visions required tremendous public acts of forgetting and future-oriented confabulations—fabricating experiences as compensation for loss of memory. *Through what fictions and forgetting must the making of a serial thing called a "national economy" proceed?*

The projected fishing industry would require a program of environmental controls for off-platform pollution, and yet as Eugenio intimated, and as others explained to me on multiple occasions, there had not been a single independent environmental review of Equatorial Guinea's offshore oil industry. Equatoguinean officials tasked with environmental assessment and regulation had no relevant training and often took their days on-platform as a vacation. In one story recounted to me, the visiting "assessor" left his room on the rig only for meals. Industry environmental reviews seemed little better. They were consistently done in-house between subcontractors, and the results were always positive, despite documented evidence on one major platform of the intentional manipulation of water quality testing. On the education of oceanographers and engineers, the country has one university—the National University of Equatorial Guinea—which, during my fieldwork, had sporadic electricity, a library of perhaps fifty books, and six computers belonging to an American study abroad program doing primate research (a group of white people known locally, in a delightful inversion of racist colonial stereotypes, as "los monkeys"). While President Obiang's regime is arguably more open to intellectual endeavor than that of Macías, the country was without a single bookstore during my fieldwork.[6]

A slightly deeper history of the fishing industry and Annobón's prospective role therein takes us back to 1975, the year in which Macías officially outlawed doctors and medical care and half of the island's inhabitants died, as recounted in the introduction. During my fieldwork, one could get to Annobón only by boat—a state-owned ferry service that was notoriously unreliable and often diverted to Bata. A new airport was under construction on Annobón, however, and although quoted airfares would be prohibitively expensive for the majority of those commuting to and from Annobón for their livelihoods, the airport was central to the imagined tourism industry, allowing access to Annobón's exquisitely beautiful, sparsely populated beaches. Annobón's future as a tourist utopia of untouched beaches, depopulated by leprosy and various forms of political violence, is overlaid not only with the island's history, but also with contemporary forms of surveillance and repression throughout Equatorial Guinea. During my fieldwork, maps were illegal, as was photography in cities or anywhere in the country in view of a policeman or soldier. Roads were interrupted by official and unofficial checkpoints guarded by armed soldiers. While simple payments of beer or a few dollars were all that was required at most stops, circulation in the country was difficult for foreigners and locals alike. Websites and tourist brochures advertised the beautiful views to be had from atop El Pico, the mountain that rises along Malabo's southwestern edge, yet the road up the mountain was guarded by soldiers, and during my fourteen months in Equatorial Guinea, I was never able to secure permission to go up.

The utopian agricultural future totters uncomfortably on dystopia as well. Recall from the introduction and chapter 4 that labor in Equatorial Guinea's colonial cocoa industry was provided largely by Liberians and then Nigerians—the latter of which left en masse when Macías stopped paying them (Daly 2013, 2017; Martino 2017). This led to Macías decreeing a compulsory labor act for Equatoguinean citizens in 1972, requiring unpaid labor from all men, sparking another mass exodus, this time of Equatoguineans. Since the precipitous decline of cocoa production that followed, Equatorial Guinea has had no industrialized agriculture. While much of the population outside the major cities is involved in subsistence farming, with some selling small surpluses to local markets, bulk agricultural products are imported from Cameroon. Less than a month after the conference's proclamations on the future of agriculture, the Minister of Agriculture and Forestry (the president's much-maligned son, Teodorín) was shown on television throwing money and basic tools to elated farmers, arms outstretched.[7] The 2007 antilogging law—requiring all felled timber to be used in-country—did not seem to disrupt Teodorín's large personal timber concession with Malaysia, later used by an American diplomat to explain how Teodorín came to have the money to fund his personal collection of mansions and luxury cars, despite his small ministerial salary (Smith 2009). To imagine that the resource curse might account for how to strengthen or diversify economic activity in the wake of these histories would be laughable, if it weren't so productive.

The surreal utopias presented at the conference were not the product of resource curse theory alone. The detached, idealized, private sector fantasies were also rooted in local experiences of a suffocating authoritarian regime and translocal imaginings of the endless riches to be had on the Equatoguinean frontier (Tsing 2005). Despite, or perhaps because of, the mirrored surrealities of a fishing sector with unimpeded offshore pollution; a tourism industry without maps or cameras; forest conservation or a flourishing agricultural sector with a minister who personally maintains the country's most lucrative illegal timber concession, I repeatedly found myself in the unexpected position of cheering for this imaginary object of desire called a "private sector," out of whose way the state claimed to want to get. As presenters narrated future utopias of microfinance, small business loans, export infrastructure, and access to credit for small businesses, even my graduate training to sniff out neoliberalism in all its forms could not overpower my visceral support of something, anything, other than a totalitarian state apparatus, strengthened beyond measure by the complicity of US oil companies. At the conference and elsewhere, I was not alone.

UTOPIA AND DYSTOPIA II: "PRIVATE" SECTOR AND "PUBLIC" ADMINISTRATION

One of the first audience members to ask a question after the surreal narration of future utopias was the head of the Equatoguinean delegation for small and medium-sized businesses. Obliquely referencing the amnesias of the current conference, he reminded the audience of a 1982 national forum on the promotion of business, and of the 1997 conference, where it was also established that the private sector would be the motor of development. After these reminders, he asked, "Where are we now? We need financing and access to credit, education, good labor conditions, access to technology, and a fiscal climate according to the law. I'm no xenophobe, but foreign companies are granted all the big contracts, and then we're hired at dismal wages as subcontractors. I have the capacity but I lack the capital. If I had the money, I too could subcontract an architect or an engineer." When he finished, much of the conference room erupted in boisterous applause and full-throated cheers, and he turned, smiled, and waved at the crowd. His question was a thinly veiled critique of the state and the repetitive conferences that came to the same conclusions, yet produced nothing—no credit, no education, no fiscal climate according to law, no disengagement of the state from its control of the private sector. Despite these serial failures, the boisterous response to his question revealed that the conference and its imagined private sector *did* offer the opportunity, otherwise unavailable, for people to gather in large numbers and cheer loudly at indirect critiques of the regime.

The desire for an idealized private sector was a visceral experience that caught me off guard more than once during my fieldwork. Indeed, during my time in Equatorial Guinea, this imagined private sector was the widely preferred realm of fantasies of freedom and opportunity for many. While these fantasies were certainly not evenly shared, neither were they the exclusive daydreams of the wealthy or well-connected. These desires stretched from those of impoverished young musicians who had to go to Gabon to find a recording studio free from government or Spanish sponsorship/censorship, to those of young Americans working in an "MBAs without Borders" program who were shocked at the extent to which, as they put it, "big brother is watching everything in Equatorial Guinea." A private sector detached from this pervasive feeling of surveillance and control felt downright radical.[8] In the pages that follow, I explore different moments of this desire for and idealization of a future private sector, thinking through its ambivalent character of subversion on the one hand and intimate subvention of state power on the other.

On weekends through June and July in 2008, Orange—the central brand of France Telecom, at that point trying to make inroads into the cell phone market in Equatorial Guinea—sponsored a free music and dance festival, setting up a stage and sound system in the public plaza around Malabo's city hall. Replete with obnoxious recorded advertisements about Orange goods and services blaring between youthful acts, the festival allowed for something that rarely happened in Equatorial Guinea: free public art and an open-air public gathering, seemingly unrelated to the regime or the Spanish or French cultural centers, at least at first glance. I attended the festival nearly every weekend, surrounded mostly by groups of teenagers with younger siblings in tow; but a variety of others lingered around the crowd's edges, curious about this unusual public spectacle. The programming was heavily focused on hip hop culture and included numerous rap and break dancing groups. Where *alabaciones* (praise songs) flooded the official music industry, whose only source of income was the state, throughout the Orange summer stage, there was not a single song praising Obiang or the regime. Indeed, there were several songs that critiqued him, directly or indirectly.[9] There was even a song about marijuana.

With labor and event management subcontracted to a small Equatoguinean media company owned by a young Equatoguinean recently returned from France, the audience was clearly seduced by what Orange and its local collaborators were able to facilitate. In a place where the outlets for dissent of any kind were curtailed to hushed conversations around dinner tables, or framed by "international community" spaces like the Spanish Cultural Center, the spaces that private enterprise seemed to open up—like this Orange stage—felt oddly radical and out of control. As with the head delegate for small and medium-sized businesses, who felt free to critique the state at a conference on the economy, on the Orange stage, too, capitalism in its own image became a *felt* space of possibility and unpredictability for the Equatoguinean youth performing, for those of us in the audience, and clearly for the mayor herself, whom I watched leave in a huff during the marijuana song. Rather than dismiss everything that happened on the Orange stage that summer—or indeed fantasies of the private sector, in general—as essentially and exclusively problematic, I follow Miyazaki (2013) and others in suggesting that to take the economy seriously is, in part, to take seriously people's fantasies about it.

The seduction of the private sector was a common trope with which to narrate what was possible, and by extension what was not possible, in Equatorial Guinea. A US development worker who had spent his entire professional life working for NGOs in Africa, and was now in Equatorial Guinea doing corporate social responsibility work for an oil company, said of his new position, "Things happen a lot quicker in the private sector." He continued:

> In the NGO world, the pace of change [was slow]. The private sector doesn't think that way. It's "We've got the money, let's do it." . . . What I see here is very unlike any African country I've been in. This is virgin territory; there haven't been NGOs. Future opportunities are linked to the private sector more than other places I've been. They need NGOs and they need more work in the development side of things, but what they have is that the private sector has parachuted in. There are few places in the world where investors or businesses looking to expand can come and it's a clean slate. If you come with the right idea and the right partners, you can pretty much do what you want to do here. Everyone wants to talk business.

Feliciano, a recently returned Equatoguinean man, was working for a software company with a contract to "modernize" the social security system by digitizing records. He noted with some pride that this work strained against ethnic and other local divisions, implying that those divisions were exacerbated by the hand-written, inconsistent record-keeping practices that his company sought to put in the past. He said his project would bring real change "if we don't get kicked out first." I smiled and said that I too was hoping not to get kicked out, to which he replied, "Oh, but we're not an NGO. We're a business in it to make money." The difference, he explained, was in the politics. NGOs represent political interests and are thus more likely to be expelled from the country. Profit, on the other hand, is apolitical. While Feliciano knew that I was a researcher affiliated with a foreign university, unattached to any NGO, my lack of a profit motive placed me on the "political" side of a line. It was only profit that offered safe haven in a place where getting expelled was a distinct possibility.

From the Orange summer concert series, to the newly privatized devel-

CHAPTER FIVE

opment worker's sense of "virgin territory," to Feliciano's conviction that the profit motive sheltered his company's presence and potential to foment social change, the inhabited feeling that the private sector was variously unencumbered and full of potential was pervasive. And yet, it is precisely because private profit is almost entirely of a piece with public power in Equatorial Guinea that it seems to operate so freely. In other words, capitalist expansion and frontier-making—while it undeniably opened up contingent and newly unstable spaces of expression, dissent, or the opportunity for private gain—was fundamentally a state-sanctioned and state-building project. (We might think again about the power of the production sharing contract here, the signing of which not only directly supports Obiang's regime, but also then proliferates a world of subcontracting arrangements.) As the development worker put it, "if you come with the right partners," you can pretty much do what you want to do, gesturing to the need to collaborate with a local *socio* in all business ventures. Orange undoubtedly lined the pockets of local power holders to bring goods and services into the country. And Feliciano too went on to explain that, in order not to get expelled, his company had cultivated local power holders to ensure their work. Decree 137, which mandated that all businesses be 35 percent locally owned, made "socio" a common role for Equatoguinean state personnel or high-powered individuals, put on retainer to serve as government liaisons. Each company paid a high-level functionary to protect their interests, push their ideas and needs ahead, and get their paperwork through. While Feliciano was excited that his company pushed *against* localisms of various kinds, it also used those very localisms (referred to as *enchufes*, or plugs into the system) to ensure success. Thus, this idealized, virgin, private sector Eden—one that shimmered in the mural of Bata, beckoned in the boundless future utopias of fishing and tourism, and lured those of us living in Malabo's boom times out into summer evenings for corporate-sponsored concerts—was deeply entangled with Equatorial Guinea's state apparatus, itself made by histories and presents of licit wealth creation at the intersection of public office and private gain.

DYSTOPIA AND PUBLIC ADMINISTRATION

In this final section of the chapter, I take up public sector finance (Elyachar 2012; see also Collier in Ong and Collier 2005) as a constitutive piece of the national economy form. Misunderstanding "the economy" as coterminous with the private sector has been one of the less-remarked effects of neoliberalism when, in fact, across Organization for Economic Cooperation and

Development (OECD) countries we see that total government output accounts for nearly half of GDP (Stiglitz et al. 2011). In Equatorial Guinea, both the private sector and the economy as a whole are exceptionally coterminous with "public sector finance," to such an extent that it would be difficult to draw the boundary in practice. Oil production blurs those lines further and, *because of that intensified blurring*, hails state and corporate actors to draw those boundaries in new ways (Mitchell 1991). In order to think about Equatoguinean public sector finance in the age of oil, and to think creatively about the national economy as an ethnographic object and the way it articulates il/licit state and corporate power, development dreams, and spectacular accumulation, we must first think about "corruption," as I started to do in chapter 3 on production sharing contracts. Here, I am interested in the intersection of Equatoguinean norms around state officials and accumulation (themselves contested and multiple), and the equally heterogeneous norms of liberal, rule-of-law bureaucracy that come with oil companies and intensified attention from the World Bank, IMF, and groups like Transparency International.

Remember, again, that Equatorial Guinea is widely ranked among the most corrupt and kleptocratic dictatorships in the world by Transparency International, the World Bank, Freedom House, the Economist Intelligence Unit, and other global "anticorruption" organizations. After being deemed "too opaque to rank" by Transparency International in 2016, Equatorial Guinea was number 171 of 180 countries on their influential Corruption Perceptions Index in 2017. In 2016, Freedom House put the country in its "worst of the worst" category, along with North Korea, Sudan, and Turkmenistan, earning an eight out of one hundred on their scale of "not free" (0) to "free" (100). Obiang himself consistently places among the richest leaders in the world. Scholarly accounts of Equatorial Guinea also often frame the country as exceptional. Even Bayart (2009), who argues that "we should not draw too hasty conclusions about the privileged relationship between power and wealth [given that] the positions of power never absorb all the channels of wealth," goes on to say that "only the political gangsterism of a Touré family in Guinea, or an Nguema family in Equatorial Guinea, approaches a de facto confiscation of the means of wealth" (91). And yet, the now approaching four-decade rule of Obiang Nguema Mbasogo is, like all state projects, a "multilayered, contradictory, translocal ensemble of institutions, practices and people" (Sharma and Gupta 2006, 6). Not least, as this book has chronicled, it owes its longevity to US oil companies, plain and simple. While Obiang, his family members, and their close associates have long benefited from absolute rule and access to positions of power, it is also important to stress the fractures and tensions at the very heart of the regime—fractures that the national budget lays bare, as I'll go on to illustrate. But first, I will briefly narrate the story of singularity.

Obiang's family is from Mongomo, a town in the continental province of Wele Nzas. He is ethnically Fang, as are the majority of Equatoguineans. Widely circulating ethnic stereotypes in Equatorial Guinea (emerging from colonial and postcolonial politics) portray Fang people as aggressive, power hungry, and fiercely loyal to family. But the "Nguema family," or the "Fang people," or "people from Mongomo," or even "Obiang's regime" are hardly the monoliths they are so often made out to be. It is widely known among Equatoguineans, for instance, that Obiang has jailed members of his own family for treachery or potential treachery; that some of his fiercest political opposition comes from Mongomo; and that Fang people are not only the demographic majority of power holders, but also the demographic majority of the poor and dispossessed in the country. Still, Bubi, Combe, and Annabonés people, let alone immigrants, are routinely the victims of ethnically based structural violence. With narrower kinship networks and fewer direct links to power, it is often harder for Bubi or Annabonés people to get jobs or to make other claims on the state often mobilized through kin-based networks. Thus, while I want to destabilize the homogenizing stories of Obiang and his family, I also want to underline the significance of kinship in politics more broadly. In such a small place, as one friend put it to me, "*la politica toca la fibra familiar*": *Politics touches (but more accurately, is made in) the fabric of families.* Equatoguineans interested in oppositional politics or contemplating a public critique of the regime are generally disciplined with familial responsibility, even if family members agree with the critique. "Your aunt works at the port" or "your cousin is minister of such and such," which is to say, keep critiques to yourself so as not to jeopardize shared income streams or create family conflict.

Understanding how power and wealth creation are made licit in a given place takes aim at assumptions about legal liberalism—a separation of "public" and "private" spheres, and then naming their commingling as corruption—as unproblematic, or the only frame within which licit wealth might be created. As I wrote earlier, Mbembe (2001) and others (Berry 1993; Guyer 2004) criticize this approach for undermining "the very possibility of understanding African economic and political facts" (Mbembe 2001, 7), let alone the realities of US politics, and instead producing a situation in which we know everything about what African states seem to lack, but nothing about what they actually *are* and how they *work*. While I agree with Mbembe and others, it has been a guiding intervention of this book that the *danger* in leaving the analysis there is the potential to overlook the effects in the world of all that lack talk. Misunderstandings of African economic and political facts—the absurd indices that rank freedom on 0–100 scales—are not simply wrong. They are incredibly productive processes in the world, not only because people in powerful positions hold them and act on them, but also because they become institutionalized in "neutral," "technical" and "liberal" instruments, such as national economies and contracts. It is one thing to understand the intimate relationship of the idealized private sector to political power on its own terms, and indeed I start here with a few "local" stories to animate certain contested Equatoguinean norms around accumulation. It is quite another thing, however, to show the new terrain on which conflicts over public wealth take shape post-oil.

PUBLIC POWER AND PRIVATE ACCUMULATION
BEFORE AND AFTER OIL

Outside Equatorial Guinea's two main cities, the small villages and towns of Bioko and Río Muni are dotted with large, cement homes that often stand out amid smaller homes made from more provisional materials. Asking about those homes would inevitably lead to a story of someone from the community who now worked in the administration and had built a house in his or her natal town postappointment. While there were always exceptions, it was incumbent upon those with higher positions in the administration, or who otherwise got rich, to construct a prestigious cement house—the bigger and more ostentatious the better—in the town of their birth. I would also hear, scathingly, that "so and so is now vice minister of such and such and he hasn't built anything here yet." In other words, if you *didn't* build one of these structures, you could be seen as selfish, as a failure, and as ungenerous. These buildings were not "philanthropic" structures in the liberal way I had come to understand philanthropy; they were neither schools nor orphanages, clinics nor libraries. Rather, they were private residences whose interiors would be enjoyed only by the immediate and extended network of the powerful persons who owned them. For the wider population, however, these buildings *were* philanthropic: material signs of ethical action, monuments to the correct moral relationship between powerful people and the rest. They showed investment in and attention to the owners' local communities, despite their wealth having been gained in the capital city.

I use a story of Luis, an Equatoguinean man in his late thirties who had lost his job at an airline office, to sketch another form of ethical obligation at stake in questions of public power and private accumulation. Beside himself and with no job prospects, Luis contacted our mutual friend Josefina—a well-connected woman from a powerful family—for help, advice, and perhaps an enchufe for a job. I was there when Josefina took Luis's call, and after the animated phone conversation to which I was half-privy, Josefina was irate, recounting the ethics of the situation in sharply ethnic terms: "Luis's father is rich and powerful! He has a fleet of cars and houses everywhere! But he has given his son nothing! This is how the Bubis are! It is horrible! If his father was Fang, even if the son had no education, he would be given a job!" Josefina, from a Fang family, saw Luis's plight as *complete kinship negligence* on the part of the father, explicable only in terms of his ethnicity. In other words, for Josefina, the only acceptable thing for a father of means to do in this situation would be to use his wealth and influence to get his son a job. In Equatorial Guinea, arguably among certain ethnic groups in particular, this behavior is expected, even *morally demanded*, and its absence may be seen as a grave moral transgression.

At the same time, nearly every Equatoguinean who I came to know, regardless of their ethnicity or relationship to the regime, was disgusted by what they perceived to be the *excesses* of influence and private accumulation committed by the most powerful at the blatant expense of public goods and services. In other words, all forms of power and privilege have limits and thresholds. Equatoguineans could demand without hypocrisy that influential fathers must find positions for their children, often in government; that important public figures must build prestigious private homes in their natal towns; *and* that those in power must stop the excessive abuse of their positions for private gain. State and nonstate actors constantly, uneasily negotiated these tensions (Sharma and Gupta 2006; Smith 2007).

While these stories of prestigious houses in rural towns and kin-based hiring practices give texture to the daily life of "public power and private gain," the stakes of these negotiations and the scales at which they ramify changed radically with the coming of oil. Recall the Riggs Bank scandal I discussed in chapter 3, in which a US Senate subcommittee concluded that everyone from the Riggs Bank employee directly responsible for the Equatoguinean accounts, to his wife, to the examiner in charge from the Office of the Comptroller of Currency (OCC), to Federal Reserve regulators, to US oil company management, to President Obiang himself were guilty of anti-money-laundering infractions (Coleman and Levin 2004). While there was brief protest in the streets of Malabo after Spanish television channels made this news quasi-public, the lasting legacy of the Riggs Bank scandal seems to have gone in several other directions, remaking the literal *landscape* in which public office and private accumulation might meet. First, as one of my friends managing a construction firm put it, "After Riggs, the government learned that they couldn't just walk around with suitcases of cash; they had to find more nuanced ways of getting money." Now, he went on to explain, the government pays radically inflated costs—on the order of 500 percent—to Chinese and Egyptian parastatal construction companies working on infrastructure projects. Subsequently, 80 percent of that inflated outlay goes to government officials as "kickbacks," or indemnification (*indemnizaciones*). After Riggs, in other words, infrastructure projects took on new importance both as a visible sign of the public investment of petro-profit, and as public/private contracts through which money could still change hands under increased scrutiny. This is no longer a question of locally powerful individuals building prestigious homes or offering jobs to family members. The post-oil landscape of public office and private gain now implicates the American Federal Reserve Bank, US Senate subcommittee investigations, parastatal companies from around the world, millions of dollars, and massive infrastructure projects.

In the wake of the Riggs Bank scandal, the Government of Equatorial Guinea released a document in their own defense: "Statement of the People and Government of Equatorial Guinea in response to the report on Riggs Bank of the Permanent Subcommittee on Governmental Affairs of the Senate of the United States of America" (República de Guinea Ecuatorial 2005). This document offers another window into the effects of the subcommittee investigation. First, arguably buttressing my informant's statement above about the role of infrastructure in post-Riggs forms of public/private accumulation, the last sixty-six pages of the document include tables of infrastructure projects, the companies responsible, and the costs, as well as "Some Investment Pictures," as Annex IV is titled, with photos descriptively captioned: "View of Urbanization and Asphalting" or "Rehabilitation of Sendje Bridge." These tables and photos are intended to document the *licit* forms of investment to which oil money has gone and to counter claims that government appointees appropriate all of the wealth. The first thirty-seven pages of the document offer a different illustration of the struggle over the licit modes of power and accumulation.

	A COMPANY	B PROJECT	C LOCATION	D TOTAL COST
1	Grupo FADEM	Construction sheet metal roof, Loza Barracks, Moganda	Bata	47,836,800
2	TECNOBAT GE	Construction Minascom Regional Offices	Bata	221,479,301
3	MANSO EXP	Land clearance	Bata	200,461,750
4	MANSO EXP	Maintenance lawn & esplanade	Bata	65,759,600
5	ACCESSEGE	Restoration of Mongomo Courthouse	Mongomo	94,110,665
6	SOGEA	Reconditioning Mbini - Nume Highway	Mbini	351,540,068
7	SOGEA	Rehab. of sidewalks & drainage	Mongomo	1,878,572,494
8	PREGUIS S.L.	Refurb. of offices of Nat. Fed. of School & Univ. Sports	Malabo	78,433,400
9	EMPOCSA	Reconditioning of Nkoa-Ntoma barracks esplanade	Bata	674,600,000
10	Const. MARIN	Refurbishment of Governor's residence	Luba	32,402,595
11	SOGECO	Reconditioning terrain La Paz Stadium	Malabo	198,535,040
12	EMPOCSA	Reconditioning terrain Nsok Nzomo Barracks	Nsok Nzomo	112,000,000
13	Bur. VERITAS	Malabo station terminal	Malabo	27,280,000
14	SOGECO	Widening Malabo-Airport highway, & turnpikes	Malabo	2,327,799,953
15	ISOLUIX WAT	Upgrading gas turbine power station to 20 megawatts	Malabo	10,053,266,513
16	FILYAL	Enlargement of Acacio Mañe School	Malabo	37,854,934
17	GRAL WORK	Enlargement of airport's aircraft parking facilities	Bata	7,262,187,500
18	BOUYGUES	Enlargement of aircraft parking facilities	Malabo	768,696,682
19	BOUYGUES	Bridge-widening on Malabo - Basacato Highway	Malabo	9,832,775,585
20	INCAT	Repair & widening of Old Bridge	Malabo	6,127,313,500
21	SOGECO	Asphalt access to garbage dump on Rebola highway	Malabo	61,898,200
22	ECOCSA	Asphalt Ngolo - Asonga Highway	Bata	4,817,772,400
23	SOGECO	Asphalt La Paz Street in Malabo & 20 meter bridge	Malabo	996,873,461
24	SOGEA	Asphalt seaside promenade Lote 1	Bata	2,085,747,815
25	SOGECO	Asphalt Rebola's main street & square	Rebola	260,535,040
26	SOGECO	Asphalt streets in Los Angeles Neighborhood	Malabo	222,004,794

Figure 5.3. Infrastructure budget details in defense of Riggs bank scandal.

Reminiscent of Nigeria's 1976 constitution, in which the authors drafted political power as "the opportunity to acquire riches and prestige, to be in a position to hand out benefits in the form of jobs, contracts, gifts of money, etc. to relations and political allies" (Bayart 2009, xvii; see also Simone 2004, 186), the Riggs Response document (2005) also seeks to sanction the conflation of public office and private accumulation, albeit with recognizably contemporary rhetoric of free markets, competition, efficiency, and "entrepreneurial flair":

> Article 29 of the Constitution of Equatorial Guinea establishes and defines the country's economy as a free market economy, which guarantees economic freedom, recognizes private initiative, and promotes competition and efficiency.... *Every single citizen, including the President of the Republic, has the right to engage in commerce and trade, because the constitution and the laws therefrom derived permit them to do so.* Foreign legislation, such as that of the United States of America, is not in force and does not apply to Equatorial Guinea, and as a result can in no way prohibit the President of the Republic, Ministers, Civil Servants, Citizens, or foreign residents in Equatorial Guinea from undertaking business initiatives.... *Persons in positions of authority are permitted to promote the creation of companies to be managed by third parties.... In the Republic of Equatorial Guinea, fellow family members of such Authorities are not prohibited from engagement in commerce* (2).... The laws of the Republic of Equatorial Guinea do not prohibit any citizen, whether national or foreign, from developing their entrepreneurial flair in the country. (5; emphasis added)

At issue here are prevailing modes of wealth creation and distribution under new scrutiny, and the negotiation of those modes in the presence of both new, fabulous amounts of money, and new mechanisms and pressures through which that money must be distributed. The post-oil terrain on which these battles are fought is radically expanded beyond rural towns and family hiring practices to include not only the OCC, the Federal Reserve Bank, and the US Senate, but also Chinese and Egyptian construction parastatals and the disposition of private property required by their projects; US-based lobbyists newly active in drafting documents like the one quoted above; and the globally circulating claims of neoliberal capitalism itself—economic freedom, private initiative, competition and efficiency, and entrepreneurial flair.

Despite the boldness evinced in this document—an unapologetic validation of the wealth of the ruling elite and an aggressive assertion of sovereignty—the daily life of Equatorial Guinea's public administration, and the administration of the national economy in particular, was far more ambiguous. Citizens and state actors alike generally acknowledged the government's pervasive failures not only in service provision, but also in the basic rituals that facilitate the provision of public services in the first place. I refer here again to the equivocal role of administrative procedures, including budgets and other forms of record-keeping, accounting, reporting, auditing, and monitoring—in general, to the ambivalent relationship of the state to procedures of administration and bureaucracy.

STATE FAILURE, PROCEDURALISM, AND A VISIT TO EDUARDO

After the futuristic confabulations of the main conference proceedings, all participants returned to the luxurious auditorium for a third day of concluding summaries. The conference participants seemed noticeably tired on this closing day. People straggled into the conference hall late, and there seemed to be fewer animated discussions in the hallways. Having spent the evenings between conference days discussing the historic event over dinners with Equatoguinean friends, our conversations too had grown tired, moving from early laughter at the impracticable goals, toward the torpor of long days and wasted time. After participants slowly filled the large hall that morning, we sat and sat, waiting for the president to appear, slowly sinking down in our chairs, wrinkling our clothing, and trying to hush our growling stomachs. I wrote in my notebook, "The future is exhausting." Finally, the voice of the president's protocol roused all of us from our slumps: "Su Excelencia, Jefe Del Gobierno," and the president's Moroccan security guards strode in before him. The audience clapped in rhythm, and Obiang joined in as he walked to his seat in the center of dignitaries and government ministers at the head table. And then, it was as if for a moment the impossibly heavy future gave way, and the present's dystopia slipped in. In front of the president and others at the head table, government presenters elaborated on, with a bluntness that surprised me, the serious and obvious problems the country would have to overcome to achieve their future goals:

> We are essentially without all basic social services. There is little to no running water, none of it potable. Electricity is sporadic in the cities, and not distributed throughout the territory. The health sector is essentially nonexistent for the majority of the country's residents, and the education sector is little better. There is no transparent access to credit for small businesses, and no

regularized process according to which one might start a business. There is a total lack of legal instruments or regulation in any and all sectors. There are serious problems with private property and contract law. There is no state contract law or set of laws regarding the quality of work done.

There is no state contract law. There is a total lack of legal instruments. These statements do not solely, or even centrally, describe a corrupt state. This is a description of the depth of imperial debris through which US oil companies profitably traipse. This is a description of the spectacular profit to be made in the not-yets of liberalism. The depth and breadth of the problems described by government presenters that morning raise the question of the relationship between so-called corruption and the daily life of administrative work. If the problems are this deep, can the conflation of public office and private gain fully account for them? What about the daily life of service provision, the making and implementation of legal instruments, or regulation? National economies are built not only at the interface of forms of licit accumulation, but also by the mundane procedures of budget-making and record-keeping. Conference documents from both 1997 and 2007 dwell at length on bureaucratic chaos, complicating a picture of simple corruption:

> More than corruption, Equatorial Guinea suffers from the absence of the procedures and tools of administration that facilitate the control of spending and limit the risk of the diversion of funds. It seems that Equatorial Guinea has not benefitted from an administrative tradition, allowing the country to construct a truly structured State. *The modern Equatoguinean administration is still to be built* (República de Guinea Ecuatorial 2007a, 42). (Parece que Guinea Ecuatorial no se haya beneficiado de una tradición administrativa, permitiéndole construir un Estado verdaderamente estructurado. *La administración Ecuatoguineana moderna queda por construir.*)

As if to drive home the extent of what exactly remains to be built, another document narrates a bafflingly vast series of required reforms. Under the heading "Implement Quality Governance at the Service of Citizens," the list includes, among other directives: (1) launch a planning state and a modern administration; (2) revise the legislative framework; (3) reform the judicial system; (4) encourage the participation and representation of citizens; and (5) secure the respect for human rights (República de Guinea Ecuatorial 2007b, 10–11). The documentation from the 1997 conference details a series of implementations agreed to by the government that are equally shocking, given the absences they acknowledge. To paraphrase: *The government will propose legislation to ensure that accounting procedures and financial and regulatory laws are followed, backed by the threat of external audit. The government will introduce new rules and procedures to control the budget process and identify the functionaries responsible for discrete actions through the process. The budgets of every agency will be identified, and the financial administration at departmental levels will be improved* (República de Guinea Ecuatorial 1997, 28; emphasis added). In other words, in 1997 there was no legislation requiring the monitoring of accounting or other financial procedures, nor were there codified processes that separated agencies by budgets, or budgets by responsible party. Thinking back again to this "first" conference, these documents show, again, how newly circulating oil wealth demanded these procedures be put in place. To give one more example from the 1997 document, an adopted reform under the heading "The Function of the Treasury" reads: "The government will seek the necessary technical assistance to develop and design the systems, processes, and procedures of control for spending and [accounting]. This will allow the Treasury to maintain accounts for all agencies collecting and spending public funds" (29). In a country where public funds are of a piece with private investment, what is a national economy in the absence of national budget accounting procedures or of national treasury accounts separated by agencies collecting and spending public funds? And what does it mean to try to make one, when suddenly billions of dollars are pouring in, and yet the treasury has long operated somewhere between Nzeng Ayong and Riggs Bank? It was in conversations with Eduardo that I began to think through these questions.

Without house numbers and with only a few main streets named, it was often a challenge to find unfamiliar destinations in Malabo. The orienting landmark Eduardo had given to find his house was a well-known pepe soup restaurant. Friends and field interlocutors had long explained that Eduardo did not talk to just anyone, so I had waited until late in my fieldwork to approach him, when I finally felt I had a list of informed questions that might be worthy of his time (and when I knew my way around the city!). A scholarly lawyer long active in opposition politics, Eduardo was understandably cagey about those to whom he granted an audience, and why. Despite all my careful preparation, approaching the pepe soup restaurant I was nervous that he would regret the invitation. When I finally found the corner store that marked his building and walked up a flight of stairs to the second-floor apartment, a young man greeted me, opening double doors to a shocking site: books. Books! Eduardo's airy, high-ceilinged flat was lined floor-to-ceiling with bookshelves, and still more books covered the floor in inclining piles. And papers! Stacks of papers and files leaned this way and that, covering and spilling off of available surfaces onto the tiled floor. That the sight of so many books—philosophy, history, political theory, law, and literature—caught me off guard has stayed with me for a long time. Beyond attention to knowledge production, as anthropologists we also must attend to the histories, uses, and relationships to specific kinds of knowledge in specific places—here, the kind to be found in books in a country with no bookstores. Indeed, in one sense, the idea of a "national economy" is but a representation of accumulated information; however, it's not merely the content of that information that matters, but also what information as a category means, and to whom, at the intersection of a country where being an intellectual was reason to be put to death for a time and an industry wherein agnotology is often the order of the day (Appel et al. 2015).

I sat amid the books to ask Eduardo questions about oil's more intimate histories in local politics, and he began by telling me an origin story that I recount various times throughout this book. The late 1980s and early 1990s were a time of deep economic and political crisis in Equatorial Guinea. External debt levels were among the highest on the continent, and Obiang's regime, in power for ten years at that point, was under mounting pressure from a strong and newly unified opposition, emboldened by pressure from the international community. Chester Norris was the American ambassador to Equatorial Guinea during these years, between 1988 and 1991. During his tenure, Norris acted openly as the intermediary between local political figures and new American oil interests in the country. Walter International was the first American production company to produce oil in Equatorial Guinea, and Chester Norris served as their representative. In 2009, Avenida Chester Norris was one of the few named streets in the city, alongside Avenida Hassan II, after the Moroccan King whose kingdom has provided Obiang with his most trusted security guards. According to Eduardo, the money from those early contracts brokered by Norris went directly, personally, to Obiang (see chapter 3).

Many in Equatorial Guinea and within oil companies themselves will readily admit that the early days of oil contracts were full of personalized dealings and "irregularities," and most will end these stories by explaining that now *all that has changed*. Whether oil companies now face more public scrutiny or have undergone meaningful reforms in the post-Enron/Sarbanes-Oxley moment is unclear, and whether local power holders have learned to deal differently with transnational corporations or were merely chastened by Riggs Bank is also unclear. Regardless, actors on both sides

CHAPTER FIVE

generally admit to early transgressions in order to demonstrate contemporary compliance, reminiscent of Baudrillard's (2001) critique of capital staging its own death: "It is always a question of proving the real by the imaginary; proving truth by scandal; proving the law by transgression. . . . Power can stage its own murder to rediscover a glimmer of existence and legitimacy" (179).

Rather than explaining to me how things had changed since Obiang, and perhaps Norris himself, pocketed Walter International's money, Eduardo took a different turn in our conversation. He walked across the room and reached into the middle of a paper mountain, returning to our seats by the window with a thick, dog-eared sheaf in hand. Shaking the stack as anticipatory punctuation, Eduardo spoke in a whisper so intense that it seemed to emanate more from the desire to keep from yelling than to not be heard. "Money from petroleum exploitation didn't appear in state budget documents until 2005! Look for the budgets from 2002, 2003, and 2004. Petroleum money does not appear!"

The stacks Eduardo held were national budget documents stretching from 2004 to 2007. As he flipped through them, he narrated his close readings of these budgets faster than I could take notes. Roughly 50 percent of annual budgets go into a category of "investments," of which the largest portion is construction and infrastructure (recalling the post-Riggs moment of public office and private gain, above). While the budget documents contained long lists of these projects and their projected costs, there was no proof or information indicating whether these projects had, in fact, been completed, nor were there numbers regarding their actual costs once completed. Eduardo's voice rose and he looked me in the eyes:

> There are paragraphs that say, "We don't know where this money goes." There are sectors of the administration where the money does not enter the treasury. They are essentially saying *here* [while stabbing the document angrily with his pointer finger] that the money made by the government from the sale of national territory does not enter the treasury! These are reflections of inner-regime conflicts, but the document is still approved! There is no follow-through or monitoring. People in the regime with personal conflicts have suffered trying to develop accurate budget documents. [These documents] are a confession of the impotence of those who work in the Ministry of Finance.

As we sat and talked our way through the documents, the story that began to emerge most forcefully was of Equatorial Guinea's parastatals—

legally separate entities created by the government to undertake commercial activities. These entities are largely funded by the government, but ostensibly operate with some degree of autonomy. (National oil companies are famous examples of parastatals, such as Saudi Arabia's ARAMCO or Brazil's Petrobras. Within the US context, Fannie Mae and Freddie Mac are other, now notorious, examples.) Starting with the story of Bata Ports, a parastatal in charge of port administration in Bata, Eduardo explained that in 2006 they received a government subsidy of 100 million CFA (about US$200,000), 100 percent of which was unaccounted for. Did they deposit it in the treasury? Where did it go? And, more pointedly, who might dare to follow up on these questions? He went on to explain that a host of institutions, including the university, the social security administration, and the national gas company, were never asked to account for the money they received from the government. But with opposition pressure on this point starting in 2006, these institutions began to bring budgets to the table. "These are like feuds," Eduardo explained. "When Obiang names someone [to lead an institution], they just use their appointment as an opportunity for private accumulation." "But surely," I said, "there's more than avarice at work here." I talked about my experiences in the Extractive Industries Transparency Initiative process (documented in chapter 6), noting that there were radically different levels of bureaucratic competence in the ministries. While a few people I worked with clearly had the training to produce complex budget documentation (training that I do not have), others definitely did not; even among those who did, many felt that crunching numbers on Excel spreadsheets was menial work compared to their daily tasks of face-to-face meetings where agreements were brokered and work actually got done. "So, how much does procedural competence have to do with it," I asked. "How much expertise is there in the formalities and procedures of budget-making?" Citing Chabal and Daloz's *Africa Works: Disorder as Political Instrument* (1999), Eduardo responded, "This disorder, this incompetence—when incompetent people are appointed to these posts it's on purpose. It is using disorder as a political instrument."

I left Eduardo's house as the sun was setting and the harmattan sky was deepening to dusky orange. I was dizzy from the hurried flipping through documents full of numbers and charts, accompanied by Eduardo's fervent narration. I was anxious about what an anthropological eye/ear might be able to bring to bear on documents so central to the national economy cosmos. Before leaving his house, I had asked if I might photocopy the documents we had gone over so that I could study them at greater length and then come back to him with further questions. He agreed, on the condition that I come back the next day to pick them up, to give him the chance to organize the documents and take note of what he had lent me. When I returned to Eduardo's house the next day, he was out, but the same young man who had so quietly ushered me into the book-filled room passed me five bursting manila envelopes of documents that I took to a small *ciber*—an establishment that rented time on computers and happened to have a copy machine, as well. The Cameroonian clerk and I laughed and joked and talked about soccer as he took out the documents. But as soon as he saw the contents of what he was about to photocopy, he grew reticent and asked me if I worked for the government. I boldly said that I did not and that I was doing a research project. Our conversation ended abruptly there, and I sat for a few hours watching the afternoon fade into night as he taciturnly copied and collated the materials.

Pouring over these documents, I indeed found the problems to which Eduardo drew my attention. The documents limn the cracks and internal tensions of an ostensibly monolithic and smoothly kleptocratic regime. Of the National University (UNGE), a 2006 document reads, "This institution lacks both accounting books and basic notions of accounting. This situation does not permit us to have reliable data" (República de Guinea Ecuatorial 2006a, 17). The same document three years later says of the university, "Despite the presentation of budgets that don't even minimally respect the established criteria, [La UNGE] has never justified the use of the funds put at their disposal by the government" (República de Guinea Ecuatorial 2009, 6). By 2009, according to the Ministry of Finance responsible for these documents, the university had yet to produce a roster detailing how many professors it had on payroll and their salaries. Despite this lack of personnel records, the document notes that in UNGE's anticipated budget for 2009, "there are 122 professors who do not appear on the remitted payroll" (República de Guinea Ecuatorial 2009, 6).

Concerning the National Institute for Business Promotion and Development (INPYDE)—a particularly relevant institution given the general thrust of the Horizon 2020 plan—the budget document contemptuously reports: "This institution that advocates business promotion and development seems to be marginal to the realization of these activities. During the fiscal year that closed on December 31st, 2005, this institution had not accomplished a single activity of their own that advocated business promotion and/or development. . . . Of the 195 million CFA subsidy INPYDE received [roughly US$400,000], 62% went to pay personnel, 10.8% went to education even though they did not hold a single seminar, and 8.7% went to travel costs" (República de Guinea Ecuatorial 2009, 18).

Concerning the Bioko Chamber of Commerce, Agriculture, and Forestry's final reported numbers, the 2005 budget document reads that "these results are fictitious and paradoxical when it is known that the Chamber is in debt to BGFI Bank for 335 million CFA" (República de Guinea Ecuatorial 2005, 19). And finally, in a document intended to review all submitted budgets of all parastatals from 2005 to 2009, it indicates that GEProyectos —the entity ostensibly responsible for all infrastructure projects nationwide—had submitted *no budget data* in four years. "This company has not remitted any information about the execution of its budgets" (República de Guinea Ecuatorial 2009, 10). Nevertheless, 2009's anticipated national budget showed that GEProyectos was to receive a *200 percent increase* in its government subsidy.

Despite Eduardo's insistence on the intentional and strategic use of disorder, even this perfunctory glance at thousands of pages of budget documents reveals both an internally fractured regime and its ambivalence about the role of procedures and techniques of administration, accounting in particular. At the very least, we see the Ministry of Finance openly critiquing responsible parties at the National University and the National Institute for Business Promotion and Development, arguably trivial institutions in terms of both budget allocations and political clout. However, when that criticism extends to the information provided by the national gas company and GEProyectos, the friction takes on increased political weight. Moreover, from both the National Economic Conference and the National Budget documentation, I want to insist on the significance of statements about bureaucratic procedures. If the National University "lacks both accounting books and basic notions of accounting," and if, by their own admission, in 2007 the government had not adopted "procedures and tools of administration that facilitate the control of spending and limit the risk of the diversion of funds," then there is more to be said than simply that "new" or "Western" forms of accounting and accountability came in with oil that were in tension with local forms that worked in the past. Rather, local political conflicts and their minimal room for maneuver are expressed through who may or may not be fluent in "basic notions of accounting."

Although, as Eduardo tells it, there is no one in the government who would dare to follow up on these arguably shocking "irregularities," the Ministry of Finance still ends the document by *daring* the Parliament to pass these budgets: "We present [the results] to this highest legislative body, to determine whether or not to validate the budgets presented by these Autonomous State Entities, despite the preliminary conclusions reached by the Ministry of Finance and Budgets" (República de Guinea Ecuatorial 2006a, 20). Looking at ethnographic data at this level alerts us to the *relationship* between broader political economic analysis—complicity among US banking and regulatory systems, transnational oil companies, and various dictatorial regimes—and the minutiae and infighting in the more localized daily practices of accounting and budget validation.

If Equatorial Guinea's national economy is made at the intersection of political infighting, insufficiency in basic accounting techniques, and spectacular accumulation, then what is this fetishized object? "Economic expertise is forced largely to overlook . . . leakage, network, energy, control, violence, and irrationality," as Mitchell (2002) writes.

> It does not take them seriously, for that is not its task. The role of economics is to help make possible the economy by articulating the rules, understandings, and equivalences out of which the economic is made. . . . This has been its impossible project. . . . This self-deception is essential, for otherwise it would have to follow these links, powers, and leakages, and admit that there could be no economy. (300)

In other words, despite all this leakage, violence, and transnational irrationality, there is an "Equatoguinean" economy in the name of which the eternal deferral of the future is justified; the government signs contracts worth billions of dollars; and carnivalesque conferences are convened and convened again.

CONCLUSION

There is an enduring tension—epitomized in the national economy form, but stretching far beyond it—between the messy daily practices of national budget-making and the authoritative work those eventual forms do in the world; between the dystopian histories and presents of life in Equatorial Guinea and the utopian futures imagined in an idealized private sector, or a disembodied and diversified national economy. In negotiating this tension, I have tried to avoid the approach of simply demonstrating the ways in which concrete national economies differ from economists' "abstract" national economies (Callon 1998; MacKenzie et al. 2007; Miyazaki 2013;

Holmes 2014; Appel 2017). Instead, I have proceeded on the understanding that "the economy is an artifactual body—a fabrication, yes, but as solid as other fabricated objects, and as incomplete" (Mitchell 2002, 301). Once we can hold both poles of the artifactual simultaneously, the solidity and the incompleteness, then we can shift analytic attention to the spaces opened up by these tensions in the world: productive spaces of misrepresentation, simplification, forgetting, and confabulation, in which powerful objects like "national economies" come to exercise much of their power. If, for instance, every country everywhere has a national economy, whose historical trajectories and potential futures can be compared, avoided, or emulated by employing authoritative economic and political tools like the resource curse, then we end up with dream spaces like the National Economic Conference I attended and the reverberations of an idealized private sector far outside conference walls. Involvement in a matchlessly lucrative sector of global trade like the oil industry compounds both the surrealism and the productivity of these spaces, as it requires intensified negotiation with new forms and geographies of legibility, licit accumulation, accounting, demography, record-keeping, the disposition of private property, and contract law, among others. These negotiations *make* Equatorial Guinea into a place from which hydrocarbons can be extracted and exported licitly, even as they retrench translocal forms of power, inequality, and rule. Central to these negotiations, then, is the *alienability* of hydrocarbon and its associated profits from Equatorial Guinea, epistemologically the same space of separation required between a national economy and the aggregate transactions and histories it can and cannot represent.

That it was the resource curse that enframed these temporal and spatial separations at 2007's National Economic Conference is particularly demonstrative, because at its core this body of theory is *about* the entanglement of oil money with local power structures. Because of the narrow focus on money, the suggestion is that this entanglement is somehow only "in" the national economy—the revenue from oil is misused; revenue streams aren't diversified. The antidotes, similarly, become located "in" national economy fixes, for which the national government is solely responsible. But given the four preceding chapters, we can see the much more complicated daily life of the entanglements of the US oil industry with local power—entanglements in contracting and subcontracting regimes, infrastructure, labor, and enclaved corporate and residential life, all of which are required *long before* oil and gas circulate back into national economies as money.

The 2007 conference literature ends with two appendices that describe other national economies worth emulating. The first explains Dubai's twenty-three-year transition away from an economy solely reliant on oil:

> In 1990, Dubai's government confronted their economic dependency on fossil fuels, and decided to begin a conversion to new technologies, commerce, and luxury commercial ventures. In fifteen years, petroleum's contribution to GNP fell from 80% to 10%. Dubai's accelerated development was possible thanks to the voluntary recourse to foreign workers at all levels . . . to cover the country's central necessities. This translated into extremely simplified procedures for business contracts and the hiring of foreign workers. In 2003, 84% of Dubai's population was immigrants, 71% of which were men. (República de Guinea Ecuatorial 2007a, 71)

The second narrates Singapore's forty-year rise from "a third world island without resources" to "a true global economic dragon":

> The development of this island without resources was based essentially in the development of strong public infrastructure and the creation of an especially favorable business climate: Policies to attract foreign investment; A reliable judicial system that protects property rights; The most efficient and least expensive manual labor among developed countries; Prestigious business schools administrated by Harvard; Perfect control of trade logistics; Logistical efficiency at air and sea ports. . . . *Singapore surely owes its success to the great visionary Lee Kwan Yew, who knew to use his authoritarianism for the singular purpose of forging a common identity from a multiethnic society. In this way, he was able to inculcate concepts of Asian values including the primacy of the collective over the individual, the benefit of consensus, work, and savings into the population.* (República de Guinea Ecuatorial 2007a, 72; emphasis added)

These disembodied pasts, offered as Equatorial Guinea's desired futures, perfectly illustrate the fantasy of something called a "national economy," an object that enables the framing of "collective growth or decline, and remedies for improvement, all in terms of what seemed a legible series of measurements, goals, and comparisons" (Mitchell 2002, 272). Dubai offers a model in which local power holders would never have to respond to the small businessman's popular plea for access to credit or capital, but could continue and even expand the "extremely simplified procedures for business contracts and the hiring of foreign workers" like body shopping and pay according to nation of origin (República de Guinea Ecuatorial 2007a, 71). Singapore offers a model in which a combination of authoritarianism and orientalist essentialism pave the way for the Harvard-approved "perfect" control of trade logistics, a judicial system that seems to protect only private property rights, and the most efficient port in the world.

These false equivalences and disavowals of historical connection enabled by the national economy form—the same in Singapore, Dubai, and Equatorial Guinea—are precisely the effects of macroeconomic tools like GDP or national growth rate when laid across postcolonial inequality. While these tools offer an expert language of comparability and potential equality, they also legitimize and dehistoricize radical inequality, rendering hierarchies of global supremacy licit in scientific language (Speich 2011; Appel 2017). The particular violence of the national economy form in the global South, and the enrolling of economic theories like the resource curse in that violence, entreat ethnographic attention. Independence-era Senegalese political leaders Mamadou Dia and Léopold Senghor were prescient in their concerns that, even as they fought for political independence from France, relational inequality between colonizers and soon-to-be nation-states would not be undone by "nominal independence" in a world they insisted was increasingly interdependent (Cooper 2015, 68; see also Pierre 2013; Ralph 2015). The sequelae of their concerns live on, here in National Economic Conferences in Equatorial Guinea, and thrumming in those moments when world and representation are not one; when development plans and economic theory occlude transnational corporate power; and when performativity is not so complete that the model becomes the world. These are the spaces of political possibility within the foreclosures of modern politics. To find them is to participate in the classic ethnographic project of making the familiar strange: "Given the centrality of the economy to modernist social representations . . . it is necessary to defamiliarize the economy as feminists have denaturalized the body, as one step toward generating alternative social conceptions and allowing new political subjectivities to be born" (Gibson-Graham 2006, 97).

CHAPTER SIX

THE *Political*

As I have shared the research on which this book is based in presentations and publications over the last five or six years, I'm often met with differing versions of these questions: What is the solution? What is the solution to the deep complicity between US oil companies and repressive regimes like Obiang's? What is the solution to the incantation—*waterschoolselectricityhealthcare* (water schools electricity healthcare)—that haunts too many resource-rich countries with the repeating specter of squandered oil revenues? And of course, far beyond this book, the serial horrors of the transnational oil industry in the global South, in particular, also provoke the question: What is the solution? For people and organizations variously invested in and effected by extractive industries, *transparency* is often the answer. Forcing open the ledgers of infamously secretive transnational corporations and the governments with whom they collude, the idea goes, will reveal the flows of oil revenue—from whom and to whom—to the public, and thus better enable that public to claim their fair share for running water, healthcare, infrastructure development, and education. This is the (mostly implicit) theory of politics embedded in the Extractive Industries Transparency Initiative (EITI) documented in this chapter and epistemologies of transparency more broadly.

This theory of politics as transparency—liberal through and through—is also a theory of social change; that is, if you disclose information, perhaps if you can show that $28 billion is missing (as the EITI process revealed to much fanfare in Nigeria), then a cascade of political change will follow. Hetherington (2011) offers a useful summary of the political theory undergirding transparency:

> If authoritarianism was built on the state's control of information, then democracy is to be built by giving citizens access to, and indeed control over and responsibility for, all information. This ultimately is what a transparent society is supposed to look like: all state knowledge is public knowledge, and citizens can therefore "see" what goes on in government and in the economy, not directly, but by receiving information about it. Transparency is information so complete that it seems unmediated; it is an access to the real through a medium so perfect as to disappear from the scene it is describing. This can never be realized, but remains always the ideal: a world of perfect information in which citizens and entrepreneurs can make fully informed decisions about how to organize their society. Capable at last of choosing rationally among political and market options, governments will be held accountable, corruption will decrease accordingly, and markets will become more efficient, leading to better growth, and hoisting developing countries out of poverty. (6–7)

This theory of political change, in turn, has embedded within *it* a series of assumptions—an already-assembled public simply waiting for information, an already sorted-out national economy, or the very fact of a nationally organized order of things separate from transnational corporations. In other words, transparency is at once an enormous industry and a gravely false conclusion. It is how liberal politics is imagined but doesn't work. After all that this book has recounted about the quotidian violence of liberal capitalism—from the segregated spaces on which it depends, to the gross inequalities of the contract form, to the chimeric yet serial "national economy"—there is a special irony to ending with, or to imagine that politics ends with, transparency.

This chapter, the last full ethnographic chapter in the book, uses the EITI process as it unfolded in Equatorial Guinea between 2006 and 2008 to return to the question of political possibility. Watching the EITI process unfold ethnographically allows a grounded critique of liberal theologies of social and political change, and shows that our epistemologies of capitalism—how we come to understand it and to *see* it—affect how we might seek to change it. I begin by explaining how I came to be so intimately involved in Equatorial Guinea's EITI process and offer a brief, emic overview of what the global EITI program aims to do. Thereafter, I move quickly through the literature on transparency before offering a detailed account of the first EITI meeting in Equatorial Guinea, eventually held in November 2007. The remainder of this chapter weaves in and out of meetings among civil society groups, oil companies, the Equatoguinean government, and foreign consultants that took place between November 2007 and June 2008, to think about politics and the political, before transitioning to the book's conclusion on the quotidian violence of liberalism, and political possibilities within and against capitalism.

Insofar as this book is an exploration of the more general forms and processes—the offshore, contracts, infrastructures, something called "the" economy—that facilitate diverse capitalist projects around the world, this chapter adds another such form. *Transparency* is often invoked as the antidote not just to the ravages of extractive industries, but also to the excesses of capitalism more generally. Indeed, like the NGOs that so often take transparency as their rallying cry, transparency-as-solution is now in dialectical relation to capitalism itself—a permanent structural feature of it (Kirsch 2014). Thus, the mandate this book attempts to fulfill—to know capitalism ethnographically—is also, today, a mandate to know transparency.

THE EXTRACTIVE
INDUSTRIES TRANSPARENCY INITIATIVE

During my first week of long-term fieldwork in Malabo, I was discussing my project over lunch with Alfredo (the Equatoguinean economist with whom I recounted a conversation about audit, corporate governance, and the Foreign Corrupt Practices Act in chapter 1) and a group of five or six other Equatoguineans, all of whom had recently returned from lives and educations abroad. Upon his return to Equatorial Guinea from postgraduate study and employment in London, Alfredo had worked first for the Major Corporation; by the time I arrived in Equatorial Guinea, he was working for Regal Energy. This was our first meeting (it would be months before our conversation on corporate governance), and it was basically a fieldwork networking lunch, in which we were getting to know one another and I was asking for other Equatoguinean contacts in and around the industry who might be willing to talk with me. As I rambled about my interest in the relationships among international capital, the state, and Equatoguinean citizens, Alfredo practically cried out that I had to work with Isabel, who was also sitting among our large group at lunch. He introduced us.

Only a few years older than me, Isabel had recently returned to Equatorial Guinea from a life abroad. When we met that day, she was working as a director—essentially a high-level technocrat—in the Ministry of Finance and Budgets. In addition to her work at the ministry, Isabel ran a consulting business, helping potential international investors navigate Equatorial Guinea's murky investment waters. It was clear why Alfredo had steered me to her. Isabel was young, open, hyper-educated, and worldly. (Over lunch, it became clear that she was fluent in Spanish, English, French, German, and Bubi.) She and I agreed to meet formally to discuss possible collaborations, and a week or so later in her consultancy office, she offered me a participant observation position at her side. She would give me access to past and present state budget and finance documents, laws, decrees, and other documentation— notionally "public" information, but practically impossible to find. She would also bring me with her to the upcoming Second National Economic Conference and offered participation (as I found it interesting) in her consultancy work, helping her design small, field-based research projects on various subsectors. (*X* company wants to invest in the health sector. What are the needs there? How does the government solicit bids? Who wins and why? What companies are already working in that sector?)

Isabel was also the government functionary recently appointed to lead Equatorial Guinea's national Extractive Industries Transparency Initiative, which was scheduled to begin the following month. Tasked with the enormous bureaucratic labor of this initiative, in addition to her functionary duties in the ministry and her consulting business responsibilities, Isabel was overwhelmed by her workload and offered me full access to the EITI process—meetings, correspondence, document drafting, and the World Bank consultancy relationship—in exchange for what turned out to be periodic full-time work from me, at least as meetings or external EITI deadlines approached. In response to Isabel's generous offer, I let her know that I was most interested in working with and for her on the EITI process, although I was willing to help out elsewhere as necessary. Ultimately, it was in part through my work for her on EITI that I gained access to oil company management, high-level Equatoguinean bureaucrats, much of the book's document-based analysis, and more. But what is it that I had signed up for? What is the Extractive Industries Transparency Initiative?

Launched in 2002 by Tony Blair, then prime minister of the UK, the Extractive Industries Transparency Initiative is an effort backed by the G8, the International Monetary Fund (IMF), and the World Bank, among other multilateral institutions, to promote what they see as good governance and fiscal transparency in resource-rich countries. The EITI process ostensibly differentiates itself from other G8, IMF, and World Bank agreements and accords—structural adjustment comes to mind—by being a *voluntary* program, nominally led by the national governments that choose to sign on to the process. Between EITI's inception in 2002 and Equatorial Guinea's entrance into the process in 2007, twenty-four countries had endorsed or were implementing EITI. Among these twenty-four, fifteen countries had met the four criteria for "candidate country" status, including Azerbaijan, Cameroon, Gabon, Ghana, Guinea, Kazakhstan, Kyrgyzstan, Liberia, Mali, Mauritania, Mongolia, Niger, Nigeria, Peru, and Yemen. The nine remaining countries of the twenty-four, including Equatorial Guinea, as well as Chad, Congo, Congo DRC, Madagascar, Sao Tome and Principe, Timor L'este, and Trinidad and Tobago, were in the process of completing basic implementation criteria. Between its 2002 inception and 2007, no country had been fully validated as EITI "compliant."

What does EITI *do*? Or more accurately for the moment, what does it intend to do? In short, the program proposes an accounting-based cure for the resource curse that the Second National Economic Conference set out to diagnose. The following description of the EITI program appeared in the sourcebook we received in Equatorial Guinea in 2007:

> In many countries, money from oil, gas, and mining is associated with poverty, conflict, and corruption. Commonly referred to as the "resource curse," this is often driven by a lack of transparency and accountability around the payments that companies are making to governments, and the revenues that governments are receiving from those companies. The Extractive Industries Transparency Initiative (EITI) seeks to create that missing transparency and accountability. It is a voluntary initiative, supported by a coalition of companies, governments, investors, and civil society organizations. Alongside other efforts to improve transparency in government budget practice, the EITI begins a process whereby citizens can hold their governments to account for the use of those revenues. (EITI Sourcebook, March 2005)

In short, the Extractive Industries Transparency Initiative is a protracted exercise in account balancing, designed to reconcile "the payments that companies are making to governments, and the revenues that governments are receiving from those companies." (Note again in this description the narrow oil-as-money approach and the same liberal myopia about national governments as sources of pathology.) Companies and governments report those payments and revenues to a mutually agreed-upon auditor, referred to as an "aggregator" in EITI lingo. The aggregator is appointed to reconcile oil companies' accounts of how much money they have paid to a government with that government's accounts of how much they have received from the companies. (Proposed aggregator names floating around Equatorial Guinea as the process began included PricewaterhouseCoopers and Ernst & Young.) As of 2007, EITI defined a compliant country as one in which "(1) an aggregator from an internationally recognized firm has been selected; (2) oil companies have disclosed their government payments to this individual; (3) governments have separately disclosed their revenue to this individual, at which point; (4) she or he will reconcile the figures, identify discrepancies, and recommend improvements. A public report of the auditor's findings will be disseminated, and the country will then undergo another external validation process to be labeled 'compliant'" (World Bank 2008).

In any version of this modular EITI process, there are three "stakeholder groups": the given national government (because it signs the hydrocarbon contracts, as detailed in chapter 3); the extractive industry companies (both foreign and national); and "civil society." In its self-descriptions, the EITI process offers each group potential benefits from successful implementation. The promise to a given government is that the better management of resource revenue will lead to economic and political stability, which in turn will prevent the conflict found in so many resource-rich regions. A successfully completed EITI process also offers the government "an improved investment climate . . . providing a clear signal to investors and international financial institutions that the government is committed to greater transparency" (EITI Sourcebook, March 2005, 5). Benefits to accrue to extraction companies include the mitigation of reputational risk, a clear and public demonstration of their contribution to a given country's economy (as their payments to the government are made public), and the mitigation of conflict, hence the greater possibility of returns on the capital-intensive, long-term investments required by many natural resource projects. Finally, EITI promises civil society more information in the public domain about the revenues that accrue to their government from extraction, which in turn may enable citizens to make more demands on those revenues and make governments more accountable to those they govern. While the EITI process does not technically encompass how mineral revenue is spent by the government (only that they do indeed receive it and in what amount), the wider scope of the program (then known as EITI ++) allows a platform for that discussion, and some countries have used the EITI process to bring issues of revenue allocation into public debate.

While I offer an analysis of the EITI program and transparency more broadly throughout this chapter, I want to note for now Strathern's (2000) insight about accountability's "dual credentials in moral reasoning and in the methods and precepts of financial accounting" (1). Mary Poovey (1998) has made similar observations about the moral histories of double-entry bookkeeping. In the EITI program, against the perception that both oil companies and the Equatoguinean state intentionally obfuscate revenue and investment information, EITI offers a platform for transparent *accounts* (in the dual sense) of how much money there is and where it is going. Accountability, this program suggests, is to be found, or at least initiated, in accounting.

ON TRANSPARENCY

Critiques of transparency are widespread in anthropology and beyond, including a few authors who have written about the EITI process specifically. Andrew Barry (2008) has called EITI's intentionally delimited and depoliticized accounting process "a form of political microscopy" which enables an assembly to examine a surface which has been prepared for it to inspect" (7). This critique of transparency as microscopy—looking at a radically delimited and prepared surface in place of politics—has its roots in Power (1997), Strathern (2000), and others who urge us to look for what the visibility allegedly provided by transparency processes *conceals* (Roberts 2009, 962). (See Ballestero 2012 for a helpful overview.) Not without humor, Barry (2013) points out that Azerbaijan was the first country to be validated in the EITI process in 2009, the same year in which the country was ranked 143rd on Transparency International's Corruption Perceptions Index. In the midst of these apparent contradictions, Shore and Wright (2015) worry that transparency processes like EITI "recast political programs as mundane administrative and technical matters to be dealt with by experts, thereby masking their ideological content and removing them from the realm of contestable politics" (421). (See also Burchell 1993; Shore and Wright 1997.)

One of the most interesting features of these critiques of transparency is how widely held they are, including by practitioners themselves. Despite transparency's staying power as development dogma, even as other tenets of hardline neoliberal reform have been abandoned,[1] Strathern notes that "what is odd about our embrace of transparency is that 'everyone knows' about its inadequacies: that transparency involves a simplistic abstraction and decontextualization from the complexity of the world" (in Roberts 2009, 968).[2] Indeed, as has been the case for many of the forms and processes of capitalism explored in this book, abstraction and decontextualization are not accidents or shortcomings in the EITI process. Rather they are, in large part, *intentional*, even *aspirational* qualities. When it comes to revenue from natural resources, the EITI process intentionally keeps "politics"— corruption, kleptocracy, violence, collusion—out of the audit frame, with the idea (often implicit) that this might create a safe space in which to talk politics by other means.

In Equatorial Guinea, a place where politics cannot be spoken in public, the delimiting that came with the EITI process—where the mandate was to talk only about revenue in and revenue out—was actually drawing *new* territory, new spaces for contestation and debate that were not there before. Nearly every ethnographic account of transparency processes has found similar effects. As Ballestero (2012) writes: "There are plenty of circumstances . . . in which the utility of indicators lies in their capacity to create a new entity, one that [was] unspeakable before the quantification exercise [was] performed. This capacity to make entities speakable . . . requires a rethinking of transparency-creation arrangements in light of their multiple productivities" (164). Barry (2013) and Hetherington (2011) came to similar conclusions. Regarding the Baku–Tbilisi–Ceyhan pipeline, Barry argues that the work of transparency, in fact, fosters new forms of dispute. Hetherington too notes that "the practices of representation that go into creating transparency are saturated at every turn with precisely that aspect of social life that they are meant to get rid of: politics. Indeed, far from stabilizing information, these larger technical networks create new spaces for disagreement and contestation" (7). The EITI process in Equatorial Guinea also illustrates this productivity, multivalence, and, indeed, unruliness of an ostensibly delimited and delimiting process.

While depoliticizing microscopy was certainly present in the EITI process as it unfolded in Equatorial Guinea during 2007 and 2008, it was only as a haunted hope, a wished-for horizon. Over more than a year of Equatorial Guinea's intensive participation in the EITI program, the process never got *anywhere close* to auditing or accounting. Rather, participants struggled through messy, embodied questions, including Who really *is* civil society in Equatorial Guinea, where it is technically illegal to meet in large groups without explicit government permission? How can we reconcile the advice and guidebooks of transnational experts with Equatoguinean daily life? Is an overtly depoliticized process the *only* method through which to openly address political questions at the crossroads of the US oil industry and Obiang's regime? In other words, EITI did not produce anything close to the outcome for which it was designed, nor did the mandate to focus on accounting successfully limit the parameters of the program to those envisioned in advance. Consider how the first meeting unfolded—literally came undone at every possible edge—as a case in point.

THE FIRST EITI MEETING

At my first official meeting with Isabel in late October 2007, she told me that the initial gathering of the EITI "civil society" stakeholder group would take place in less than two weeks. She asked if I would be willing to give a presentation at that meeting on the concept of civil society—its history as a term, its evolution, its use within the context of the global South, and its current popularity in the international development world. This presentation was the first work Isabel requested of me, and while the thought of giving it repelled me (I was, after all, a judgmental graduate student steeped in critical development studies and deeply skeptical about "civil society" as a political category), accepting this assignment felt like both a gesture of good faith and my ticket into the EITI process. I agreed to do the presentation and offered to help plan the seminar as well, given that ten days before the event, Isabel had no venue, no agenda, no speakers, and no invitations. In talking more about her expectations for my presentation, Isabel let me know that many of the people I would be speaking to (none of whom I had met at that point) didn't have very "high qualifications" and would be a diverse group, including parliamentarians, members of voluntary associations, NGO representatives, and beyond. Isabel encouraged me to make liberal use of diagrams so that people would have visual representations of the material, and she reminded me repeatedly that this would be a "capacity-building" program.

The evening after Isabel talked about the incapacities of this civil society group, I worried over pages and pages of fieldnotes. I wrote that I struggled to even imagine who might show up to the meeting, in a country where people were not allowed to gather in groups, and where there was no press separate from government-controlled television or radio. I railed about the performative shortcomings of the civil society category, how it creates another *not yet* (Chakrabarty 2000) for Africa: civil society must be located, developed, their capacity built and organized before they might emerge as some kind of political force. Indeed, this is Mehta's (1997) argument about the teleological exclusions of liberalism itself, in which, far from a universal rubric, liberalism proceeds via a specific set of cultural norms, "a constellation of social practices riddled with a hierarchical and exclusionary density." He goes on to explain:

> [These norms] draw on and encourage conceptions of human beings that are far from abstract and universal and in which the anthropological minimum is buried under a thick set of social inscriptions and signals. They chart a terrain full of social credentials. It is a terrain that the natural individual equipped with universal capacities must negotiate before these capacities can assume the form necessary for political inclusion. In this, they circumscribe and order the particular form that the universalistic foundations of Lockean liberalism assume. It is a form that can and historically has left an exclusionary imprint in the concrete instantiation of liberal practices. (70)

The people to whom I would present, according to Isabel, who was of course addressing me and perhaps tailoring her descriptions accordingly, were not yet suitable as civil society, but had to achieve certain social credentials before their ostensibly universal capacities could "assume the form necessary for political inclusion." Both Barry (2013) and Hetherington (2011) note this kind of interpellation or subject-making capacity of transparency processes, in which novel forms of civil society or publics are *made* by projects like EITI and do not predate them in any neat way. "It is common to assume," Barry (2013) writes, "that the public has an immanent existence, waiting to be addressed and activated, only constrained by the absence of appropriate liberal democratic safeguards" (97). But transparency in practice shows that the assembly of publics is itself a disputed process, and that was certainly the case in Equatorial Guinea.

At our next meeting, I asked Isabel for more detail on who was to make up this civil society group, and she passed me a set of World Bank guidelines on how to select civil society for the EITI process, noting that the Ministry of the Interior would make sure the World Bank guidelines were respected and could be reconciled with government definitions. The World Bank criteria stated that (1) civil society participants must be officially registered with the state; (2) they must be compliant with government regulations; and (3) no one should be excluded. Now, clearly, the first and second criteria are completely contradictory to the third, because certainly not everyone who might be considered civil society is officially registered with the state or compliant with government regulations, so all those people are automatically excluded. Moreover, in Equatorial Guinea, the process *to become* an officially recognized "civil society" group is arduous; and most applicants are rejected on the explicitly vague, but implicitly obvious, grounds that they are somehow threatening to the dictatorship. The process is "long and full of obstacles," as Okenve (2017) summarizes:

> It requires the organisation's promoters to submit to the Ministry of the Interior an affidavit certifying that it will submit to its control on a quarterly basis, plus a favourable report from the Ministry of the area in which the organisation wishes to work, and another report from the governor or provincial government delegate. It also requires [the organization] to formalise the constitution of the entity before a notary public, who in turn must obtain an authorisation from the Ministry of the Interior to validate this act. No legally constituted association is allowed to receive any donation, whether local or foreign, private or public, above US$100 without prior authorisation from the Interior Minister [and] no legally constituted organisation, that is, no organisation that has been allowed by the government to function, can deal directly with a beneficiary community without an additional authorisation or credential; this is not what the law says, but it is "customary."

During my time in Equatorial Guinea, a friend was going through this process to get permission to start an association that would show films (there was no movie theater in the country at that point), hold conferences, host speakers, and plan art exhibitions. Her application was rejected, implicitly because the idea was too subversive. So who, then, was to count as Equatoguinean civil society for the purposes of EITI? Who among the as-yet undefined, incapacitated, not-yet of Guinean civil society would be the right ones to participate in this effort? Isabel told me that she would sit down with people from the Ministry of the Interior and the World Bank technical expert, and together they would identify shared guidelines on which groups might participate.

This meeting to-be among the national EITI coordinator, the Minister of the Interior, the World Bank technical expert, and unknown others to *define* civil society for the purposes of EITI can easily seem like a conspiracy at the highest levels of power to ensure an innocuous "participatory" process. And yet, despite intentions to the contrary, this meeting and others like it never happened. The ingredients for the conspiracy were there, but the willful, organized, strategic, planned, and plotted execution was not. This is not to say that the same *outcomes*—exclusion, depoliticization, the general defanging of EITI—are not possible in the absence of tactical and deliberate conspiracy, but the processes, methods, and intentions via which those outcomes are reached are not nearly as organized or intentional as it may be convenient (for analytical and critical purposes) to think, even in an internationally notorious dictatorship. In fact, over the year that I participated in nearly every detail of the EITI process, there was no enforcement of either the government's or the World Bank's rules of inclusion/exclusion in defining civil society for the purpose of EITI participation. At each meeting, there were new civil society participants who had not participated in previous meetings, and no one checked if they were registered officially as civil society or not. It seemed enforcement would only come into effect if someone stepped dramatically out of line.

The larger planning process of the first civil society meeting had similar contours. When viewed from the outside, it had all the makings of a depoliticizing conspiracy; from the inside, however, it was a bewildering tour through the daily life of a large bureaucracy. Originally to be held on November 9 and 10, the meeting went through an almost farcical series of postponements and shape-shifts, which again, from the outside, could easily be attributed to an intentional, conspiratorial effort on the part of the government to confuse, exclude, and nearly guarantee the failure of the meeting and hence civil society's participation in the process. But, having witnessed the postponements and shape-shifts quite literally alongside the organizer herself, I know that they were, in fact—while admittedly staggering in their regularity— a series of logistical conflicts and timing issues. Isabel made it clear to me that by the end of October (nine days ahead of the scheduled meeting), she still did not have the appropriate signatures from her superiors to *approve* the meeting, and technically we could not go ahead with publicity, invitations (to take place over the radio), or planning without that permission.

In the meantime, the Major Corporation had taken the lead in planning this meeting by arranging to host it on their compound and flying in an Azerbaijani expert on civil society participation in EITI, now working for them out of Houston. In the months leading up to the meeting, and before my involvement, Major representatives told Isabel that they had experience with civil society participation and with the EITI process, and that they were happy to host. Although, as noted above, EITI differentiates itself from other multilateral programs by being a government-led process, where possible, the Equatoguinean government deferred that responsibility to US-based companies. Isabel recounted a meeting to me between a minister and a company representative in which, having been given the files on the EITI effort, the minister turned to the company representative and asked, "So, what do we have to do?" Again here, we see the implicit assumptions on which liberalism and EITI are built—for instance, a clear line between state and company; already-assembled publics—as mere fictions, but *productive* fictions.

As the last days of October flew by, Isabel was still without official permission to hold the meeting. Nevertheless, Major went ahead with their hosting plans, and Isabel and I began planning as well, although we couldn't air anything on the radio or even tell anyone that this meeting would be held. I was somewhat incredulous and said repeatedly to Isabel that no one would come to the meeting if they didn't know about it. She assured me that *it's always like this here*, and people would listen to the radio the day before, if need be, and know that they were invited. Then she would call people on their cell phones and let them know again that they were invited. Over the next few days, the planned seminar changed from two days to one, and from a series of speakers to only me on civil society and an oil company representative on *What Is EITI and Civil Society's Role in EITI?* Isabel was to present and go through the country work plan. Note, of course, the potential conflict of interest in an oil company not only hosting the seminar, but also delivering the instruction on what EITI is and how Equatoguinean civil society is meant to participate, let alone my own conflicted role in presenting this dreaded civil society talk.

At some point, it became clear to Isabel that she was not getting permission for the meeting because all the ministers were too busy organizing for the upcoming Second National Economic Conference on the mainland, and no minister would be available to open the EITI meeting proceedings on the currently scheduled day because they would all have already left for Bata, where the economic conference was to take place. At that point, only three days before the meeting was scheduled to begin, Isabel postponed and relocated it—now to be held in Bata on the day before the economic conference was set to begin. Radio announcements in Malabo started five days ahead of the rescheduled event, now to be held on the continent, calling registered civil society members. Radio announcements started in Bata seven days ahead of the rescheduled event.

I flew into Bata the night before the rescheduled meeting (and two days before the Second National Economic Conference I chronicled in chapter 5), arriving at 10:30 p.m. I had been working for the previous four days on my civil society presentation—three days writing in English, and then one day translating into Spanish and editing with the help of the same friend whose film club was deemed too subversive to be approved as "civil society." Isabel had not had time to look at the developing presentation earlier, so when I arrived in Bata late that night, she called me into a government office (where she was still working at 11:00 p.m.) to have a look.

Isabel read the document in a speed-reading, out-loud voice and made a series of changes to take out any reference to the potential political (let alone revolutionary) dimensions of civil society. In a sentence on de Tocqueville, she crossed out "The independent associational life of civil society can act as protection against the domination of society by the state." She revised a sentence about the way in which dissident intellectuals fighting totalitarianism in Eastern and Central Europe in the 1970s and 1980s used the concept of civil society to read: "The concept of civil society has influenced civic participation in Eastern European populations and has played a primordial role in the changes in that region during the 1990s." But by far the most excruciating change she made, in an already excruciating document, was an edit to a section in which I discussed the role of civil society in South Africa during the antiapartheid struggle. She deleted a sentence in which I had written that civil society "organized strikes and boycotts, and were not simple protesters but approached a revolutionary force against the apartheid regime." I knew why she edited the previous examples above—Equatorial Guinea is a dictatorship, and those participating in EITI would not be allowed to "politicize" this process—so I said nothing when she made those changes. But in the case of the South African sentence, I asked her, "Why do I have to cut this?" Her response was that *"the rich will never say anything bad against the rich."* It shocked me that what she viewed as the Equatoguinean government's class position aligned them more closely with the former white South African regime than with those mostly Black South Africans who fought against it. I was awake until 3:30 the next morning making these changes in a dizzying exhaustion crowded with anxiety about complicity, depoliticization, and methodology.

I arrived early to the opulent conference hall where the meeting was to take place, eager to make sure my PowerPoint presentation worked and to calm my under-slept, over-caffeinated nerves. Major Corporation personnel were there as well, checking the technology they provided for the event, including a laptop for all presenters, a projector, and simultaneous translation technology. (Most of them did not speak Spanish.) Two hours after the indicated start time of 9:00 a.m., some members of civil society had trickled in, but the local coordinators and functionaries, with the exception of Isabel, were nowhere to be found. Isabel was in constant contact with the relevant government personnel via text, and she explained to me that if not enough people showed up for the meeting, they wouldn't come (following a widely accepted practice among highly ranked government personnel to only attend functions deemed important enough for their presence). Finally, at 11:30 a.m., Isabel relayed a message from the prime minister, who would have been the functionary to officially open the conference. He felt, she said, that the meeting was so important that he wanted a quorum of civil society people present. Thus, in his name, Isabel postponed the meeting again, now to be held on November 15, the day *following* the national conference. The Major personnel immediately began muttering to themselves, and the Azerbaijani expert was clearly upset, having already postponed her trip once and now not at all sure she'd be able to attend the meeting a week hence.

While announcing the postponement, Isabel explained EITI in very general terms for those civil society members who were present, and she let them know what they could expect of the upcoming meeting. She then asked for questions. The first and only question came from his Excellency the Representative of Muslims in Equatorial Guinea, who wanted all civil society members present to be formally introduced. In a paranoid dictatorship, especially around a de/politicized political process, participants always wanted to know who was present to avoid a situation analogous to my own disastrous evening with Elena and "the entrepreneur" who was also a security agent. After the group of roughly ten people introduced themselves, Isabel invited everyone to breakfast, and as we settled ourselves around the table in the restaurant, an animated discussion started among them about civil society and Equatorial Guinea in general. "How big is our population?" someone asked, making the point that even basic statistics were impossible to come by. Another demanded basic statistics on oil revenue. Isabel replied that this information was available and that it was the obligation of civil society to find it. The man responded that he's sick of obligations and that he wants rights; he has a right to information. Another added, elliptically, that "everyone has the right to dream in their country." He had heard of someone who wanted to start a car factory, he explained. It wouldn't work, "but let him dream." Food came, and the conversation slowed but continued. A clearly outspoken member of the group, whom I would later come to know well, explained that people "are sick of the same shit, participating in something that they know will go nowhere, or they know exactly where it will go." As the conversation about demands and rights continued, I was increasingly mortified about the civil society presentation I had narrowly escaped that day, but for which I was still on the hook in a week's time. Far from a group of not-yet capacitated potential liberal subjects, the people I was sitting among were a diverse group of Equatoguineans who were generally suspicious of one another, and rightly suspicious not only of the EITI process, but also of the US corporations who seemed to be behind it, as well as their own government, and yet I couldn't address any of that openly in the presentation I still had to give.

One week later, after the National Economic Conference had passed, and after more statistical information (of questionable origin) about Equatorial Guinea had been released into the public than perhaps ever before, we all re-assembled in the opulent conference hall. Three ministers strode in to open the meeting, and we stood as the prime minister called the meeting to order, flanked by the Minister of Finance and the Minister of Planning. The prime minister's opening speech that day was an almost verbatim enactment of Hetherington's account of the politics of transparency (2011), in which "democracy [is] formally similar to capitalism—a rule-based playing field on which the rational choices of citizen-consumers equipped with transparent information were to discover and elect optimal governments" (4), while also optimizing markets. In the wake of the National Economic Conference (which most of us in the room had attended), the prime minister explained that the EITI process represented "the first time that civil society will involve itself in important economic efforts at the national level," and that this must be understood "in the context of calling the population to participate in the larger economic plan presented at the conference." He said that EITI represented a restructuring of the relationship between state and society, with society now as the protagonist: *"el pueblo en acción,"* or the people in action. He explained the importance of administrative transparency to this restructuring, which would lead to authentic participation, effective information, and access to administrative documentation "so that the world of petroleum won't be secret." The prime minister finished his speech by invoking the uneven geographies of globalization, noting that "competition is what Equatorial Guinea needs to occupy the center of globalization, not the periphery. For those on the periphery, globalization offers more threats than benefits. What we're doing now is to move to the center from the periphery. Civil society is central in this move." Meeting attendees—who represented an assorted crew of political parties, rural community initiatives, youth organizations, an organization of people with disabilities, and a representative from the National Association of Musicians of Equatorial Guinea—applauded politely, at which point the prime minister introduced Susan, the corporate social responsibility representative for the Major Corporation, by saying, "You all know Major, those that handle our petroleum, and sometimes give us something for it."

On the agenda, Susan (representing the Major Corporation more broadly) was slated to introduce EITI—what it is, why it is important, and what civil society's role would be therein. She started her presentation with a description of the resource curse, which, she explained, emanated from a lack of transparency regarding the payment of oil revenues. Conveniently for Susan and her corporation, she specified that EITI was a process in which citizens hold their *governments* accountable for spending from oil and mining. The corporation disappeared from her description. After explaining what the EITI process was specifically designed to do—identify a third-party aggregator familiar with oil finance and Equatoguinean contractual regimes who will collect revenue reporting from the government and payment information from companies, and then reconcile them—Susan received an amazing series of questions from the civil society members in attendance. Below, I translate and transcribe several of them, interspersed with brief, parenthetical clarifying commentary of my own. Each question was asked by a separate enquirer, and Susan answered all of them, unless I note otherwise.

Roughly half the questions to Susan addressed Major's role in recent Equatoguinean history, notably, the company's complicity with the government and their role in corruption. (It is important to remember that among all the major US-based companies in Equatorial Guinea, Major had been there the longest.) Some of these questions also brought in the complicity of multilateral partners—the World Bank and the IMF. Read together, I interpret these questions less as a defense of the Equatoguinean state than as a polyvocal indictment of the companies in the wake of Susan's presentation, which absolved them completely and pointed all the blame at the government. Note that Susan's answer to nearly every question was that the question's content was not relevant to the specific (political microscopy) of the EITI process.

Q *About the curse (la maldición, referring to the role of the resource curse in her presentation): Your company has been here for many years, and 90 percent of local people don't have access to work for you. If we want to work for you, we have to pay an agency. Most people can't pay this. Guineanos should be able to have access to work for the companies. The people just see oil platforms and don't know anything about the businesses. Maybe you can talk to the government or the companies about these problems.*

A I will answer this question at lunch. I want to keep these questions about the process of EITI. But to answer quickly, some people sit for an entrance exam, and if they pass, Major hires them. There are no third parties in that process. We do hire people through subcontractors, and they comply with EG laws. (Note that the entrance exam process is just for higher levels of work. This isn't true for maids or drivers, for example, who are hired through Voxa or other subcontracting agencies, as discussed in chapter 4.)

Q   *Your people live in beautiful estates and we have no access to them. Why is that?*

A   This too is not a part of EITI. But we do have outreach in communities. But the EITI process is intended to give you access. In other words, it's a way in.

Q   *Can you give us a better definition of how we should define "corruption"? Only economic? Political as well? And good governance: to whom does this refer? To your companies and your leadership, or to our state?*

A   In EITI, "corruption" refers to misuse of government revenues. (Note that this is factually untrue in EITI's own terms, which specify the relationship between company payments and government receipt of those payment. How the government chooses to use those revenues is beyond the primary scope of the auditing process, as explained above.)

Q   *The businesses have now been in our country for a while, and they don't have trade unions. When will they have trade unions? (Everyone laughs.)*

A   This is not a question I can answer. It's beyond the scope of EITI. (Isabel joins in to say that unionization is up to the workers. As far as she knows, "unions are legal here." They are not, as discussed in chapter 4.)

Q   *If I buy a car for 20.000 CFA but the receipt says 15.000 (in other words, if the receipt is falsified), how can civil society moderate in this process where everyone is taking their little bit and putting in false receipts? Where will we get our information from?*

A   The aggregator: PricewaterhouseCoopers, Ernst & Young, someone who is completely objective. Everyone takes part in selecting the aggregator. It will be someone that everyone trusts and that everyone agrees is objective.

Q   *We have heard that the World Bank, the IMF, and others are partners in this effort. In the economic chaos of 1995, the government followed IMF suggestions to fire two thousand people from public administration. But now that we have money, can't EITI help, give advice, tell them to hire these people back? We can't get rid of poverty if people don't have jobs. (Everyone applauds.)*

A   This is not what EITI does. The World Bank and IMF continue to give advice, but not EITI.

Isabel speaks up again to say that the advice of the structural adjustment era is not relevant to this time. Now, she says, we're looking at diversifying the economy; the IMF and World Bank were at the conference, and this will make jobs.

There was another series of questions about civil society itself. Some participants openly or tacitly agreed with the capacity-building framework, and they actively requested their own tutelage to "become" civil society members that might be capable of understanding the accounting complexities at issue. Others noted the historic difficulties of civil society groups in Equatorial Guinea and wondered pointedly how this process would (or would not) change that history.

Q   *For civil society to commit to this process, we still need more training, more details. The concepts are new to most of us. I would like to request more seminars.*

A   Please speak up when you need more information. Again, remember that EITI is not an event, but the beginning of a process.

Q   *If we don't know who we are as civil society, how will we know what to do with the information we are given? We don't know what to do with the information.*

A   (Isabel responded directly to say that the civil society groups will be given more education on finance, payments, and other aspects of the program to enable them to follow the process.)

Q   *The European Union (in Equatorial Guinea) did a study on the difficulties of civil society in the country. That study concluded that civil society was not organized; that the actors who should participate don't have training; and that there was a lack of economic resources to support civil society. These problems persist. The EU has tried to put together an overview organization of civil society, but they couldn't get it ratified. Now EITI is requesting our participation, but we haven't resolved the original issue.*

A   Equatorial Guinea is in a difficult situation. But perhaps we can use this opportunity to build capacity, to solve some of the problems.

We can see in this question-and-answer session what Ballestero (2012), Hetherington (2011), and Barry (2013) found in their respective ethnographic accounts of transparency processes. The EITI process was making entities *speakable*—labor unions, corporate and government collusion, the question of local employment, critiques of international financial institutions—that were *not speakable* before the process began, or certainly not in a more or less public forum like this one. We see evidence of both Barry's "new forms of dispute" (2013, 5) and Hetherington's argument that "the practices of representation that go into creating transparency are saturated at every turn with precisely that aspect of social life that they are meant to get rid of: politics" (2011, 7). In other words, the ostensible delimitation of the EITI process *was*, in fact, creating new kinds of space for political speech in front of high-level oil company and government functionaries, not to mention fellow Equatoguineans relatively unknown to one another. And the substantive content of this political speech was nowhere near accounting. Indeed, for the most part, the questions weren't even near oil-as-money, but much closer to the material in this book's preceding chapters on the conditions of possibility for oil-as-money in the first place—luxurious enclaves for foreign management, lack of local employment, contracting and subcontracting regimes, trade unions and their absence, and the difficulties of finding any kind of information in Equatorial Guinea. And this question content persisted despite Susan's repeated protestations that these questions "weren't relevant to EITI." The one question that notionally addressed accounting—about how falsified receipts can possibly be accurately accounted for—was, of course, a jab at the companies and the Equatoguinean state alike, believed by most people to be equally complicit in falsifying financial information; thus, a process to look at already-falsified receipts (i.e., the EITI process) would go nowhere.

While the EITI process undeniably opened up new spaces for political speech, these spaces remained within a persistent subtext of paranoia, mutual suspicion, and hesitation to participate, meaningfully shared by all but one civil society participant—already a well-known and often-arrested member of an opposition party—whom I will call Sonrisa. This mutual suspicion was evident at the first canceled meeting, when the representative for Muslim groups asked everyone to introduce themselves. It was even more evident when—after long hours that included my ultimately uneventful presentation, during which people just nodded politely at the weird white woman, and Isabel's discussion of the action plan for civil society that they were all asked to sign—the EITI meeting concluded with a long a discussion of *who would count as civil society* for the purposes of EITI.

Isabel introduced the day's final discussion topic by explaining that "all who are here now will be participants. Everyone who is here will be called for every event. But when there are national committee meetings, only the selected committee will be able to participate." With that, she posed a question to the group: "Who do you think should make up that selected committee? Given the complicated level of discourse around EITI, who should do it?" The answer she was fishing for, and the answer she was given, was that the selected committee should have education and relevant work experience, which ruled out roughly four-fifths of the people present. Nevertheless, Isabel asked those present to discuss the membership of the fifteen-member national committee and initiated that discussion by reading aloud Government Decree 42/2007, which regulated the participation of civil society in EITI. After reading the Decree, Isabel explained, "This law has a trap (or a trick) [*la ley tiene trampa*]. It says that five of the fifteen members should be from PDGE [the ruling party]. Over time, we should work to improve this law."

Isabel's statement—inviting the assembled members of civil society to discuss the content of a law, and not just any content, but content that specified a clear over-representation of the ruling party (which Isabel represented to many people present)—was, plainly, a radical one. Assembled members of civil society were incredulous. One representative asked, "Why did we have all that discussion before if there's already a law!?" Others chimed in to the same effect, stating unequivocally that the law cannot be changed because it's been published officially already. But Isabel insisted to everyone that we could interpret the document. She explained that the government legislated under time pressure from an external EITI deadline, but the law does not necessarily reflect the needs of the EITI process. She proposed that the assembled group make a recommendation to the government of what would work best for EITI, and went on to suggest: "For instance, on the issue of political parties, perhaps one representative should be from PDGE and another from the opposition." She said that a recommendation of this kind wouldn't mean that we were disrespecting the document (metonymically, the government), but that we were involved in finding the best application of Equatoguinean law to the EITI process. Having known and worked intensively with Isabel for a month at this point, I believe her request for the assembled people to suggest a change in the law was genuine. It was not mere bait to sniff out dissent, as some people present clearly feared.

As Isabel insisted on the pliability of this law and the opportunity meeting participants had to change it for the better, civil society members interrupted her repeatedly to insist that they couldn't change the decree, with a clear subtext of fear of government reprisal. And yet, Isabel herself was a government functionary telling them they could. To me, in that moment, this debate seemed like such an important chance to potentially transcend the political microscopy of the EITI process through actually debating and improving an existing law. I found myself increasingly incensed and impassioned with each interruption of Isabel's insistence. I wanted to interject, and I drafted the following in Spanish in my notebook: "Let me see if I can explain myself. Ms. Isabel, as the national coordinator of EITI and as a member of the public administration, is giving you, as civil society, an incredible opportunity to participate in the best implementation of this decree and to deepen your participation in the EITI process. But it seems no one is willing or interested. Shouldn't we take this opportunity?" But I didn't get a chance to say anything. It was 5:30 p.m., and we had been working since eight o'clock that morning without food, having taken only one thirty-minute break. Isabel brought the meeting to a close, and afterward, I stood with Susan and her Equatoguinean translator, to whom I voiced my incredulity about what had just happened. Susan replied: "There's no real debate. EITI is only to decide the amount of money that the government gets. Civil society will have input into who is the aggregator, but the aggregator is chosen by a bid process. So, it's just for civil society to go and approve." While Susan's response was textbook transparency-as-depoliticization, having been in Equatorial Guinea on and off for a decade at that point, she continued: "That said, if you had told me ten years ago that people would be in a group debating a decree, saying it's not valid, *in a government building*, I would never have believed you. I can't believe how much things have changed." In Susan's brief recap, we see both the radically depoliticizing potential of transparency processes—"there's no real debate"—*and* the out-of-bounds political potentials these processes create, in which a group (hesitates to) debate a decree in a government building.

With both of these truths held in tension in my own experience of the meeting, I returned to my friend Josefina's family home in Bata, where I'd been staying. I told her about the meeting: how the people had an amazing opportunity and they didn't take it, and how I thought it was because they didn't understand what was actually being offered to them, or maybe because they were scared. Isabel herself said it was because people still thought they were living under a dictatorship. But Josefina disagreed. She said they did understand, but that no one wanted to be perceived as being in the opposition. "People always feel like people will talk [as in, inform], so even if inside they wanted to say something, they won't because of others. That's why security works so well in this country, because the masses talk just to fuck over others. Not even because they get paid or receive anything in return."

TRANSPARENCY'S MESS, ETHNOGRAPHY'S SEDUCTION

Over the next three months, there was a series of EITI meetings with various combinations of the stakeholder groups—civil society, foreign and national oil companies, and the Equatoguinean government. In this section, I offer a sequence of brief vignettes and descriptions of these meetings, in part to give a sense of the arc of the EITI process over the year I participated. I also offer these accounts to show how that process was perhaps not an arc at all, but tenuously connected fits and starts of processes, conversations, and priorities that stretched not only between Houston-based companies and the Equatoguinean government with whom they had signed lucrative contracts, but also into the worlds of transnational development work and nonprofit organizations.

Shortly after the meeting chronicled above, Isabel traveled to Oslo, Norway, where the EITI International Secretariat was holding its annual meeting. She was greeted in Oslo with the news that Sonrisa—who, as an active member of an opposition party, had fairly extensive contacts in the international NGO world—had submitted a complaint letter to Publish What You Pay (an NGO affiliated with EITI) about the November meeting chronicled above. In the letter, Sonrisa wrote that the notice civil society members were given before the meeting was too short; those who eventually showed up to the meeting were not sufficiently independent from the government; and he had signed the civil society action plan under pressure. Publish What You Pay representatives confronted Isabel when she arrived in Oslo, Sonrisa's letter in hand, and requested a meeting with her, which she refused, after which she was asked to meet with "some other expert" who told her that the process in Equatorial Guinea "is not transparent enough." Recounting this humiliating experience to me first over the phone from Oslo and later upon her return, Isabel said that she had tried to explain to the foreign "experts" and NGO representatives in Oslo that in Equatorial Guinea, the short notice was sufficient; none of the NGOs were explicitly affiliated with the government; and civil society members were in no way pressured to sign the action plan. She was then told that there needed to be another meeting and that Peter Eigen (at that time the head of EITI's International Secretariat) should come. She questioned this assertion both in her Oslo interactions and again to me: "So it will be taken more seriously if *he* does it!? It always feels like people don't trust the government here. But in the [EITI] Sourcebook it says this is a government-driven process! All I had was three days of fighting in

Oslo. They don't want to listen to what the government says. They think that they are right and the government is wrong."

In response to her experience in Oslo (in mid-December), Isabel called a civil society meeting after the winter holidays, in late January. She began that meeting by stating that when she got to Oslo, she was greeted with the accusation that the government forces NGOs to agree to things against their will. Everyone at the table shook their heads, and I found myself in the odd position of nearly agreeing with them. Having participated in every twist and turn of the November planning meeting, and of course the meeting itself, I knew full well that no one was forced to do anything. I wrote in my notebook, "It's good that Sonrisa hears this and understands the consequences of his actions, but then of course it may turn others [in the civil society group] against him." Isabel told assembled civil society members that they needed to call their own meeting, ideally by February 4, so that they could participate meaningfully, and on their own terms, in the meeting with companies and the government scheduled for February 5. She said that *they*, as civil society, had to get the announcement on the radio, and that *they* would be responsible for letting everyone know with sufficient lead time. "And there will be international people here as observers on the fifth, so maybe then they'll believe our efforts." Sonrisa, knowing that all of this was in response to his letter, said he wanted to help call the meeting on February 4, but couldn't do it himself. He requested administrative support from Isabel who responded, not without sarcasm, that if the government (read: her) organizes and announces the meeting on the fourth, then maybe civil society will feel that they can't speak openly. Sonrisa responded, "Don't tell me a black pen is a blue pen! I can't go to the Ministry of the Interior and ask for an NGO list! It's just not that easy. We need your coordination office to facilitate this." Isabel responded, "Go with a copy of the decrees, go with the acts to the Minister of the Interior. We can't try to change the functioning of the whole administration; we are just working in EITI."

And this, of course, was a central problem with the EITI process in Equatorial Guinea, in general. How do you do it in a dictatorship with no independent press? How do you do it where civil society is essentially illegal? How do you do it between the demands for liberalism from Publish What You Pay and the exigencies of what would actually happen if Sonrisa were to go to the Ministry of the Interior, or try by himself to get an announcement put on the radio? How do you do it between histories of violence and repression that pester, even as recently returned government functionaries like Isabel promise (and seem to genuinely believe) that something else is possible? Having become friends at that point, Sonrisa and I conferred after that meeting and agreed to work with whomever else was willing to organize an independent meeting. I would serve as the institutional go-between, using my relationship with Isabel to request needed information.

Also in late January, Isabel called a meeting for the corporate and government EITI stakeholders. Held in the Major compound, this meeting brought together country managers of all the major US-based companies (Major, Smith, Endurance), along with Regal Energy, two Chinese-owned oil companies, some smaller US-based exploration companies and representatives from both Sonagas and GEPetrol—the national gas and oil companies— as well as high-level government representatives from the Ministry of Finance and the Ministry of Mines. The preoccupations at this meeting were remarkably different from those expressed at the civil society meeting. After presentations by Susan and Isabel on the EITI process, nearly the whole meeting was taken up with accounting; again, however, not accounting in a narrow, depoliticized sense, but in a tense, accounting-as-politics and as information-potentially-full-of-liability sense. After the initial bland presentations on the EITI process, a representative from the Ministry of Finance asked, "Which are the different concepts of revenue that we want the aggregator to look at? What will oil companies share? We will share the same." The country manager for the Major Corporation responded: "We will report lease bonus payments, lease rental payments, any sale of hydrocarbon that created revenue, any take in kind provided to government or another company. For example, GEPetrol is lifting half of our crude. GEPetrol would have to report that they received that and sold it. It's that level of definition: product sales and takes in kind, plus other major fees paid to governments for our contracts: royalties, taxes. The government will have to take a view [then he interrupted himself], well, I would *recommend* that they take a view that we won't take small registration fees: port fees, etc."

We can see from this exchange that what actually counted as revenue for the sake of the EITI process was not self-evident. Clearly, from the country manager's description, "revenue" in the world of oil and gas contracts is a radically disaggregated and distributed category. We are not simply talking about double-entry bookkeeping here. Note, however, that at least at this point in the process, the government and the companies seem happy to work together on what will count as revenue, with the government representative effectively asking the oil companies, "What should count as revenue? Whatever you decide, that's what we'll report as well." In this question of what counts, or what will and will not be counted in an accounting process, the various complicities between companies and the government grow starker, including the question of confidentiality. The Major country manager again: "Here's what we should do, what I think we should do." (He constantly corrected himself from a command voice to a suggestion voice.) "We have to maintain the confidentiality of our production sharing contracts through this process. We should try to have a model confidentiality agreement where we could all sign the same thing. We need to bind the aggregator to this agreement also, bind him [*sic*] individually back to the companies and the government. Fully back and forth confidentiality provision." This preoccupation with confidentiality brings us back to the exploitative production sharing contract terms I discussed in chapter 3. Here, Major's country manager is worried that public disclosures of payments and revenue through the EITI process could inadvertently disclose the specificities of contract terms. Royalties, taxation rates, takes-in-kind, profit sharing percentages—all of these are not merely revenue categories but negotiated political relationships. And Major, as the company with the oldest standing contracts in Equatorial Guinea, undoubtedly had exceptionally lucrative (for the company) / exploitative (for the government) contract terms that they did not want revealed either to other companies or to a more general public. Here again, a process focused narrowly on accounting aims at political microscopy but splatters over a wide swath of political relationships in practice.

The final preoccupation at this more-than-accounting meeting was a question at the intersection of audit, aggregation, and temporality. Having made our way slowly through discussions of what would count as revenue and how confidentiality would be maintained, the question of revenue *from when* arose. How many years back does the EITI process need to go? How current does it need to be? Isabel suggested that at least for Equatorial Guinea's first foray into the process, they report perhaps one to two years of payments and revenue. But *which* one to two years was a surprisingly thorny problem. The country manager of the Regal Corporation explained the temporality conundrum as follows: "The information we share as a company will come from audited financial reports. This means that we will be publishing with years of delay. Does that meet [the EITI] Secretariat requirements? If information can only be submitted following an audit process, then we won't have the data until a year later—after our internal auditing process." Major's country manager picked up the thread: "We are subject to an internal audit as a company. I have audited books. When I report [to EITI], I report what we have. It's just that as a business my books do get audited by an outside entity because I'm a public company and I have to do that. My auditor doesn't close out my books until early the following year. If we're in '08, we could submit '05 and '06."

As these two country managers went back and forth with one another, there was a clear subtext about the Equatoguinean government's bookkeeping. After all, nearly everyone in that room had just been through the National Economic Conference together and knew full well the admitted shortcomings of the government's bookkeeping practices. Isabel finally addressed the palpable tension, now in her capacity as technician in the Ministry of Finance and Budgets: "The government doesn't have formal auditing as such. We have a yearly IMF mission [Article IV] that has a look in our accounts. So, the government will submit IMF-reviewed data as our audited data." In this meeting, we can see that even where political microscopy and ostensibly depoliticized accounting are the goals, and where participants are addressing EITI *on its own terms*, those terms are deeply politicized and controversial. They are saturated with potential liability for companies and governments alike. Thus, Shore and Wright's (2015, 421) anxiety that transparency programs "recast political programs as mundane administrative and technical matters to be dealt with by experts, thereby masking their ideological content and removing them from the realm of contestable politics" may be misplaced, certainly in the case of Equatorial Guinea. Seemingly mundane administrative questions—What counts as revenue? What can we keep confidential?—asked by and among experts who are both in competition and collusion with one another, do not mask ideological content, but beckon it. Force it to be spoken. Certainly, in the case of EITI in Equatorial Guinea, if the companies and the government would collude successfully, then this removal from the realm of contestable politics would be achieved; but this would happen only through great effort that included wading through imperial debris and confidentiality tensions in shareholder-owned companies, all of which shape the deep politics of accounting itself.

Less than a week later, on February 4, Sonrisa and I spent the entire day in an NGO office in Caracolas, drafting a proposal from civil society to the National Coordinator (Isabel) that outlined adequate civil society participation in the EITI process. Sonrisa had made cell phone calls to other civil society members (on a prepaid card provided by the government), and we were joined for a time by two of the more senior members of the civil society group, in addition to two foreign experts who had flown in for the next day's meeting. One of these foreign experts was an Italian man I will call Piero, who was working in Cameroon as the regional director of Publish What You Pay, and one was a Venezuelan man—Luis—who was also an EITI board member. I had picked them up from the airport the evening before, and Luis marveled at the good quality of the roads driving from the airport, illustrative, in my experience, of people's apocalyptic imaginations of Equatorial Guinea, so often contravened by lived experience. Piero had had visa problems because the government didn't send him a letter of invitation in time, and he paid 30.000 CFA in the airport (roughly US$50) as an entry fee. After paying, Piero asked me loudly in the airport if this was corruption, and if so, stated that this was the first time in his ten years in Africa that he had paid a bribe. (I personally thought that 30.000 CFA was a reasonable, if unposted, fee for letting a foreign national in without a visa.) As I drove them from the airport to their hotel, they asked me what I did, and among other things I mentioned my research and relationships with the migrant wives of oil company managers. Piero made the analogy to Italian mafia wives—taking care of kids and having potlucks while their husbands decide whose leg to cut off. It was in this interesting company that Sonrisa and I drafted the proposal that he would present at the first meeting among all three stakeholder groups, to be held the following day.

That meeting felt both momentous and uneventful. It was perhaps the first time that civil society—by which I mean, essentially, regular Equatoguineans—high-level government personnel, and high-level oil company personnel were all in one small room together, with the idea that they should address one another. So that seemed momentous, as did the ten minutes during which Sonrisa read aloud the proposal we had drafted the day before. But in other ways, the meeting was quotidian and felt purposeless. Oil company people don't speak Spanish, and there wasn't simultaneous translation, so many sat there with their arms folded across their chests as the Minister of Mines opened the meeting or as Sonrisa presented the proposal; so too when it was the companies' turn to speak, and one of the Equatoguinean ministers was visibly asleep at the head of the table. But two short weeks later, with the blessing of the foreign visitors who came to witness the process, the International Committee of EITI met in Accra, Ghana, and Equatorial Guinea was officially recognized as a candidate country. This recognition gave them two years (until February 2010) to complete all EITI requirements and to move from a candidate country to a compliant country.

After this flurry of meetings in early 2008, clearly scheduled to meet the Accra deadline, we didn't have another EITI meeting until June, this time on the occasion of the one-month residency of a Peruvian World Bank/EITI consultant I will call Carlos. Carlos did not work for the World Bank, but was subcontracted by them given his central role in the EITI process in Peru.

A Cameroonian expert was also supposed to attend this meeting, but Isabel explained to me that he refused to come because he didn't get his invitation letter from the Equatoguinean government in time. Subsequently, he didn't trust Isabel's advice to just go to Douala and that she would get him into the country. I heard her, exasperated on the phone with him, explaining, "You know how African governments work! If you want to participate in this conference according to European norms, fine. But you are an African and we are an African government, and you have to trust me that this is how this will work."

In June, at the final official meeting I attended, Carlos gave what was, for Equatorial Guinea, a wildly radical political speech. He began by repeating, indeed hammering, what I have called "the incantation" in this book—the contradiction between so much oil wealth and so much poverty. If the resource curse was the acceptable way to tell this story in Equatorial Guinea, Carlos's speech offered no such niceties. Rather than relying on the comfortable distance of economic theory, he was directly critical of the Equatoguinean government. I wondered how government representatives in the audience were taking it. Civil society people I had come to know well over the last many months were visibly uncomfortable and squirmed in their chairs. "All Equatoguineans are owners of these [hydrocarbon] resources" Carlos explained "and, organized as civil society, they can fulfill functions that the government should fulfill."

Again, in Equatorial Guinea, it was difficult if not impossible to say directly, in public, and still less in front of government ministers, that there were things that the government did not do. Most basically, this was because you could not criticize them, and more specifically because even the imagined state was not a service-providing state. It was a patronage state. Services, when provided, were provided through personal connections—*enchufes* into the system, or kin networks. But Carlos carried on about Sendero Luminoso and the trial of Fujimori for corruption and human rights abuses. Carlos's speech again highlights the ways in which the EITI process did not stay within the lines of political microscopy, but rather spat up politics in the most unanticipated places. Who knew, for instance, that the World Bank subcontractor would be the most radical show in town?

When Carlos finished and opened the floor for questions, the first to speak was a lower level government functionary, who offered a rhetorical question that was in fact a warning to those considering asking actual questions. He said, in brief, that this EITI meeting was not an opportunity to issue personal political complaints. But rather than use the opportunity of Carlos's politicized speech to do that, members of civil society chose to illustrate the chasm between what Carlos narrated of his experience in Peru, and their experiences in Equatorial Guinea:

Q   *You talk about your experience in Peru, but I don't have anywhere near the capacity that you talk about. You say in the Peru case they only invited NGOs [to participate] whose institutional work already had to do with the themes of EITI. EITI was an extension of what they were already doing. If this were to be the case here, obviously there would be no EITI (pointing to the fact that there are no NGOs monitoring government budgets or corruption in Equatorial Guinea).*

Q   *I feel like you have given us a very interesting case, and one that is very different from ours. For example, if here only those members of civil society who already worked in transparency or fiscal vigilance were to participate, we wouldn't have anyone. I want to say that here we find ourselves in a very different situation.*

Rather than mustering a substantive response to these grounded and genuine questions, Carlos was basically empathetically dismissive of Equatoguinean civil society in his exit report for the World Bank and the Equatoguinean government:

It is not difficult to understand that a large number of representatives of civil society organizations have seen, in this initiative, an opportunity to obtain financial resources to carry out their particular projects. These can range from hospital care to artistic events but have nothing to do with issues of transparency and accountability. No wonder that some have felt themselves "misplaced" when confronted with the real definition of the EITI. . . . They are more concerned about what their participation in the EITI-EG can bring to them, rather than what they can bring to the country through their participation in the Commission.

And indeed, many civil society participants were consistently concerned not only with potential revenue for their resource-starved organizations, but also with the potential of *desplacamientos* (small payments, similar to per diems) for their personal participation in the process. Just like the practice of a second salary for government ministers, who receive a desplacamiento for agreeing to speak here or participate there, it seemed perfectly reasonable for civil society participants to demand payments for participation, confident that this was perhaps the only guaranteed good that could come from such a problematic process. Finally, Carlos also reported that although Isabel seemed to be genuinely dedicated to her job and to the EITI process, she was radically overextended, and he suggested that the government name a full-time EITI director. My own role came up in Carlos's final report as well:

> This consultant finds that the person in charge of the National Coordinating Office should have a full-time dedication to her duties. The progress of the EITI process in Equatorial Guinea has been favored by the collaboration of an American anthropologist, Ms. Hannah Appel, who is in Malabo carrying out the field work for her Ph.D. program. Ms. Appel arrived six months ago and in six months she will return to her country; her departure will make clearer the limitations of the Coordination Office to fulfill its responsibilities.

Over my year of participation in the EITI process, during which I grew to be close friends with both Isabel and Sonrisa and gained access to more documents, ministries, and US oil company bureaucracies than I could ever have anticipated, what I saw more than anything was the messiness and unpredictability of the EITI process. Certainly, there is government and corporate collusion, and at the same time there are unanticipated spaces of politicization and resistance. This messiness is precisely what ethnography, especially in recent decades, is accustomed to finding. But what is all this messiness *doing*? Is it creating interstitial political spaces for dissent and friction? Sure, to an extent. Is the Extractive Industries Transparency Initiative a simple stage for the reproduction of the power of transnational liberal political practice? I would say, given the above account, *no*, to an extent. Not only was the process itself messy and ever-far from EITI's stated goals, but by February 2010—the date set in the Accra meeting for Equatorial Guinea to demonstrate sufficient progress to move from a candidate to a compliant country—the country was officially delisted from the EITI process. It had failed.

CONCLUSION

Transparency—the ontology in the licit life of capitalism that was supposed to be about politics—failed, on its own terms. I want to note the contrast here with, for example, the offshore, the contract, or the national economy, which despite their own forms of messiness, *did not fail*, or at least did not in Equatorial Guinea. As I wrote of the offshore, it is not the capitalist utopia of placeless economic interaction. Rather, it is a teeming and situated social space: men from twenty different countries and seventeen different companies, consequentially divided by nationality and race; Equatoguineans underpaid and held indefinitely at the level of trainee; and a corporate form so multiple and attenuated that, paradoxically, it can seem to disintegrate altogether. But nor is the power of the offshore, or its effects, undone by attention to this teeming and contentious sociality. So too with contracts, which, I contended, render licit blatant forms of neocolonialism; frame multiplicities into legally recognized and politically consequential singularities; and change contested political regimes into petro-powerhouses. We can point to contracts' reliance on and manipulation of postcolonial inequality and sovereignty without imagining that these "social explanations" or "historical contexts" somehow undo their power. On the contrary, I argued that they are constitutive of it. But transparency, that which was meant to address or redress or hold *accountable* some of the excesses to be found in these earlier forms, did not *work* in this same way. As I will argue in the book's brief afterword, liberalism (here in the form of transparency, civil society participation, and an improved investment climate), especially when mobilized as the moral architecture for resistance to transnational corporations, is too much in the service of capitalism to work as a trenchant form of politics.[3]

In 2017, a new General Secretary of EITI Equatorial Guinea was named, and the three stakeholder groups—the government, civil society, and the oil and gas companies—began to meet again. Indeed, as of June 2017, roughly a decade after the above year of work, the government's website reported that "EITI members from Gabon and Equatorial Guinea will receive . . . a training seminar which will allow their Civil Society Organisations to learn the tools and mechanisms for internal handling, and how to carry out fully and effectively their missions within the extractive sector, and boost sustainable governance in this promising strategic sector."[4]

———————————

As the decades wear on, it seems that Equatoguineans remain candidates for liberalism. Equatorial Guinea is, in fact, a candidate country, in EITI's language, hoping still to become *compliant* liberal subjects (again, EITI's language). And this iterative deferral itself reproduces fertile ground for the licit life of capitalism in Equatorial Guinea.

AFTERWORD

This book has engaged a specific capitalist project—US oil companies working off the shores of Equatorial Guinea—to make an argument about global capitalism more broadly. Refusing both totalizing theories that attribute to capitalism an intrinsic systematicity or logic, *and* arguments for an endlessly varied, specific, and fractured form, this book traces the work required to make Equatorial Guinea into an oil-exporting place. In so doing, it attempts to show the relationship between capitalism's coherence and power and the radically heterogeneous sites through which those qualities are made—and made again. Methodologically, this approach asks us to take the "as ifs" on which capitalism has so long relied—abstraction, decontextualization, and standardization—*themselves* as ethnographic objects, always-haunted aspirational processes and political projects that we can follow in the field. Rather than an attempt to recover the complexity and friction that those concepts famously elide, this is an ethnography of how things come to *seem* smooth, and of how the US oil and gas industry works to *seem* distanced from, and even outside of, Equatoguinean life. It is an ethnography of the effects of economic theory, or transparency rankings, or the contract form. It is an ethnography that traces the "real world effects of the phantoms" (Povinelli 2006, 13).

Each chapter—"The Offshore," "The Enclave," "The Contract," "The Subcontract," "The Economy," "The Political"—chronicles a site where capitalism's apparent smoothness is made, where systematicity is built, and where local complexity and heterogeneity are more or less successfully mustered into legibly, and *licitly*, capitalist practices. Indeed, it is the *licit life of capitalism*—contracts and subcontracts, infrastructures, economic theory, corporate enclaves, "transparency"—and the forms of racialized and gendered liberalism on which it relies that allow oil and gas to move from subsea deposit to futures price with both mundane reliability and spectacular accumulation. A supple form of vernacular liberalism—most often in the mouths of migrant managers, although also present in the EITI process and national economy documentation—gives the licit life of capitalism its moral architecture. Law, on the one hand, and densely historical forms of white supremacy and heteronormative conjugal intimacy, on the other, offered the US petro-project in Equatorial Guinea a performative stage on which to enact distance and tutelage, and to peddle standardization and market rationality.

First, a closing thought on law. Each site chronicled in this book—from the offshore to transparency—is meaningfully subtended by legal liberalism. The law of the sea, international tax law, contract law, labor law, regulatory takings, Sarbanes-Oxley, FCPA—each weaves in and out of the industry's daily practice in Equatorial Guinea, not only in the straightforward sense as a law to be followed, but also much more circuitously as invocation ("tax planning," "local law"), absence (the archive cemetery and imperial debris), and future (revise the legislative framework; reform the judicial system; secure the respect for human rights). The relationship of capitalism to law, and to legal liberalism more broadly, is central to the licit. Plainly, many of capitalism's most egregious excesses are lawful, or proceed *dans le vrai* of the law (Pistor 2019). This is precisely what Cheryl Harris (1993) means when she writes that "whiteness as property retains its core characteristic: the *legal legitimation* of expectations of power and control that enshrine the status quo as a neutral baseline, while masking the maintenance of white privilege and domination" (1715; emphasis added). It is also de Tocqueville's point, whom she quotes: "The United States has accomplished this twofold purpose of extermination of Indians and deprivation of rights, legally, philanthropically, and without violating a single great principle of morality in the eyes of the world. *It is impossible to destroy men with more respect for the laws of humanity*" (in Harris 1993, 1723; emphasis added).

Second, then, is the intertwined role of race—white supremacy, in particular—and gender in this supple vernacular liberalism. In Equatorial Guinea's oil industry, select postcolonial meanings attributed to whiteness, including expertise, technology, meritocracy, and philanthropy, worked together with the apparent standardization of rigs, subcontracts, economic theory, and globe-trotting transparency programs to produce a world in which racial discrimination and spatial segregation did not detract from, but *added to*, licit practice; white : nonwhite was semiotically mapped onto standard : corrupt :: global : local. Here, we see the relationship between production sharing contracts and heterosexual white marriage contracts, where the sanctity of the latter seems to validate the sovereign violations of the former. All three contract forms—pscs, subcontracts, and marriage contracts—disturb fantasies of liberal equality, showing how "the liberal" is made in and by radical power imbalances always-already available to it. We see that law, gender, race, and capitalism are intimately knotted: "Contracts about property in the person constitute relations of subordination, even when entry into the contracts is voluntary . . . [and] the global racial contract underpins the stark disparities of the contemporary world" (Pateman and Mills 2007, 3).

The licit life of capitalism, then, is made at the intersection of technology, race, law, gender, materials, markets, and phantom philosophies of liberalism. The global labor market for oil is *made* in the colonial relationship of the Philippines to US shipping and military industries, or Chavez's firing of unionized workers in Venezuela. Supply and demand are *made* by the mobility of Jim Crow segregation, apparent in the ability to licitly categorize workers as Third Country Nationals and in the fungibility of ten Filipinos for one American. Here, race, gender, empire, and capitalism are co-produced; one is not epiphenomenal to the other. Markets do not merely deepen postcolonial inequality, they are *made by* that inequality. It is in this argument that this book is most clearly indebted both to feminist approaches to capitalism and to the Black radical tradition, which have long argued (in their own ways) that our bodies, histories, socialities, and conscriptions are not epiphenomenal, peripheral, or merely affected by something called capitalism. Rather, those histories and embodiments of inequality, exploitation, and difference are the grounds for arbitrage; the grounds for profit-seeking; the grounds for ownership, property, and dispossession; and the grounds, of course, for resistance.

Neither the Black radical tradition nor feminist political economy can be narrowly defined as bodies of scholarship. Both traditions emanate from "the modern project of emancipation" (Hudson 2016; see also Kelley 2003). I end, then, with a note on my own commitments to emancipation and the question of this book's contribution. If anthropological knowing has been a mode of power, then we must know more about that which we need more power over. Consequently, as I wrote in the introduction, it was *capitalism*—its ideologies and institutions, people and dreams, ecologies and erasures—that I took as my ethnos. Methodologically, the book's chapters suggest ethnographic thresholds for the anthropological study of capitalism. In some, I tread well-worn ground—on transparency, for instance—but in others, like the contract or the national economy form, there is much work still to do. So there are hopeful programmatics here—that more richly ethnographic accounts of the daily life of capitalism will help us reimagine it. I am also aware, however, of my ambivalent belief in anthropological knowing—and scholarship more broadly—as a form of power. In the field, I saw again and again how the resource curse or social science on corrupt African states had powerful effects in the world, and powerful teeth in the mouths of transnational corporations, to invoke Simpson's (2014) framework that I used in the introduction. But, to be frank, the types of anthropology and critical theory that have long insisted on complexity, contingency, critiques of patriarchy, or racism did not seem to have the same liveliness in the field. They did not wield the same power. Thus, on the one hand, I insist that we trace the work of simplifications and systematization as themselves ethnographic objects. On the other hand, I want to harbor no illusions about the work theory does in the world, that is, the work it does and does not do when confronted with the world. Simply because capitalism is a project, for instance, does not mean that it can be undone simply. As I wrote in the introduction, bringing capitalism's otherwises into being is a profound challenge that requires much more than simply calling it a project.

I have committed myself increasingly to this particular challenge in the thirteen years since I began this research (see Appel 2012a, 2012b, 2015, 2019). With each passing year I am, on the one hand, more aware of the difficulties of this work—that it can only be realized in immanent and incremental worldly action, that it both demands collectivity and exposes the excruciating fractures of solidarity. And thus, on the other hand, I am more thankful for the creative refuge of paid and insured intellectual life. But living and working in these worlds simultaneously, I often worry that anthropological analysis, which often tacitly (and sometimes explicitly) presents itself as radical, seems to suggest that we know the answers—that we know how radical social change might proceed. My ongoing experiences as an activist have destabilized that conviction for me. Intellectual endeavor is one place where we can bask in the fullness of radical visions and radical critique but we must never forget the limits they meet beyond the page and, in my opinion, always commit ourselves to pushing those limits by putting ourselves beyond the page as well.

Beyond the page, we live in a world where the nexus of capitalism, vernacular liberalism, and white supremacy explored in this book—market rationality, legal regimes, contracts and those they privilege—is hegemonic. This means that "no strategy credibly poses a direct threat to the system in the sense that there are good grounds for believing that adopting it will generate effects in the near future that would really threaten capitalism. This is what it means to live in a hegemonic capitalist system: capitalism is sufficiently secure and flexible in its basic structures that there is no strategy possible that immediately threatens it" (Wright 2010, 332). Thus, when people ask me, wouldn't a *liberal* Equatorial Guinea, with free and fair elections, an end to dictatorship and impunity, respect for human rights, be better than today's illiberal (or antiliberal) Equatorial Guinea? Wouldn't oil companies that follow environmental laws and desegregate workforces be better? The answer to these questions is, *of course*. Of course we succumb to the banal seduction of liberal projects. But for me, those commitments should be made warily and partially, not least because of the deep betrayals of justice that subtend the liberal orders within which these reforms make their demands. We might also commit ourselves warily to liberal reforms because liberalism is felicitous in a liberal world. Liberal demands allow legible victories, like changing laws and changing regimes and voting for a better goddamn representative. And yet, at the same time, we can also commit ourselves to the fullness of radical projects: the scope of their vision, the depth of their analysis, their slow simmer and occasional explosion into public consciousness. These projects are antiracist and antipatriarchal and, for me, anticapitalist and antiliberal, in the historical sense that liberalism as exclusion and dispossession presented itself to me, in Equatorial Guinea. The legible victories here are fewer and farther between, and when they are legible, it is often because they have been yoked to liberalism—a law changed, a candidate defeated. But the space they offer, which is to imagine otherwise, to articulate, enact, and embody it slowly and stutteringly, is as expansive as the open ocean, seen from above.

NOTES

INTRODUCTION

1. This sentence, like the rest of the book, rests on the work of many people. To start, I will simply point out three. First, I refer you to Karen Ho's (2014) discussion of Andrew Orta's 2014 article on business students abroad, in which she asserts that capitalism is not a context. Second, in Anna Tsing's work on projects, she defines them as "organized packages of ideas and practices that assume an at least tentative stability through their social enactment, whether as custom, convention, trend, clubbish or professional training, institutional mandate, or government policy. A project is an institutionalized discourse with social and material effects" (Tsing 2001, 4; see also Tsing 2000a, 2000b). Finally, Édouard Glissant (1989) refers to the West as a project, not a place. Insofar as capitalism has been central to the project of the West, the material in this book demonstrates the same.

2. This argument is in direct dialogue with Robinson (1983), who shows how race, in particular, is a form of difference that long predated capitalism in European society and, thus, was widely available to it as a commonsense way to differentiate and (de)value—to the point of enslavement—labor, especially. As he writes: "The tendency of European civilization through capitalism was thus not to homogenize but to differentiate—to exaggerate regional, subcultural, and dialectical differences into 'racial' ones. As the Slavs became the natural slaves, the racially inferior stock for domination and exploitation during the early middle ages, as the Tartars came to oc-

cupy a similar position in the Italian cities of the late middle ages, so at the systemic interlocking of capitalism in the sixteenth century, the peoples of the Third World began to fill this expanding category of a civilization reproduced by capitalism" (26). Robinson uses the term "racial capitalism" (2) to capture this ongoing history and agency of racial differentiation as a material force. This book aims to contribute to the intellectual and political project of racial capitalism by showing how many of the general forms and processes on which capitalism relies—the offshore, contracts, infrastructures, something called "the" economy—are made by various forms of (de)valued difference including, but not limited to, race and gender.

3. Butler (1993) writes that performativity should focus on "the process of materialization that stabilizes over time to produce the effect of boundary, fixity, and surface we call matter" (9). See also Callon 2007 on "performation."

4. The people I refer to in this paragraph as "itinerant oil company management" are high-level managers for the overseas subsidiaries of the US-based companies on which this book focuses. As I will go on to detail, during my fieldwork in Equatorial Guinea, these were exclusively white men from the US or Western Europe. In earlier drafts of this manuscript I referred to them as *expatriate* managers, which is a term they use, and a term that I, in turn, had used uncritically. Now, however, I have chosen to use the terms "itinerant" and "migrant" to describe them, only using "expat" or "expatriate" where it was someone's actual usage (including my own) in Equatorial Guinea. I have made this shift because "expatriate" is a racialized term that essentially refers to the relative ease of white global mobility, based on the ongoing colonial advantage secured by European nations. Migrant, by contrast, is also a racialized term but refers to the relative difficulty of nonwhite mobility. Thus I refer to these managers as migrants as opposed to expatriates in order to denaturalize both terms and draw attention to their racialized constitution. In this choice, I draw on the work of Mongia (1999); Neumayer (2006); Andrucki (2010); and Mau et al. (2015). I have also relied heavily on Achiume (2019), who writes: "First World citizens have far greater capacity for lawful international mobility relative to their Third World counterparts, even setting aside questions of personal financial means. One's nationality determines the range of one's freedom of movement in a way that completely belies frequent claims that assert or imply that all persons are equally without the right of freedom of international movement in our global order. This is because of the robust web of multilateral and bilateral visa agreements that privilege First World passport holders and pre-authorize their movement across the globe. . . . Freedom of movement is, in effect, politically determined and racially differentiated. . . . And because of the persisting racial demographics that distinguish the First World from the Third—demographics that are a significant product of passports, national borders, and other successful institutions partially originated as technologies of racialized exclusion—most whites enjoy dramatically greater rights to freedom of international movement, by which I mean travel across borders, than most nonwhites."

5. I use "Guinean" and "Equatoguinean" interchangeably to refer to people from Equatorial Guinea. "Guinean" is closer to local usage, where Guineano/a refers to an Equatoguinean national; however, because there is a separate country called Guinea

(often referred to as Guinea-Conakry), this usage can be confusing outside national borders. Thus, I interchange it with Equatoguinean.

6. More recent histories characterize Río Muni as having a greater part in global trade and colonial connections, with active timber and rubber trade by German and British firms from the 1890s forward, and a vast labor-recruiting network with Liberian, Portuguese, and coastal Ndowe agents (Nerín 2010; Martino 2016a, 2018b).

7. See Mamdani 1996 on the colonial and postcolonial history of forced labor on the continent. See Martino 2016a for more details on the Equatoguinean/Nigerian situation.

8. *Dictadura* is dictatorship in Spanish. *Dura* means hard, where *blanda* means soft. Thus, *dictadura* to *dictablanda* refers to the softening of the dictatorship.

9. Kirsch (2014) offers a powerful account of this moment in the transnational mining industry, arguing that, from this moment, "the dialectical relationship between corporations and their critics has become a permanent structural feature of neoliberal capitalism" (3).

10. See Kirsch (2014), who writes about "Not another OK Tedi," and Bond's (2013) work about change in the oil industry as an ongoing response to disasters of their own making.

11. For more recent and detailed accounts of the political situation in Equatorial Guinea, including the six-month detention of political cartoonist Ramón Esono Ebalé, or the internet blockage and oppositions arrests leading up to the November 2017 elections, see resources on the EG Justice website: https://www.egjustice.org/.

12. In their "Feminist Manifesto for the Study of Capitalism," Bear et al. (2015) call these assemblages formalizations or conversions through which "diverse social and economic projects come to appear coherent despite the heterogeneous, disaggregated practices from which they are constituted."

13. For resonant approaches see Appadurai 2003, and the Miller/Callon debate: Callon 1998, Miller 2002, Callon 2005, and Miller response 2005; see also Bear et al. 2015.

14. Quantitatively, flaring is measured in million cubic feet per day; for the largest producer in Equatorial Guinea, it averaged around 75 million cubic feet/day between January and April 2007 off of one platform (internal document).

15. On modularity as an ethnographic object, see Appel 2012c.

16. Ansley (1989) defines white supremacy as "a political, economic, and cultural system in which whites overwhelmingly control power and material resources, conscious and unconscious ideas of white superiority are widespread, and relations of white dominance and non-white subordination are daily reenacted across a broad array of institutions and social settings" (1024). This is the definition I use throughout the book.

1. THE OFFSHORE

1. All company and personal names are pseudonyms, here and throughout the book.

2. In general, access to offshore infrastructure for any amount of time was quite difficult for me to arrange and required months of relationship building and anticipa-

tory research with the company in question. Once I gained hard-won access for my twelve-hour visit to the FIPCO, I sought to organize return trips. Significant to the larger argument I develop in this book, by the time I did so, that rig had already been contracted by a different company and had changed locations at sea.

3. See Nixon 2011 on slow violence, and Hughes 2017 on the argument that the oil industry is most dangerous when it is working normally.

4. Of course, government officials in Equatorial Guinea and elsewhere are often complicit in this vast transfer of funds. Ghana—the next major African offshore oil exporter—has chosen a different approach to this issue and has been working with Barclays Bank since 2005 to establish itself as a tax haven (Mathiason 2009).

5. Thanks to the dissertation writers' seminar led by Sylvia Yanagisako in 2009 for pushing me in this direction.

6. Transfer pricing refers to the practices by which legally related entities (a parent corporation and its subsidiaries, for example) negotiate cost and payment for goods and services among subsidiaries. Transfer pricing becomes a consequential practice because taxable income is determined based on net profit, giving companies an incentive to inflate costs and pay exorbitant rates to their subsidiaries in order to minimize their tax burdens.

7. The incessant finger-pointing among BP, Halliburton, and TransOcean in the wake of the Deepwater Horizon conflagration demonstrates this insight in practice.

8. MEND is the Movement for the Emancipation of the Niger Delta (see Adunbi 2015; Kashi and Watts 2008). As a northern neighbor of Equatorial Guinea, and effectively sharing productive offshore waters, every time there was a serious uptick of violence in Nigeria, Equatorial Guinea would go on high alert. The "intel" grapevine would start buzzing with rumors that an attack was also planned on EG, the strategy being that the Nigerian government had stopped paying attention to militant action within their borders, so fighters were going to carry out attacks outside of Nigeria to regrab the attention of their own nation.

9. Thanks to Ramah McKay (personal communication) for encouraging me to explore this line of thinking. I only wish that my data bore out the generosity her comments suggested was part of these managers' statements.

10. During eighteen months of fieldwork, I only encountered one female rig worker. The offshore oil industry is an exceptionally patriarchal space.

11. Although my rig visit in Equatorial Guinea predated the 2010 Deepwater Horizon conflagration, that disaster forcefully brought these potentialities into the consciousness of a wider American, and arguably international, public. See Bond 2011, 2013.

2. THE ENCLAVE

1. "The wives" is an emic label that these women used to refer collectively to themselves. The great majority of the women who played cards were in Equatorial Guinea because they were married to migrant male managers in the oil and gas industry. Others who played were married to male diplomats or men in oil services companies.

Thus, "wives" made sense to them as a collective category. Among the regular group of women—a rotating cast of approximately fifty at any given time—there was only one whose presence in the country wasn't primarily defined (by her or others around her) as an accompaniment to her husband. This was "Joyce," a Chinese migrant who had come to Equatorial Guinea in the mid-1990s to open a Chinese food restaurant. While her husband was indeed with her, as he was the cook, she was the social and commercial face of the business. When referring to this group of women in this chapter, when appropriate, I use their term, "the wives," although it is one of the purposes of this chapter to think through their modes of being—gendered, secondary, and otherwise—in Equatorial Guinea.

2. The three large enclaves I discuss here are by no means the only oil or oil-related compounds in the country. As the industry exploded in Equatorial Guinea and more companies came in every day, small walled complexes began springing up everywhere. But these companies are smaller, often related to Smith, Major, or Endurance through subcontracting relationships discussed at length in chapter 4. With fewer in-country employees, their compounds were also noticeably different. They were generally located in affluent residential neighborhoods of Malabo, often recognizable by high walls, razor wire, and uniformed security. These complexes housed between three and fifteen migrant employees at any given time, and were often more incorporated into the communities around them, both architecturally and in terms of the circulation of their personnel. The men and occasional woman in these companies went into town more often and certainly fraternized outside their walls more, even if simply to go to another compound, usually the big ones that I deal with here, which became migrant management social centers. While these smaller developments are interesting and were an increasing presence in Malabo during my time there, I don't discuss them explicitly in this chapter except to note here that they exist, and that the lifestyles of the people who work in them can be somewhat different than what I describe. In terms of business practices, because they are subcontractors or consortia investors for Endurance, Major, or Smith, their ring-fencing is largely taken care of by being under the wing of a larger operating company.

3. See Pierre 2013 for whiteness as a form of increased mobility through securitized spaces in postcolonial Ghana.

4. Given the centrality of state-sponsored violence to trade in this era, Beckert (2015) encourages us to replace "mercantilism" with War Capitalism, which "better expresses [the period's] rawness and violence as well as its intimate connection to European imperial expansion." His rereading of Marx's original accumulation and rethinking of mercantilism relies on the scholarship of the black radical tradition, including Willams 1944. See Hudson 2017b for a critique.

5. On the centrality of Africa and the African diaspora to capitalism (and critical rereadings of Marx on primitive accumulation), see James 1963; Robinson 1983; and Johnson and Kelley 2017.

6. See Martino 2017 for resistance among nonunionized, colonial recruited labor in Equatorial Guinea.

7. After Obiang's overthrow of Macías in 1979, the bienes abandonados decree

stipulated that during a specified time period, Spaniards who had abandoned "their" land under Macías could return to recuperate it, although they were no longer recognized as owners and were asked to buy the land back from the new regime. Some Spaniards returned, purchased land, and resumed work, mostly in agriculture. But in many (if not most) cases, the Spanish were not able to repossess their land, even if they were willing to pay for it, especially those with property in or near the city, or with particularly valuable holdings like Punta Europa. In general, those who returned slowly realized that there was no juridical guarantee of stability for their land tenure, even having successfully reclaimed ownership. Many began returning to Spain. The decree was widely understood as having multiple motives, including not only revenue for a new regime whose treasury had been raided by departing settlers, but also the official transfer of many of the most valuable holdings into the hands of Obiang and his new cohort of leaders. Punta Europa has belonged to Obiang as private property since this time.

8. Working for the Clinton White House, J. Paul Bremer served as the leader of the Coalition Provisional Authority (CPA) in Iraq from May 2003 until June 2004, following the 2003 invasion of Iraq by the United States.

9. See Introduction note 4 for the meaningful distinction between *migrant and expatriate* and how I use these terms in the book.

10. On white ambivalence in charity and development work, see Elisha 2008 and Kowal 2015. On the thinness of white ambivalence as a form of narcissism that only buttresses white privilege and supremacy, see Heron 2007 and Goudge 2003. On race, whiteness, and blackness in the development industry and Africa more broadly, see Pierre 2013.

11. Prohibitions on public transport were company policy for all migrant workers, though perhaps only enforced by and among managerial levels. These policies were not specific to women. Prohibitions on or permissions for driving private/company cars for men varied with their position in the company.

12. On histories of intra-European or white racialization, see Robinson 1983; Brodkin 1998; Ignatiev 1995; Roediger 2018.

3. THE CONTRACT

1. Zalik (2009) writes of subcontracted Mexican oil workers that they too "are often denied severance pay supposedly guaranteed under contract" (573).

2. On concessionary companies, see Bouteillier 1903; Coquery-Vidrovitch 1972; Cantournet 1991; and Hardin 2011.

3. The opening four "Recitals" of the sample production sharing contract (República de Guinea Ecuatorial 2006b) specify this ownership regime, starting with the first: "WHEREAS all Hydrocarbons existing within the territory of the Republic of Equatorial Guinea, as set forth in the Hydrocarbons Law, are national resources owned exclusively by the state" (1). The following three recitals go on to say that the state wants to develop hydrocarbon deposits within a specific contract area; the contractor has the financial, technical, and professional ability to do so; thus, the state

and the contractor enter into this agreement. This ownership regime is also con-
secrated in Equatorial Guinea's Hydrocarbon Law (República de Guinea Ecuatorial
2006a), which begins: "The fundamental Law of the Republic of Equatorial Guinea
consecrates and designates as the property of the people of Equatorial Guinea all resources found in our national territory, including the subsoil, continental shelf, islands, and the Exclusive Economic Zone of our seas. It is by the mandate and delegation of the people, to whom these resources legitimately belong, that the Government
undertakes to manage them."

4. For this reason, as Latour (2005) argues, we cannot offer a "social" explanation,
but must instead offer an account of "reassembling the social" itself.

5. See Hale 1923 and Mnookin and Kornhauser 1979 for classic accounts of this
argument. See Banaji 2003 for a historical materialist account.

6. The company-as-party listing in the sample PSC reads: "[INSERT NAME], a
company organized and existing under the laws of [INSERT JURISDICTION], under
a company registration number [INSERT NUMBER], and having its registered office
at [INSERT ADDRESS], (hereinafter referred to as [the Company]), represented for the
purposes of this Contract by [INSERT NAME], in his capacity as [INSERT POSITION]"
(República de Guinea Ecuatorial 2006b, 1).

7. See Sawyer 2006 on this dual potential of corporate personhood.

8. Until its merger with PNC Financial Services in 2005, Riggs Bank was a venera-
ble Washington, DC, financial firm. Its two major lines of business were Private Banking, "financial services provided exclusively to wealthy individuals," and a specialized
area known as Embassy Banking. Riggs opened and administered accounts to "more
than 95% of the foreign missions and embassies located throughout the Washington
metropolitan area" (Coleman and Levin 2004, 13).

9. In another egregious regional example of the contract as a vehicle for liability
denial, in 2009, a group of Nigerian citizens took Chevron to court in an attempt to
hold the company responsible for the murder of their relatives on an offshore platform. "The Company" was easily able to claim that it was a nonsovereign entity operating under legal and contractual conditions, which absolved them from security
outcomes even on their own platforms (Michael Watts, personal communication).

10. See Harris 1993 for the ways in which transformative liberal legal precedent—
Brown v. Board of Education—enshrined the white supremacist status quo as the neutral baseline under the sign of newfound equality.

11. "To include a country on the index, Transparency International analyzes at
least three reliable data sources from credible organizations. This year, the anticorruption organization was unable to find a third source of information for Equatorial
Guinea. The absence of one single source such as the African Development Bank
made it impossible to get the necessary three sources to be ranked compared to last
year's available sources. Compare this to neighboring Cameroon (ranked toward the
bottom of the index at 136th), where Transparency International was able to identify
eight reliable data sources" (http://www.egjustice.org/post/eg-too-opaque-rank-0).

12. "The antibribery provisions of the FCPA make it unlawful for a U.S. person,
and certain foreign issuers of securities, to make a corrupt payment to a foreign offi-

cial for the purpose of obtaining or retaining business for or with, or directing business to, any person. Since 1998, they also apply to foreign firms and persons who take any act in furtherance of such a corrupt payment while in the United States. The FCPA also requires companies whose securities are listed in the United States to meet its accounting provisions. See 15 U.S.C. § 78m. These accounting provisions, which were designed to operate in tandem with the antibribery provisions of the FCPA, require corporations covered by the provisions to make and keep books and records that accurately and fairly reflect the transactions of the corporation and to devise and maintain an adequate system of internal accounting controls" (http://www.justice .gov/criminal/fraud/fcpa/docs/lay-persons-guide.pdf).

13. The brief legal history in this paragraph is based on conversations with this judge, a series of lawyers, and the scant legal history available, including Liniger-Goumaz 2000, and Campos Serrano and Micó Abogo 2006. On archives in Equatorial Guinea, see Martino 2014 and Enrique Martino's incredible Open Archives Project: opensourceguinea.org.

4. THE SUBCONTRACT

1. On contemporary body shopping practices, see Aneesh 2006; McKay 2007; 2014; Biao 2006; and Parreñas 2008, 2015.

2. On racial capitalism in particular, and a summary of the Black Radical Tradition, see Robinson 1983. For more recent work that explores the relationship between racialization and transnational capitalist processes, see Hoang 2015; Hudson 2017a and b; and Beckert 2015.

3. Again, back to chapter 1, note here that the dispersed corporate geographies and liability-dissemination practices are definitive not only of the large operating companies, but of the oil services companies as well. Laurel's finance manager described this setup as "the 'legal' way to do whatever they want with their money. They can declare some. They cannot declare some. It's not double accounting, but handling things the way they want, creating their own fiscal paradises."

4. Here "the party" refers to PDGE—the Democratic Party of Equatorial Guinea, or Partido Democrático de Guinea Ecuatorial. This is the president's political party, and membership therein was widely considered a requisite for any kind of gainful employment.

5. THE ECONOMY

1. As of 2018, credit and ATM cards were far more useful and (with still-declining oil prices and production) petty theft more common.

2. See Mitchell 2002 on this phenomenon with Egyptian cotton.

3. In Equatorial Guinea, the issue of where the administration is—Malabo or Bata—is for many people tied up with the Fang/non-Fang (biloblob) question. While the colonial capitol is Malabo, many suggest that Obiang privileges administrative presence and infrastructural development in Bata, and on the continent more gener-

ally, out of fealty to Fang people. While the Fang remain the demographic majority in Malabo as well, the island's rural inhabitants are mostly Bubi.

4. On whiteness, racial meaning-making, and development, see Pierre 2013; Crewe and Fernando 2006; Leonard 2010; Kothari 2006; and Kowal 2015.

5. The first paragraph of the 2007 conference's printed material starts with a summary of these looming futures. After noting that Equatorial Guinea has experienced unprecedented growth over the last ten years, a caution: "This growth is fragile, and based entirely on petroleum, whose productive peak will be reached in less than five years, while agriculture, which flourished in the past, is in decline and new sectors are not emerging" (República de Guinea Ecuatorial 2007a, 9). Even in the first National Economic Conference in 1997, when oil was the unanticipated solution to Equatorial Guinea's two decades of independent economic disaster, the resource curse cast its shadow: "Today a brilliant star has risen over Equatorial Guinea, that lights the way for all of our aspirations toward the progress and development of the nation. . . . The current economic moment invites us to examine the negative experiences of other countries whose goals have been frustrated in the use of their natural resources, and those who have been successful" (República de Guinea Ecuatorial 1997, 2).

6. The Spanish and French Cultural Centers sold a handful of books, and each had a small library accessible free to students. There were no private bookstores or public libraries. See also Williams 2011.

7. Throwing money, like "making it rain" in parts of US hip-hop culture or "shower money" at Nigerian weddings, is a recognized way to publicly display relations of patronage and hierarchy in Equatorial Guinea and beyond. Despite this recognizable ritual character, however, the Equatoguinean friends with whom I was watching the news that day were enraged by Teodorín's gesture and turned off the television in disgust.

8. Indeed, it had the MBAs without Borders fantasizing about something so separate from state control as to sound like the "informal economy": "Here, it's hard for a business to be legitimate, even if they want to be. It just means they get taxed and abused more. It's almost better to be under the radar, almost better not to have those connections. Our idea now is just to build small companies which will specialize in something, and, from that, find a model that the other companies will follow. We're trying to help build a market, generate some competition, good old capitalism."

9. Koz2Rim song translated from French: "In our days, honor is dead / and only the CFA is left standing. / To sell a homeland to the highest bidder / banks defrauded as ministers and the state fill their pockets and leave us nothing. / They cruise around, all of them, in Porsches."

6. THE POLITICAL

1. "Even as the World Bank and others turned away from hardline neoliberalism in the wake of structural adjustment, information problems remained the key to guiding the developing world out of poverty" (Hetherington 2011, 5).

2. The widespread dissatisfaction, even among economists, with GDP as a mea-

sure of national economy is a compelling analogue here, and perhaps more than analogue insofar as the abstractions in each case are substantively the same.

3. See James 1963; Sartori 2014; and Robinson 1983, chapter 9, for contrasting conclusions and discussions about the radical potentials of liberalism within capitalism.

4. "Inicio del seminario sobre Democracia y Buena Gobernabilidad de la EITI," Equatorial Guinea government website, July 6, 2017, https://www.guineaecuatorial press.com/noticia.php?id=9856.

REFERENCES

Achiume, Tendayi. 2019. "Migration as Decolonization." *Stanford Law Review* 71; UCLA School of Law, Public Law Research Paper No. 19-05. https://ssrn.com /abstract=3330353.

Adunbi, Omolade. 2015. *Oil Wealth and Insurgency in Nigeria*. Bloomington: Indiana University Press.

Alicante, Tutu. 2017. "How Our Incoming Secretary of State Helped to Enrich Africa's Nastiest Dictatorship." *Washington Post*, February 1. https://www .washingtonpost.com/news/global-opinions/wp/2017/02/01/how-our-incoming -secretary-of-state-helped-to-enrich-africas-nastiest-dictatorship/?utm_term =.0e1c010dd053.

Andrucki, Max J. 2010. "The Visa Whiteness Machine: Transnational Motility in Post-apartheid South Africa." *Ethnicities* 10, no. 3: 358–70.

Aneesh, A. 2006. *Virtual Migration: The Programming of Globalization*. Durham, NC: Duke University Press.

Ansley, Frances Lee. 1989. "Stirring the Ashes: Race, Class, and the Future of Civil-Rights Scholarship." *Cornell Law Review* 74, no. 6: 993–1077.

Appadurai, Arjun. 2003. *The Social Life of Things: Commodities in Cultural Perspective*. Cambridge: Cambridge University Press.

Appadurai, Arjun. 2013. *The Future as Cultural Fact*. London: Verso.

Appel, Hannah C. 2012a. "The Bureaucracies of Anarchy and the People's Microphone." In *Dreaming in Public: Building the Occupy Movement*, edited by

Amy Schrager Lang, Daniel Lang/Levitsky. Oxford: New Internationalist Publications.

Appel, Hannah C. 2012b. "Dispatches from an Occupation: Ethnographic Notes from Occupied Wall Street." *Social Text* blog. http://www.socialtextjournal.org /blog/topics/dispatches-from-an-occupation/.

Appel, Hannah. 2012c. "Offshore Work: Oil, Modularity, and the How of Capitalism in Equatorial Guinea." *American Ethnologist* 39, no. 4: 692–709. https://doi .org/10.1111/j.1548-1425.2012.01389.x.

Appel, Hannah C. 2012d. "Walls and White Elephants: Oil Extraction, Responsibility, and Infrastructural Violence in Equatorial Guinea." *Ethnography* 13, no. 4.: 439–65. https://doi.org/10.1177/1466138111435741.

Appel, Hannah. 2014. "Occupy Wall Street and the Economic Imagination." *Cultural Anthropology* 29, no. 4: 602–25.

Appel, Hannah C. 2015. "Possible Futures: Finance Capitalism and the Potential of Debtors' Unions." ROAR *Magazine*. https://roarmag.org/magazine/debt -collective-debtors-union/.

Appel, Hannah C. 2017. "Toward an Ethnography of the National Economy." *Cultural Anthropology* 32, no. 2: 294–322.

Appel, Hannah C. 2018a. "Infrastructural Time." In *The Promise of Infrastructure*, edited by Nikhil Anand, Akhil Gupta, and Hannah Appel. Durham, NC: Duke University Press.

Appel, Hannah C. 2018b. "Race Makes Markets: Subcontracting in the Transnational Oil Industry." SSRC *Items: Race and Capitalism*. https://items.ssrc.org /race-makes-markets-subcontracting-in-the-transnational-oil-industry/.

Appel, Hannah, Arthur Mason, Michael Watts, and Matthew T. Huber. 2015. *Subterranean Estates: Life Worlds of Oil and Gas*. Ithaca, NY: Cornell University Press.

Appel, Hannah C., Sa Whitley, and Caitlin Kline. 2019. "The Power of Debt: Identity and Collective Action in the Age of Finance." White paper available: https:// challengeinequality.luskin.ucla.edu/wp-content/uploads/sites/16/2019/03/Appel -Hannah-THE-POWER-OF-DEBT.pdf.

Aranzadi, Juan. Forthcoming. "Una propuesta de replanteamiento radical y revisión crítica de los estudios sobre guinea ecuatorial." In *Guinea Ecuatorial (des)conocida: Lo que sabemos, ignoramos, inventamos y deformamos acerca de su pasado y su presente*, edited by Gonzalo Álvarez Chillida and Juan Aranzadi. Madrid: UNED.

Artucio, Alejandro, International Commission of Jurists, and International University Exchange Fund. 1979. *The Trial of Macías in Equatorial Guinea: The Story of a Dictatorship*. International Commission of Jurists and International University Exchange Fund. Geneva, NY: AAICJ.

Asad, Talal. 1979. "Anthropology and the Analysis of Ideology." *Man* 14, no. 4: 607–27.

Ávila Laurel, Juan T. 2011. *Diccionario básico, y aleatorio, de la dictadura*

guineana. http://www.buala.org/pt/mukanda/diccionario-basico-y-aleatorio
-de-la-dictadura-guineana.

Ballestero, Andrea S. 2012. "Transparency in Triads." *PoLAR: Political and Legal Anthropology Review* 35, no. 2: 160–66. https://doi.org/10.1111/j.1555-2934.2012 .01196.x.

Banaji, Jaira. 2003. "The Fictions of Free Labour: Contract, Coercion, and So-Called Unfree Labour." *Historical Materialism* 11, no. 3: 69–95.

Banta, Martha. 1993. *Taylored Lives: Narrative Productions in the Age of Taylor, Veblen, and Ford.* Chicago: University of Chicago Press.

Barry, Andrew. 2006. "Technological Zones." *European Journal of Social Theory* 9, no. 2: 239–53. https://doi.org/10.1177/1368431006063343.

Barry, Andrew. 2008. "The Curse of Economics: Oil, Conflict and the Law." Paper prepared for Legal Knowledge and Anthropological Engagement conference, CRASSH, University of Cambridge, October.

Barry, Andrew. 2013. *Material Politics: Disputes along the Pipeline.* Hoboken: Wiley-Blackwell. http://catalogimages.wiley.com/images/db/jimages/9781118529119 .jpg.

Baudrillard, Jean. 2001. *Selected Writings.* Translated by Mark Poster. Cambridge: Polity.

Bayart, Jean-François. 2009. *The State in Africa: The Politics of the Belly.* Cambridge: Polity.

Bear, Laura, Karen Ho, Anna Tsing, and Sylvia Yanagisako. 2015. "Gens: A Feminist Manifesto for the Study of Capitalism." *Cultural Anthropology.* https://culanth .org/fieldsights/gens-a-feminist-manifesto-for-the-study-of-capitalism.

Beckert, Sven. 2015. *Empire of Cotton: A Global History.* New York: Vintage Books.

Benson, Peter. 2008. "El Campo: Faciality and Structural Violence in Farm Labor Camps." *Cultural Anthropology* 23, no. 4: 589–629.

Berry, Sara. 1993. *No Condition Is Permanent: The Social Dynamics of Agrarian Change in Sub-Saharan Africa.* Madison: University of Wisconsin Press.

Biao, Xiang. 2006. *Global "Body Shopping": An Indian Labor System in the Information Technology Industry.* Princeton, NJ: Princeton University Press.

Bishara, Fahad A. 2017. *A Sea of Debt: Law and Economic Life in the Western Indian Ocean, 1780–1950.* Cambridge: Cambridge University Press.

Bohannan, Paul, and George Dalton. 1962. *Markets in Africa.* Evanston, IL: Northwestern University Press.

Bond, David. 2011. "The Science of Catastrophe: Making Sense of the BP Oil Spill." *Anthropology Now* 3, no. 1: 36–46.

Bond, David. 2013. "Governing Disaster: The Political Life of the Environment during the BP Oil Spill." *Cultural Anthropology* 28, no. 4: 694–715. https://doi .org/10.1111/cuan.12033.

Bond, Patrick, and Khadija Sharife. 2009. "Apartheid Reparations and the Contestation of Corporate Power in Africa." *Review of African Political Economy* 36, no. 119: 115–25.

Bonvillain, Nancy. 2009. *Cultural Anthropology*. 2nd ed. Upper Saddle River, NJ: Pearson.

Bouteillier, G. 1903. *Les concessions et le Congo francais*. Albi: Pezous.

Bowker, Geoffrey C. 1994. *Science on the Run: Information Management and Industrial Geophysics at Schlumberger, 1920–1940*. Cambridge, MA: MIT Press.

Brodkin, Karen. 1998. *How Jews Became White Folks and What That Tells Us about Race in America*. New Brunswick, NJ: Rutgers University Press.

Brown, Wendy. 2009. *Edgework: Critical Essays on Knowledge and Politics*. Princeton, NJ: Princeton University Press.

Bryan, Dick, and Michael Rafferty. 2011. "Deriving Capital's (and Labor's) Future." In *The Crisis This Time*, vol. 47. Socialist Register. London: Merlin Press.

Bryan, Dick, and Michael Rafferty. 2018. *Risking Together: How Finance Is Dominating Everyday Life in Australia*. Sydney: Sydney University Press.

Burchell, Graham. 1993. "Liberal Government and Techniques of the Self." *Economy and Society* 22, no. 3: 267–82. https://doi.org/10.1080/03085149300000018.

Bureau of Democracy, Human Rights and Labor. 2004. "Equatorial Guinea." *International Religious Freedom Report* 2004. Washington, DC. http://www.state.gov/j/drl/rls/irf/2004/35353.htm.

Butler, Judith. 1993. *Bodies That Matter: On the Discursive Limits of "Sex."* New York: Routledge.

Butler, Judith. 2010. "Performative Agency." *Journal of Cultural Economy* 3, no. 2: 147–61. https://doi.org/10.1080/17530350.2010.494117.

Butler, Paula. 2015. *Colonial Extractions: Race and Canadian Mining in Contemporary Africa*. Toronto: University of Toronto Press.

Byrd, Jodi A. 2011. *The Transit of Empire: Indigenous Critiques of Colonialism*. Minneapolis: University of Minnesota Press.

Caldwell, Meg, Mark Zoback, and Roland Horne. 2010. "The Deepwater Horizon Disaster and the Future of Drilling in the Gulf of Mexico." Panel presentation, Stanford University, November 2010. https://www.youtube.com/watch?time_continue=12&v=aN2TIWomahQ.

Çalışkan, Koray, and Michel Callon. 2009. "Economization, Part 1: Shifting Attention from the Economy towards Processes of Economization." *Economy and Society* 38, no. 3: 369–98. https://doi.org/10.1080/03085140903020580.

Çalışkan, Koray. 2010. *Market Threads: How Cotton Farmers and Traders Create a Global Commodity*. Princeton: Princeton University Press.

Callon, Michel. 1998. "The Embeddedness of Economic Markets in Economics." In *The Laws of the Markets*. Sociological Review Monograph. Oxford: Blackwell/Sociological Review.

Callon, Michel. 2005. "Why Virtualism Paves the Way to Political Impotence: A Reply to Daniel Miller's Critique of 'The Laws of the Market.'" *Economic Sociology: European Electronic Newsletter* 6, no. 2: 3–20. https://www.econstor.eu/bitstream/10419/155843/1/vol06-no02-a2.pdf.

Callon, Michel. 2007. "What Does It Mean to Say That Economics Is Performative?" In *Do Economists Make Markets? On the Performativity of Economics*, edited by

Donald A. MacKenzie, Fabian Muniesa, and Lucia Siu. Princeton, NJ: Princeton University Press.

Cameron, Angus, and Ronen Palan. 2004. *The Imagined Economies of Globalization*. London: Sage, 2004.

Campos Serrano, Alicia, and Plácido Micó Abogo. 2006. *Labour and Trade Union Freedom in Equatorial Guinea*. International Confederation of Free Trade Unions (ICFTU). Madrid: Fundación Paz y Solidaridad "Serafín Aliaga" (CCOO).

Cantournet, Jean. 1991. *Des affaires et des hommes: Noirs et blancs, commercants et fonctionnaires dans l'Oubangui du debut du siècle*. Paris: Société d'ethnologie.

Cattelino, Jessica R. 2008. *High Stakes: Florida Seminole Gaming and Sovereignty*. Durham, NC: Duke University Press.

Cattelino, Jessica R. 2011. "'One Hamburger at a Time': Revisiting the State-Society Divide with the Seminole Tribe of Florida and Hard Rock International." *Current Anthropology* 52, no. S3: S137–49.

Césaire, Aimé. 1962. *Discours sur le colonialisme*. Paris: Presence Africaine.

Chabal, Patrick, and Jean-Pascal Daloz. 1999. *Africa Works: Disorder as Political Instrument*. International African Institute in association with James Currey, Oxford, UK. Bloomington: Indiana University Press.

Chakrabarty, Dipesh. 2000. *Provincializing Europe: Postcolonial Thought and Historical Difference*. Princeton, NJ: Princeton University Press.

Chakrabarty, Dipesh. 2002. *Habitations of Modernity: Essays in the Wake of Subaltern Studies*. Chicago: University of Chicago Press.

Chatterjee, Partha. 1993. *The Nation and Its Fragments: Colonial and Postcolonial Histories*. Princeton, NJ: Princeton University Press.

Cho, Sumi. 2008. "Embedded Whiteness: Theorizing Exclusion in Public Contracting." *Berkeley La Raza Law Journal* 19: 5–26.

Coleman, Norm, and Carl Levin. 2004. *Money Laundering and Foreign Corruption: Enforcement and Effectiveness of the PATRIOT Act*. Hearing before the Permanent Subcommittee on Investigations of the Committee on Governmental Affairs, United States Senate, 108th Congress, 2nd sess., US GPO, July 15.

Comaroff, Jean, and John L. Comaroff. 2001. *Millennial Capitalism and the Culture of Neoliberalism*. Durham, NC: Duke University Press.

Comaroff, Jean, and David Kyuman Kim. 2011. "Anthropology, Theology, Critical Pedagogy: A Conversation with Jean Comaroff and David Kyuman Kim." *Cultural Anthropology* 26, no. 2: 158–78. https://doi.org/10.1111/j.1548-1360.2011 .01093.x.

Conaway, Charles F. 1999. *The Petroleum Industry: A Nontechnical Guide*. Tulsa, OK: PennWell. http://site.ebrary.com/id/10344552.

Cooper, Frederick. 2015. *Africa in the World: Capitalism, Empire, Nation-State*. Cambridge, MA: Harvard University Press.

Cooper, Frederick, and Ann Laura Stoler. 1997. *Tensions of Empire: Colonial Cultures in a Bourgeois World*. Berkeley: University of California Press.

Coquery-Vidrovitch, Catherine. 1972. *Le Congo au temps des grandes companies concessionnaires 1898–1930*. Paris: Mouton.

Coronil, Fernando. 1997. *The Magical State: Nature, Money, and Modernity in Venezuela*. Chicago: University of Chicago Press.

Crawford, Margaret. 1995. *Building the Workingman's Paradise: The Design of American Company Towns*. London: Verso.

Creus, Jacint. 1997. *Identidad y conflicto: Aproximación a la tradición oral en Guinea Ecuatorial*. Madrid: Los Libros de la Catarata.

Crewe, Emma, and Priyanthi Fernando. 2006. "The Elephant in the Room: Racism in Representations, Relationships, and Rituals." *Progress in Development Studies* 6, no. 1: 40–54.

Crumley, Carole L. 2002. *New Directions in Anthropology and Environment: Intersections*. Lanham, MD: AltaMira Press.

Daly, Samuel F. 2013. "Dropped Subjects: Igbo Labor Migration to Fernando Po, 1940–1974." *Igbo Studies Review* 1, no. 1: 1–17.

Daly, Samuel F. 2017. "De trabajadores a soldados: Trabajo forzado y conscripción en la Guinea Española y la Nigeria oriental, 1930–1970." *Millars: Espai i historia* 43, no. 2: 219–41.

Davis, Angela Y. 1983. *Women, Race, and Class*. New York: Vintage Books.

De Genova, Nicholas. 2005. *Working the Boundaries: Race, Space, and "Illegality" in Mexican Chicago*. Durham, NC: Duke University Press.

Dua, Jatin. 2019. *Captured at Sea: Piracy and Protection in the Indian Ocean*. Berkeley: University of California Press.

Durkheim, Émile. (1893) 1997. *The Division of Labor in Society*. Translated by W. D. Halls. New York: Free Press. http://catalog.hathitrust.org/api/volumes/oclc/37575382.html.

Ebrahim-Zadeh, Christine. 2003. "Back to Basics: Dutch Disease. Too Much Wealth Managed Unwisely." *Finance and Development* 40, no. 1: 50–51. http://www.imf.org/external/pubs/ft/fandd/2003/03/ebra.htm.

EG Justice. 2014. "Too Opaque to Rank." http://www.egjustice.org/post/eg-too-opaque-rank-0.

EITI. 2005. *EITI Sourcebook*. https://eiti.org/document/eiti-source-book.

Ejituwu, N. C. 1995. "Anglo-Spanish Employment Agency: Its Role in the Mobilization of Nigerian Labour for the Island of Fernando Po." In *The Nigeria-Equatorial Guinea Transborder Cooperation*, edited by A. I. Asiwaju, B. M. Barkindo, and R. E. Mabale. Lagos: University of Nigeria Press.

Elisha, Omri. 2008. "Moral Ambitions of Grace: The Paradox of Compassion and Accountability in Evangelical Faith-Based Activism." *Cultural Anthropology* 23, no. 1: 154–89. https://doi.org/10.1111/j.1548-1360.2008.00006.x.

Elyachar, Julia. 2005. *Markets of Dispossession: NGOs, Economic Development, and the State in Cairo*. Durham, NC: Duke University Press.

Elyachar, Julia. 2012. "Before (and after) Neoliberalism: Tacit Knowledge, Secrets of the Trade, and the Public Sector in Egypt." *Cultural Anthropology* 27, no. 1: 76–96. https://doi.org/10.1111/j.1548-1360.2012.01127.x.

Enloe, Cynthia H. 1990. *Bananas, Beaches and Bases: Making Feminist Sense of International Politics*. Berkeley: University of California Press.

Escobar, Arturo. 1995. *Encountering Development: The Making and Unmaking of the Third World*. Princeton, NJ: Princeton University Press.

Falola, Toyin, and Ann Genova. 2005. *The Politics of the Global Oil Industry: An Introduction*. Westport, CT: Praeger.

Fanon, Frantz. 1991. *The Wretched of the Earth*. New York: Grove Weidenfeld.

Federici, Silvia. 1998. *Caliban and the Witch: The Body and Primitive Accumulation*. Brooklyn: Autonomedia.

Fegley, Randall. 1989. *Equatorial Guinea: An African Tragedy*. New York: P. Lang.

Ferguson, James. 1994. *The Anti-politics Machine: "Development," Depoliticization, and Bureaucratic Power in Lesotho*. Minneapolis: University of Minnesota Press.

Ferguson, James. 1999. *Expectations of Modernity: Myths and Meanings of Urban Life on the Zambian Copperbelt*. Berkeley: University of California Press. http://public.eblib.com/choice/publicfullrecord.aspx?p=858756.

Ferguson, James. 2006. *Global Shadows: Africa in the Neoliberal World Order*. Durham, NC: Duke University Press.

Ferguson, James. 2013. "Declarations of Dependence: Labor, Personhood, and Welfare in Southern Africa." *Journal of the Royal Anthropological Institute* 19. 223–242.

Ferguson, James, and Akhil Gupta. 2002. "Spatializing States: Toward an Ethnography of Neoliberal Governmentality." *American Ethnologist* 29, no. 4: 981–1002.

Fligstein, Neil, and Taekjin Shin. 2007. "Shareholder Value and the Transformation of the U.S. Economy, 1984–2000." *Sociological Forum* 22, no. 4: 399–424.

Fortes, Meyer, E. E. Evans-Pritchard, and the International Institute of African Languages and Cultures (IIALC). 1940. *African Political Systems*. London: Oxford University Press.

Foucault, Michel. 1980. *Power/Knowledge: Selected Interviews and Other Writings, 1972–1977*. New York: Pantheon Books.

Ghazvinian, John H. 2007. *Untapped: The Scramble for Africa's Oil*. Orlando, FL: Harcourt.

Gibson-Graham, J. K. 1996. *The End of Capitalism (as We Knew It): A Feminist Critique of Political Economy*. Minneapolis: University of Minnesota Press.

Gibson-Graham, J. K. 2006. *A Postcapitalist Politics*. Minneapolis: University of Minnesota Press.

Glissant, Édouard. 1989. *Caribbean Discourse: Selected Essays*. Charlottesville, VA: University Press of Virginia.

Goede, Marieke de. 2005. *Virtue, Fortune and Faith: A Genealogy of Finance*. Minneapolis: University of Minnesota Press.

Goudge, Paulette. 2003. *The Power of Whiteness: Racism in Third World Development and Aid*. London: Lawrence & Wishart.

Granovetter, Mark S. 1985. "Economic Action and Social Structure: The Problem of Embeddedness." *The American Journal of Sociology* 91, no. 3. https://doi.org/10.1086/228311

Granovetter, Mark S., and Richard Swedberg. 2001. *The Sociology of Economic Life*. Boulder, CO: Westview.

Granovetter, Mark S., and Richard Swedberg. 2011. *The Sociology of Economic Life.* Boulder, CO: Westview.

Grovegui, Siba N'Zatioula. 1996. *Sovereigns, Quasi Sovereigns, and Africans: Race and Self-Determination in International Law.* Minneapolis: University of Minnesota Press.

Gupta, Akhil. 1995. "Blurred Boundaries: The Discourse of Corruption, the Culture of Politics, and the Imagined State." *American Ethnologist* 22, no. 2: 375–402.

Gupta, Akhil. 2012. *Red Tape: Bureaucracy, Structural Violence, and Poverty in India.* Durham, NC: Duke University Press.

Guyer, Jane I. 1995. *Money Matters: Instability, Values, and Social Payments in the Modern History of West African Communities.* Portsmouth, NH: Heinemann.

Guyer, Jane I. 2004. *Marginal Gains: Monetary Transactions in Atlantic Africa.* Chicago: University of Chicago Press.

Guyer, Jane I. 2009. "Composites, Fictions, and Risk: Towards an Ethnography of Price." In *Market and Society: the Great Transformation Today*, edited by Chris Hann and Keith Hart. Cambridge: Cambridge University Press.

Hale, Robert L. 1923. "Coercion and Distribution in a Supposedly Non-coercive State." *Political Science Quarterly* 38: 470–78.

Hampton, Mark, and Jason Abbott. 1999. *Offshore Finance Centers and Tax Havens: The Rise of Global Capital.* West Lafayette, IN: Ichor Business Books.

Hann, Chris, and Keith Hart. 2011. *Economic Anthropology.* Cambridge: Polity.

Hardin, R., and P. Auzel. 2001. "Colonial History, Concessionary Politics, and Collaborative Management of Equatorial African Rain Forests." In "Hunting and Bushmeat Utilization in the African Rain Forest: Perspectives toward a Blueprint for Conservative Action," edited by M. Bakarr, G. D. A. Fonseca, R. Mittermeier, A. B. Rylands, and K. Walker Painemilla, special issue, *Advances in Applied Biodiversity Science* 2: 21–38.

Hardin, Rebecca. 2011. "Concessionary Politics: Property, Patronage, and Political Rivalry in Central African Forest Management." In "Corporate Lives: New Perspectives on the Social Life of the Corporate Form," edited by Damani J. Partridge, Marina Welker, and Rebecca Hardin, special issue, *Current Anthropology* 52, no. S3: S113–S125.

Harney, Stefano, and Fred Moten. 2013. *The Undercommons: Fugitive Planning and Black Study.* Brooklyn: Autonomedia.

Harris, Cheryl I. 1993. "Whiteness as Property." *Harvard Law Review* 106, no. 8: 1707–91. https://doi.org/10.2307/1341787.

Harrison, Virginia. 2013. "World's Best Economies in 2013." CNNMoney, December 27. http://money.cnn.com/gallery/news/economy/2013/12/27/best-economies/.

Hart, Jeff, Niels Phaf, and Koen Vermeltfoort. 2013. "Saving Time and Money on Major Projects." McKinsey & Company, December. https://www.mckinsey.com/industries/oil-and-gas/our-insights/saving-time-and-money-on-major-projects.

Hart, Keith, Jean-Louis Laville, and Antonio David Cattani. 2010. *The Human Economy: A Citizen's Guide.* Cambridge, MA: Polity.

Hartman, Saidiya V. 1997. *Scenes of Subjection: Terror, Slavery, and Self-Making in Nineteenth-Century America*. New York: Oxford University Press.

Harvey, David. 1990. *The Condition of Postmodernity: An Enquiry into the Origins of Cultural Change*. Oxford: Blackwell.

Heron, Barbara. 2007. *Desire for Development: Whiteness, Gender, and the Helping Imperative*. Waterloo, ON: Wilfrid Laurier University Press.

Hetherington, Kregg. 2011. *Guerrilla Auditors: The Politics of Transparency in Neoliberal Paraguay*. Durham, NC: Duke University Press.

Hirschman, Albert O. 1961. *The Strategy of Economic Development*. New Haven, CT: Yale University Press.

Ho, Karen. 2009. *Liquidated: An Ethnography of Wall Street*. Durham, NC: Duke University Press.

Ho, Karen. 2014. "Commentary on Andrew Orta's 'Managing the Margins': The Anthropology of Transnational Capitalism, Neoliberalism, and Risk." *American Ethnologist* 41, no. 1: 31–37. https://doi.org/10.1111/amet.12057.

Ho, Karen. 2016. "Racializing Normative Markets: Whiteness, Masculinity, and the 'Efficiency' of Networks." Public lecture presented at Culture, Power, and Social Change (CPSC), UCLA (May 12).

Hoang, Kimberly Kay. 2015. *Dealing in Desire: Asian Ascendancy, Western Decline, and the Hidden Currencies of Global Sex Work*. Berkeley: University of California Press. http://public.eblib.com/choice/publicfullrecord.aspx?p=1766690.

Hobbes, Thomas, and Aloysius Martinich. 2002. *Leviathan*. Peterborough, ON: Broadview Press.

Holmes, Douglas R. 2014. *Economy of Words: Communicative Imperatives in Central Banks*. Chicago: University of Chicago Press.

Hudson, Peter James. 2016. "The Racist Dawn of Capitalism: Unearthing the Economy of Bondage." *Boston Review*. http://bostonreview.net/books-ideas/peter-james-hudson-slavery-capitalism.

Hudson, Peter James. 2017a. *Bankers and Empire: How Wall Street Colonized the Caribbean*. Chicago: University of Chicago Press.

Hudson, Peter James. 2017b. "Racial Capitalism and the Dark Proletariat." In "Race, Capitalism, Justice," edited by Walter Johnson and Robin D. G. Kelley, *Boston Review* Forum 1.

Hughes, David M. 2017. *Energy without Conscience: Oil, Climate Change, and Complicity*. Durham, NC: Duke University Press.

Humphreys, Macartan, Jeffrey Sachs, and Joseph E. Stiglitz. 2007. *Escaping the Resource Curse*. New York: Columbia University Press.

Ignatiev, Noel. 1995. *How the Irish Became White*. London: Routledge.

International Chamber of Commerce. 2010. *Incoterms® rules 2010*. https://iccwbo.org/resources-for-business/incoterms-rules/incoterms-rules-2010/.

International Monetary Fund. 2009. "Republic of Equatorial Guinea: Selected Issues." Country Report 09/99. https://www.imf.org/en/Publications/CR/Issues/2016/12/31/Republic-of-Equatorial-Guinea-Selected-Issues-22815.

International Monetary Fund. 2010. "Republic of Equatorial Guinea: 2010 Article IV Consultation." Staff Report for the 2010 Article IV Consultation 10/103. Washington, DC.

International Monetary Fund. 2015. "Republic of Equatorial Guinea: 2015 Article IV Consultation—Press Release; Staff Report; and Statement by the Executive Director for the Republic of Equatorial Guinea." Staff Report for the 2015 Article IV Consultation 15/260. Washington, DC.

James, C. L. R. 1963. *The Black Jacobins: Toussaint L'Ouverture and the San Domingo Revolution.* New York: Vintage Books.

Jameson, Fredric. 1992. *Postmodernism, or, The Cultural Logic of Late Capitalism.* Durham, NC: Duke University Press.

Johnson, Leigh. 2015. "Near Futures and Perfect Hedges in the Gulf of Mexico." In *Subterranean Estates: Life Worlds of Oil and Gas,* edited by M. Watts, A. Mason, and H. Appel. Ithaca, NY: Cornell University Press.

Johnson, Walter, and Robin D. G. Kelley, eds. 2017. "Race, Capitalism, Justice." *Boston Review* Forum 1. http://bostonreview.net/race/race-capitalism-justice.

Johnston, David. 2007. "How to Evaluate the Fiscal Terms of Oil Contracts." In *Escaping the Resource Curse,* edited by Macartan Humphreys et al. New York: Columbia University Press.

Kashi, Ed, and Michael Watts. 2008. *Curse of the Black Gold: 50 Years of Oil in the Niger Delta.* Brooklyn, NY: PowerHouse Books.

Kelley, Robin D. G. 2003. *Freedom Dreams: The Black Radical Imagination.* Boston: Beacon.

Keynes, John Maynard. 1913. *Indian Currency and Finance.* London: Macmillan and Company.

Khurana, Rakesh. 2007. *From Higher Aims to Hired Hands: The Social Transformation of American Business Schools and the Unfulfilled Promise of Management as a Profession.* Princeton, NJ: Princeton University Press. http://public.eblib.com/choice/publicfullrecord.aspx?p=457842.

Kimball, Solon T. 1956. "Section of Anthropology: American Culture in Saudi Arabia." *Transactions of the New York Academy of Sciences* 2, 18, no. 5: 415–503.

Kirsch, Stuart. 2014. *Mining Capitalism: The Relationship between Corporations and Their Critics.* Berkeley: University of California Press. http://public.eblib.com/choice/publicfullrecord.aspx?p=1693155.

Kothari, Uma. 2006. "Critiquing 'Race' and Racism in Development Discourse and Practice." *Progress in Development Studies* 6, no. 1: 1–7.

Kowal, Emma. 2015. *Trapped in the Gap: Doing Good in Indigenous Australia.* New York: Berghahn Books.

Lampland, Martha, and Susan Leigh Star, eds. 2008. *Standards and Their Stories: How Quantifying, Classifying, and Formalizing Practices Shape Everyday Life.* Ithaca, NY: Cornell University Press.

Larkin, Brian. 2013. "The Politics and Poetics of Infrastructure." *Annual Review of Anthropology* 42, no. 1: 327–43. https://doi.org/10.1146/annurev-anthro-092412-155522.

Latour, Bruno. 2005. *Reassembling the Social: An Introduction to Actor-Network-Theory*. Oxford: Oxford University Press.

Leonard, Lori. 2016. *Life in the Time of Oil: A Pipeline and Poverty in Chad*. Bloomington: Indiana University Press.

Leonard, Pauline. 2010. *Expatriate Identities in Postcolonial Organizations: Working Whiteness*. Burlington, VT: Ashgate.

Limbert, Mandana E. 2010. *In the Time of Oil: Piety, Memory, and Social Life in an Omani Town*. Stanford, CA: Stanford University Press.

Liniger-Goumaz, Max. 1989. *Small Is Not Always Beautiful: The Story of Equatorial Guinea*. Totowa, NJ: Barnes & Noble Books.

Liniger-Goumaz, Max. 2000. *Historical Dictionary of Equatorial Guinea*. Lanham, MD: Scarecrow.

LiPuma, Edward, and Benjamin Lee. 2004. *Financial Derivatives and the Globalization of Risk*. Durham, NC: Duke University Press.

Lovejoy, Paul E., and Toyin Falola. 2003. *Pawnship, Slavery, and Colonialism in Africa*. Trenton, NJ: Africa World Press.

Lowe, Lisa. 2015. *The Intimacies of Four Continents*. Durham, NC: Duke University Press.

MacKenzie, Donald A., Fabian Muniesa, and Lucia Siu, eds. 2007. *Do Economists Make Markets? On the Performativity of Economics*. Princeton, NJ: Princeton University Press.

MacLachlan, Ian, and Adrian Guillermo Aguilar. 1998. "Maquiladora Myths: Locational and Structural Change in Mexico's Export Manufacturing Industry." *Professional Geographer* 50, no. 3: 315–31.

Maine, Henry Sumner. 1861. *Ancient Law: Its Connection with the Early History of Society and Its Relation to Modern Ideas*. London: John Murray.

Makdisi, Saree. 1998. *Romantic Imperialism: Universal Empire and the Culture of Modernity*. Cambridge: Cambridge University Press.

Mamdani, Mahmood. 1996. *Citizen and Subject: Contemporary Africa and the Legacy of Late Colonialism*. Princeton, NJ: Princeton University Press.

Mann, Kristin. 2007. *Slavery and the Birth of an African City: Lagos, 1760–1900*. Bloomington: Indiana University Press. http://public.eblib.com/choice/publicfullrecord.aspx?p=339114.

Mann, Kristin. 2013. "A Tale of Slavery and Beyond in a British Colonial Court: West Africa and Brazil." In *African Voices on Slavery and the Slave Trade*, edited by Alice Bellagambe, Sandra E. Greene, and Martin A. Klein. Cambridge: Cambridge University Press.

Maples, Susan, Peter Rosenblum, and Revenue Watch Institute. 2009. *Contracts Confidential: Ending Secret Deals in the Extractive Industries*. New York: Revenue Watch Institute. http://www.opensocietyfoundations.org/reports/contracts-confidential-ending-secret-deals-extractive-industries.

Martino, Enrique. 2012. "Clandestine Recruitment Networks in the Bight of Biafra: Fernando Pó's Answer to the Labour Question, 1926–1945." *International Review of Social History* 57 (Special Issue): 39–72.

Martino, Enrique. 2014. "Las fuentes abiertas de Guinea Ecuatorial o Cómo descolonizar el archivo colonial." *Revista Trimestral Editada por la Institució Alfons el Magnànim* 123: 42–47.

Martino, Enrique. 2016a. "Panya: Economies of Deception and the Discontinuities of Indentured Labour Recruitment and the Slave Trade, Nigeria and Fernando Pó, 1890s–1940s." *African Economic History* 44: 99–129.

Martino, Enrique. 2016b. "Nsoa ('Dote'), Dinero, Deuda y Peonaje: Cómo el Parentesco Fang Tejió y Destejió la Economía Colonial de la Guinea Española." *ENDOXA* 0, no. 37: 337–62. https://doi.org/10.5944/endoxa.37.2016.16616.

Martino, Enrique. 2017. "Dash-Peonage: The Contradictions of Debt Bondage in the Colonial Plantations of Fernando Pó." *Africa: Journal of the International African Institute* 87, no. 1: 53–78.

Martino, Enrique. 2018a. "Corrupción y contrabando: Funcionarios españoles y traficantes nigerianos en la economía de Fernando Poo, 1936–1968." *Ayer: Asociación de Historia Contemporánea*, 109, no. 1: 169–95.

Martino, Enrique. 2018b. "The Political Economy of Free and Unfree Labour in Twentieth Century Colonial Río Muni: The Dialectics of La Recluta and Prestación Personal." In *Guinea Ecuatorial (des)conocida lo que sabemos, ignoramos, inventamos y deformamos acerca de su pasado y su presente*. Madrid: Universidad Nacional de Educación a Distancia.

Martino, Enrique. 2019. http://www.opensourceguinea.org/.

Massumi, Brian. 2002. *Parables of the Virtual: Movement, Affect, Sensation.* Durham, NC: Duke University Press.

Mathiason, Nick. 2009. "Barclays and Ghana Plan Tax Haven." *The Guardian.* https://www.theguardian.com/business/2009/may/03/barclay-tax-avoidance-ghana.

Mau, Steffen, Fabian Gülzau, Lena Laube, and Natascha Zaun. 2015. "The Global Mobility Divide: How Visa Policies Have Evolved over Time." *Journal of Ethnic and Migration Studies* 41, no. 8: 1192–1213.

Maurer, Bill. 1998. "Cyberspatial Sovereignties, Offshore Finance, Digital Cash, and the Limits of Liberalism." *Indiana Journal of Global Legal Studies* 5, no. 2: 493–520.

Maurer, Bill. 2005. "Due Diligence and 'Reasonable Man,' Offshore." *Cultural Anthropology* 20, no. 4: 474–505.

Maurer, Bill. 2008. "Re-regulating Offshore Finance?" *Geography Compass* 2, no. 1: 155–75. https://doi.org/10.1111/j.1749-8198.2007.00076.x.

Maurer, Bill. 2010. "Orderly Families for the New Economic Order: Belonging and Citizenship in the British Virgin Islands." *Identities* 2, no. 1–2: 149–71.

Maurer, Bill. 2013. "Ungrounding Knowledges Offshore Caribbean Studies, Disciplinarity and Critique." *Comparative American Studies: An International Journal* 2, no. 3: 324–41.

Mazzarella, William. 2012. "Affect: What Is it Good For?" In *Enchantments of Modernity: Empire, Nation, Globalization*, edited by Saurabh Dube. London: Routledge.

Mba, César A. 2011. *(La construcción de) la memoria del petróleo*. Edited by Dulcinea Tomás Cámara. Alicante: Biblioteca Virtual de Miguel de Cervantes. http://www.cervantesvirtual.com/obra/la-construccion-de-la-memoria-del -petroleo/.

Mbembe, Achille. 1992a. "Provisional Notes on the Postcolony." *Africa* 62, no. 1: 3–37.

Mbembe, Achille. 1992b. "The Banality of Power and the Aesthetics of Vulgarity in the Postcolony." *Public Culture* 4, no. 2: 1–30. https://doi.org/10.1215 /08992363-4-2-1.

Mbembe, Achille. 2001. *On the Postcolony*. Berkeley: University of California Press. http://hdl.handle.net/2027/heb.02640.

Mbembe, Achille, and Janet Roitman. 1995. "Figures of the Subject in Times of Crisis." *Public Culture* 7, no. 2: 323–52. https://doi.org/10.1215/08992363-7-2-323.

McBride, James. 2016. "Understanding the LIBOR Scandal." Council on Foreign Relations. https://www.cfr.org/backgrounder/understanding-libor-scandal.

McKay, Steven C. 2007. "Filipino Sea Men: Constructing Masculinities in an Ethnic Labour Niche." *Journal of Ethnic and Migration Studies* 33, no. 4: 617–33. https:// doi.org/10.1080/13691830701265461.

McKay, Steven C. 2014. "Racializing the High Seas: Filipino Migrants and Global Shipping." In *The Nation and Its Peoples: Citizens, Denizens, Migrants*, edited by John Park and Shannon Gleeson. New York: Routledge.

McKittrick, Katherine. 2013. "Plantation Futures." *Small Axe* 17, no. 3: 1–15.

McKittrick, Katherine. 2016. "Rebellion/Invention/Groove." *Small Axe* 20, no. 1(49): 79–91.

McSherry, Brendan. 2006. "The Political Economy of Oil in Equatorial Guinea." *African Studies Quarterly* 8, no. 3: 23–45.

Mehta, Uday Singh. 1997. "Liberal Strategies of Exclusions." In *Tensions of Empire: Colonial Cultures in a Bourgeois World*, edited by Frederick Cooper and Ann Laura Stoler, 59–86. Berkeley: University of California Press.

Mehta, Uday Singh. 1999. *Liberalism and Empire: A Study in Nineteenth-Century British Liberal Thought*. Chicago: University of Chicago Press.

Meillassoux, Claude. 1981. *Maidens, Meal, and Money: Capitalism and the Domestic Community*. Cambridge: Cambridge University Press.

Miller, Daniel. 2002. "Turning Callon the Right Way Up." *Economy and Society* 31, no. 2: 218–33. https://doi.org/10.1080/03085140220123135.

Miller, Daniel. 2005. "Reply to Michel Callon." *Economic Sociology: European Electronic Newsletter* 6, no. 3: 3–13. https://www.econstor.eu/bitstream/10419 /155849/1/vol06-no03-a2.pdf.

Mills, Charles W. 1997. *The Racial Contract*. Ithaca, NY: Cornell University Press.

Mills, Charles W. 1998. *Blackness Visible: Essays on Philosophy and Race*. Ithaca, NY: Cornell University Press.

Mills, Charles W. 2003. *From Class to Race: Essays in White Marxism and Black Radicalism*. Lanham, MD: Rowman & Littlefield.

Mills, Charles W. 2017. *Black Rights / White Wrongs: The Critique of Racial Liberalism*. Oxford: Oxford University Press.

Mitchell, Timothy. 1988. *Colonising Egypt*. Berkeley: University of California Press.
Mitchell, Timothy. 1991. "The Limits of the State: Beyond Statist Approaches and Their Critics." *American Political Science Review* 85, no. 1: 77–96. https://doi.org/10.1017/S0003055400271451.
Mitchell, Timothy. 2002. *Rule of Experts: Egypt, Techno-Politics, Modernity*. Berkeley: University of California Press.
Mitchell, Timothy. 2005. "The Work of Economics: How a Discipline Makes Its World." *European Journal of Sociology / Archives Européennes de Sociologie* 46, no. 2: 297–320. https://doi.org/10.1017/S000397560500010X.
Mitchell, Timothy. 2009. "Carbon Democracy." *Economy and Society* 38, no. 3: 399–432. https://doi.org/10.1080/03085140903020598.
Mitchell, Timothy. 2011. *Carbon Democracy: Political Power in the Age of Oil*. London: Verso.
Miyazaki, Hirokazu. 2003. "The Temporalities of the Market." *American Anthropologist* 105, no. 2: 255–65.
Miyazaki, Hirokazu. 2013. *Arbitraging Japan: Dreams of Capitalism at the End of Finance*. Berkeley: University of California Press.
Mnookin, Robert, and Lewis Kornhauser. 1979. "Bargaining in the Shadow of the Law: The Case of Divorce." *Yale Law Journal: Dispute Resolution* 88, no. 5: 950–97.
Mol, Annemarie. 2002. *The Body Multiple: Ontology in Medical Practice*. Durham, NC: Duke University Press.
Mongia, Radhika Vivas. 1999. "Race, Nationality, Mobility: A History of the Passport." *Public Culture* 11, no. 3: 527–55.
Moodie, T. Dunbar, and Vivienne Ndatshe. 1994. *Going for Gold: Men, Mines, and Migration*. Berkeley: University of California Press. http://search.ebscohost.com/login.aspx?direct=true&scope=site&db=nlebk&db=nlabk&AN=42027.
Moore, Jason W. 2015. *Capitalism in the Web of Life: Ecology and the Accumulation of Capital*. London: Verso.
Mrázek, Rudolf. 2002. *Engineers of Happy Land: Technology and Nationalism in a Colony*. Princeton, NJ: Princeton University Press.
Munif, Abdel Rahman. 1989. *Cities of Salt*. Translated by Peter Theroux. New York: Vintage Books.
Nandy, Ashis. 1988. *The Intimate Enemy: Loss and Recovery of Self under Colonialism*. Delhi, India: Oxford University Press.
Navarro, Tami. 2010. "Offshore Banking within the Nation: Economic Development in the United States Virgin Islands." *Global South* 4.2.
Ndongo-Bidyogo, Donato. 1977. *Historia y tragedia de Guinea Ecuatorial*. Madrid: Editorial Cambio.
Nerín, Gustau. 2010. *La última selva de España: Antropófagos, misioneros y guardias civiles. Crónica de la conquista de los Fang de la Guinea Española, 1914–1930*. Madrid: Los Libros de la Catarata.
Neumayer, Eric. 2006. "Unequal Access to Foreign Spaces: How States Use Visa Restrictions to Regulate Mobility in a Globalized World." *Transactions of the Institute of British Geographers* 31, no. 1: 72–84.

Nixon, Rob. 2011. *Slow Violence and the Environmentalism of the Poor.* Cambridge, MA: Harvard University Press.

Okenve, Alfredo. 2017. "State of Civil Society in Equatorial Guinea: An Interview with CEID." February. https://www.civicus.org/index.php/media-resources /news/2731-state-of-civil-society-in-equatorial-guinea-an-interview-with-ceid.

Okonta, Ike, and Oronto Douglas. 2001. *Where Vultures Feast: Shell, Human Rights, and Oil in the Niger Delta.* San Francisco: Sierra Club Books.

Ong, Aihwa. 1999. *Flexible Citizenship: The Cultural Logics of Transnationality.* Durham, NC: Duke University Press.

Ong, Aihwa. 2006. *Neoliberalism as Exception: Mutations in Citizenship and Sovereignty.* Durham, NC: Duke University Press.

Ong, Aihwa. 2017. "Autonomous Zones and the Unbundling of Territorial Sovereignty." In *The Urbanism of Exception: The Dynamics of Global City Building in the Twenty-First Century*, edited by M. Murray. Cambridge: Cambridge University Press.

Ong, Aihwa, and Stephen J. Collier. 2005. *Global Assemblages: Technology, Politics, and Ethics as Anthropological Problems.* Malden, MA: Blackwell.

Orta, Andrew. 2013. "Managing the Margins: MBA Training, International Business, and 'the Value Chain of Culture.'" *American Ethnologist* 40, no. 4: 689–703. https://doi.org/10.1111/amet.12048.

Orta, Andrew. 2014. "Commentary: Response to Karen Ho on Cultures of Capitalism, Contexts of Capitalism." *American Ethnologist* 41, no. 1: 38–39. https://doi .org/10.1111/amet.12058.

Ortner, Sherry B. 1995. "Resistance and the Problem of Ethnographic Refusal." *Comparative Studies in Society and History: An International Quarterly* 37: 173–93.

Oyono, Ferdinand. 1966. *Houseboy.* London: Heinemann.

Palan, Ronen. 2006. *The Offshore World: Sovereign Markets, Virtual Places, and Nomad Millionaires.* Ithaca, NY: Cornell University Press.

Parreñas, Rhacel Salazar. 2008. *The Force of Domesticity: Filipina Migrants and Globalization.* New York: New York University Press.

Parreñas, Rhacel Salazar. 2015. *Servants of Globalization: Migration and Domestic Work.* Stanford, CA: Stanford University Press.

Pateman, Carole, and Charles W. Mills. 2007. *Contract and Domination.* Cambridge: Polity.

Peet, Richard, and Michael Watts. 2004. *Liberation Ecologies: Environment, Development, Social Movements.* London: Routledge.

Peterson, Kristin. 2014. *Speculative Markets: Drug Circuits and Derivative Life in Nigeria.* Durham, NC: Duke University Press.

Petryna, Adriana. 2005. "Ethical Variability: Drug Development and Globalizing Clinical Trials." *American Ethnologist* 32, no. 2: 183–97.

Picciotto, Sol, and Jason Haines. 1999. "Regulating Global Financial Markets." *Journal of Law and Society* 26, no. 3: 351–68. https://doi.org/10.1111/1467-6478.00129.

Pierre, Jemima. 2013. *The Predicament of Blackness: Postcolonial Ghana and the Politics of Race.* Chicago: University of Chicago Press.

Pistor, Katharina. 2019. *The Code of Capital: How the Law Creates Wealth and Inequality*. Princeton, NJ: Princeton University Press.

Platts Marketwire. 2010. "Platts Introduces World's First Daily Alumina Price Ass." Press release. n.d. Accessed December 21, 2017. https://www.platts.com/press releases/2010/081610.

Polanyi, Karl. 1957. *Trade and Market in the Early Empires: Economies in History and Theory*. Glencoe, IL: Free Press.

Polanyi, Karl. 2001. *The Great Transformation: The Political and Economic Origins of Our Time*. Boston: Beacon.

Poon, Martha. 2009. "From New Deal Institutions to Capital Markets: Commercial Consumer Risk Scores and the Making of Subprime Mortgage Finance." *Accounting, Organizations, and Society* 34, no. 5: 654–74.

Poovey, Mary. 1998. *The History of the Modern Fact: Problems of Knowledge in the Sciences of Wealth and Society*. Chicago: University of Chicago Press.

Porteous, J. Douglas. 1974. "Social Class in Atacama Company Towns." *Annals of the Association of American Geographers* 64, no. 3: 409–17. https://doi.org/10.1111/j.1467-8306.1974.tb00989.x.

Potts, Shaina. 2016. "Reterritorializing Economic Governance: Contracts, Space, and Law in Transborder Economic Geographies." *Environment and Planning A* 48, no. 3: 523–39.

Potts, Shaina. 2018. "(Re-)Writing Markets: Law and Contested Payment Geographies." *Environment and Planning A: Economy and Space*. 0308518X18768286.

Povinelli, Elizabeth A. 2006. *The Empire of Love: Toward a Theory of Intimacy, Genealogy, and Carnality*. Durham, NC: Duke University Press.

Povinelli, Elizabeth A. 2014. "Geontologies of the Otherwise." *Cultural Anthropology Website*, Theorizing the Contemporary (January).

Power, Michael. 1997. *The Audit Society: Rituals of Verification*. Oxford: Oxford University Press.

Power, Michael. 2007. *Organized Uncertainty: Designing a World of Risk Management*. Oxford: Oxford University Press.

Pratt, Mary Louise. 1992. *Imperial Eyes: Travel Writing and Transculturation*. London: Routledge.

Prescott, J. R. V. 1975. *The Political Geography of the Oceans*. New York: Wiley.

Radon, Jenik. 2007. "How to Negotiate an Oil Agreement." In *Escaping the Resource Curse*, edited by Macartan Humphreys et al. New York: Columbia University Press.

Rajak, Dinah. 2011. *In Good Company: An Anatomy of Corporate Social Responsibility*. Palo Alto: Stanford University Press.

Ralph, Michael. 2015. *Forensics of Capital*. Chicago: University of Chicago Press.

Reed, Kristin. 2009. *Crude Existence: Environment and the Politics of Oil in Northern Angola*. Berkeley: University of California Press. http://public.eblib.com/choice/publicfullrecord.aspx?p=837188.

Reno, William. 1998. *Warlord Politics and African States*. Boulder, CO: Lynne Rienner.

Reno, William. 2001. *Foreign Firms, Natural Resources and Violent Political Econo-*

mies. Institut für Afrikanistik, Universität Leipzig. University of Leipzig Papers on Africa. Politics and Economics Series no. 46.

República de Guinea Ecuatorial. 1997. "Documento final de la Primera Conferencia Economica Nacional." 8. Bata, Equatorial Guinea.

República de Guinea Ecuatorial. 2005. "Statement of the People and Government of Equatorial Guinea in Response to the Report on Riggs Bank of the Permanent Subcommittee on Governmental Affairs of the Senate of the United States of America."

República de Guinea Ecuatorial. 2006a. "Ley de Hidrocarburos de la República de Guinea Ecuatorial." http://www.equatorialoil.com/PDFs%20for%20download /EG%20Hydrocarbons%20Law%20(Spanish).pdf.

República de Guinea Ecuatorial. 2006b. "Model Production Sharing Contract." http://www.ecuatorialoil.com/pdfs/Model%20PSC_English.pdf.

República de Guinea Ecuatorial. 2006c. "Resumen Liquidaciones de Presupuestos de Las Entidades Autonomas." Ejercicio Economico, República de Guinea Ecuatorial.

República de Guinea Ecuatorial. 2007a. "Plan Nacional de Desarollo Econónomico y Social. Guinea Ecuatorial 2020 Agenda Para La Diversificación de Las Fuentes de Crecimiento. Tomo I: Diagnostico Estratégico."

República de Guinea Ecuatorial. 2007b. "Plan Nacional de Desarollo Econónomico y Social. Guinea Ecuatorial 2020 Agenda Para La Diversificación de Las Fuentes de Crecimiento. Tomo II: Visión y Ejes Estragégicos 2020."

República de Guinea Ecuatorial. 2009. "Resumen de La Situación de Los Presupuestos de Las Entidades Autónomas."

República de Guinea Ecuatorial. 2010. "Primer Informe de La Iniciativa Para La Transparencia en Las Industrias Extractivas 2007–2008." Malabo, Guinea Ecuatorial: Oficina Nacional Autónoma de Coordinación de EITI.

Riles, Annelise. 2004. "Real Time: Unwinding Technocratic and Anthropological Knowledge." *American Ethnologist* 31, no. 3: 392–405.

Riles, Annelise. 2011. *Collateral Knowledge: Legal Reasoning in the Global Financial Markets.* Chicago: University of Chicago Press.

Roberts, A. D. 1986. *The Cambridge History of Africa.* Vol. 7. Cambridge: Cambridge University Press. http://dx.doi.org/10.1017/CHOL9780521225052.

Roberts, Adam. 2006. *The Wonga Coup: Guns, Thugs, and a Ruthless Determination to Create Mayhem in an Oil-Rich Corner of Africa.* New York: PublicAffairs.

Roberts, Adam. 2009. *The Wonga Coup: Simon Mann's Plot to Seize Oil Billions in Africa.* London: Profile.

Robinson, Cedric J. 1983. *Black Marxism: The Making of the Black Radical Tradition.* Chapel Hill: University of North Carolina Press.

Rodney, Walter. 1972. *How Europe Underdeveloped Africa.* Washington, DC: Howard University Press.

Roediger, David. 2018. *Working toward Whiteness: How America's Immigrants Became White: The Strange Journey from Ellis Island to the Suburbs.* New York: Basic Books.

Rofel, Lisa. 1999. *Other Modernities: Gendered Yearnings in China after Socialism.* Berkeley: University of California Press. http://hdl.handle.net/2027/heb.04242.

Roitman, Janet L. 2005. *Fiscal Disobedience: An Anthropology of Economic Regulation in Central Africa.* Princeton, NJ: Princeton University Press.

Roitman, Janet L. 2014. *Anti-crisis.* Durham, NC: Duke University Press.

Rousseau, Jean-Jacques, and Donald A. Cress. 1987. *On the Social Contract.* Indianapolis: Hackett.

Roy, Ananya, Genevieve Negrón-Gonzales, Kweku Opoku-Agyemang, and Clare Vineeta Talwalker. 2016. *Encountering Poverty: Thinking and Acting in an Unequal World.* Berkeley: University of California Press.

Sachs, Jeffrey D., and Andrew M. Warner. 1995. "Natural Resource Abundance and Economic Growth." Working Paper 5398. National Bureau of Economic Research. https://doi.org/10.3386/w5398.

Sachs, Jeffrey D., and Andrew M. Warner. 2001. "The Curse of Natural Resources." *European Economic Review* 45, no. 4: 827–38. Fifteenth Annual Congress of the European Economic Association. https://doi.org/10.1016/S0014-2921(01)00125-8.

Sahlins, Marshall. 1974. *Stone Age Economics.* New York: Aldine.

Sahlins, Marshall. 1976. *Culture and Practical Reason.* Chicago: University of Chicago Press.

Said, Edward W. 1978. *Orientalism.* New York: Vintage Books.

Said, Edward W. 1989. "Representing the Colonized: Anthropology's Interlocutors." *Critical Inquiry* 15, no. 2: 205–225.

Saro-Wiwa, Ken. 1992. *Genocide in Nigeria: The Ogoni Tragedy.* London: Saros International Publishers.

Sartori, Andrew S. 2014. *Liberalism in Empire: An Alternative History.* Berkeley: University of California Press.

Satsuka, Shiho. 2015. *Nature in Translation: Japanese Tourism Encounters the Canadian Rockies.* Durham, NC: Duke University Press.

Sawyer, Suzana. 2004. *Crude Chronicles: Indigenous Politics, Multinational Oil, and Neoliberalism in Ecuador.* Durham, NC: Duke University Press.

Sawyer, Suzana. 2006. "Disabling Corporate Sovereignty in a Transnational Lawsuit." *Political and Legal Anthropology Review* 29, no. 1: 23–43.

Sawyer, Suzana. 2012. "Commentary: The Corporation, Oil, and the Financialization of Risk." *American Ethnologist* 39, no. 4: 710–15.

Schmitt, Carl, and G. L. Ulmen. 2003. *The Nomos of the Earth in the International Law of the Jus Publicum Europaeum.* New York: Telos.

Sharma, Aradhana, and Akhil Gupta. 2006. *The Anthropology of the State: A Reader.* Hoboken, NJ: Wiley.

Shaxson, Nicholas. 2008. *Poisoned Wells: The Dirty Politics of African Oil.* New York: Griffin.

Shever, Elana. 2012. *Resources for Reform: Oil and Neoliberalism in Argentina.* Stanford, CA: Stanford University Press. http://site.ebrary.com/id/10571088.

Shore, Cris, and Susan Wright. 1997. *Anthropology of Policy: Critical Perspectives on Governance and Power.* London: Routledge.

Shore, Cris, and Susan Wright. 2015. "Audit Culture Revisited: Rankings, Ratings, and the Reassembling of Society." *Current Anthropology* 56, no. 3: 421–44.
https://doi.org/10.1086/681534.

Silverstein, Ken. 2014. *The Secret World of Oil*. London: Verso.

Simmel, Georg. 2011. *The Philosophy of Money*. Abingdon, UK: Routledge. http://
public.eblib.com/choice/publicfullrecord.aspx?p=683949.

Simone, AbdouMaliq. 2004. *For the City Yet to Come: Changing African Life in Four
Cities*. Durham, NC: Duke University Press.

Simone, AbdouMaliq. 2012. "Infrastructure: Introductory Commentary by Abdou-
Maliq Simone." *Curated Collections, Cultural Anthropology*: Infrastructure
(November). https://journal.culanth.org/index.php/ca/infrastructure
-abdoumaliq-simone.

Simpson, Audra. 2014. *Mohawk Interruptus: Political Life across the Borders of Settler States*. Durham, NC: Duke University Press.

Sklair, Leslie, University of California, San Diego, and Center for U.S.-Mexican
Studies. 1993. *Assembling for Development: The Maquila Industry in Mexico and
the United States*. San Diego: UCSD Center for U.S.-Mexican Studies.

Smith, Adam, and Andrew S. Skinner. 1986. *The Wealth of Nations: Books I–III*.
London: Penguin Group, .

Smith, Anton. 2009. "Subject: Equatorial Guinea Raw 5: How It Works—Why It
Sometimes Doesn't." Wikileaks Cable 09MALABO031.

Smith, Daniel Jordan. 2007. *A Culture of Corruption: Everyday Deception and Popular Discontent in Nigeria*. Princeton, NJ: Princeton University Press.

"Speculative Markets." n.d. Duke University Press. Accessed December 20, 2017.
www.dukeupress.edu/speculative-markets.

Speich, Daniel. 2011. "The Use of Global Abstractions: National Income Accounting in the Period of Imperial Decline." *Journal of Global History* 6, no. 1: 7–28.
https://doi.org/10.1017/S1740022811000027.

Star, Susan Leigh. 1999. "The Ethnography of Infrastructure." *American Behavioral
Scientist* 43, no. 3: 377–91.

Stegner, Wallace. 1971. *Discovery: The Search for Arabian Oil*. Beirut, Lebanon: Middle East Export Press.

Stiglitz, Joseph E. 2010. *Freefall: America, Free Markets, and the Sinking of the World
Economy*. New York: W. W. Norton & Company.

Stiglitz, Joseph E. 2014. "On the Wrong Side of Globalization." *New York Times*,
March 15. //opinionator.blogs.nytimes.com/2014/03/15/on-the-wrong-side
-of-globalization/.

Stiglitz, Joseph E., Patrick Bolton, and Frederic Samama. 2011. *Sovereign Wealth
Funds and Long-Term Investing*. New York: Columbia University Press. http://
public.eblib.com/choice/publicfullrecord.aspx?p=908923_0.

Stoler, Ann Laura. 1989a. "Making Empire Respectable: The Politics of Race and
Sexual Morality in 20th-Century Colonial Cultures." *AMET American Ethnologist* 16, no 4: 634–60.

Stoler, Ann Laura. 1989b. "Rethinking Colonial Categories: European Communities and the Boundaries of Rule." *Comparative Studies in Society and History* 31, no. 1: 134–61.

Stoler, Ann Laura. 1995. *Capitalism and Confrontation in Sumatra's Plantation Belt, 1870–1979*. Ann Arbor: University of Michigan Press.

Stoler, Ann Laura. 2008. "Imperial Debris: Reflections on Ruins and Ruination." *CUAN Cultural Anthropology* 23, no. 2: 191–219.

Stoler, Ann Laura. 2010. *Carnal Knowledge and Imperial Power: Race and the Intimate in Colonial Rule*. Berkeley: University of California Press.

Strange, Susan. 1986. *Casino Capitalism*. Oxford: Blackwell.

Strathern, Marilyn. 2000. "The Tyranny of Transparency." *British Educational Research Journal* 26, no. 3: 309–21.

Strathern, Marilyn. 2003. *Audit Cultures: Anthropological Studies in Accountability, Ethics and the Academy*. London: Routledge.

Sundiata, I. K. 1983. "Equatorial Guinea: The Structure of Terror in a Small State." In *African Islands and Enclaves*, edited by Robin Cohen, 81–100. Beverly Hills, CA: Sage.

Sundiata, I. K. 1990. *Equatorial Guinea: Colonialism, State Terror, and the Search for Stability*. Boulder, CO: Westview.

Taussig, Michael T. 1980. *The Devil and Commodity Fetishism in South America*. Chapel Hill: University of North Carolina Press.

Thrift, N. J. 2005. *Knowing Capitalism*. London: Sage. http://public.eblib.com /choice/publicfullrecord.aspx?p=254820.

Tsing, Anna. 2000a. "The Global Situation." *Cultural Anthropology* 15, no. 3: 327–60.

Tsing, Anna Lowenhaupt. 2000b. "Inside the Economy of Appearances." *Public Culture* 12, no. 1: 115–44.

Tsing, Anna Lowenhaupt. 2001. "Nature in the Making." In *New Directions in Anthropology and Environment: Intersections*, edited by Carole L. Crumley. Lanham, MD: AltaMira.

Tsing, Anna Lowenhaupt. 2005. *Friction: An Ethnography of Global Connection*. Princeton, NJ: Princeton University Press.

Tsing, Anna Lowenhaupt. 2009. "Supply Chains and the Human Condition." *Rethinking Marxism* 21, no. 2: 148–76. https://doi.org/10.1080/08935690902743088.

Tsing, Anna Lowenhaupt. 2015. *The Mushroom at the End of the World: On the Possibility of Life in Capitalist Ruins*. Princeton, NJ: Princeton University Press.

US Energy Information Administration. 2016. "International Energy Outlook 2016." US Department of Energy, Independent Statistics and Analysis. Washington, DC: Office of Energy Analysis.

Van Onselen, Charles, University of London, and Institute of Commonwealth Studies, eds. 1986. *Kas Maine: Sharecropping in the South-Western Transvaal*. London: Institute of Commonwealth Studies.

Vanoli, André. 2005. *A History of National Accounting*. Amsterdam: IOS Press.

Vitalis, Robert. 2007. *America's Kingdom: Mythmaking on the Saudi Oil Frontier*. Stanford, CA: Stanford University Press.

Watts, Michael. 2004. "Resource Curse? Governmentality, Oil and Power in the Niger Delta, Nigeria." *Geopolitics* 9, no. 1: 50–80. https://doi.org/10.1080/14650040412331307832.

Welker, Marina. 2014. *Enacting the Corporation: An American Mining Firm in Postauthoritarian Indonesia.* Berkeley: University of California Press.

Weszkalnys, Gisa. 2015. "Geology, Potentiality, Speculation: On the Indeterminacy of First Oil." *Cultural Anthropology* 30, no. 4: 611–39.

Wikipedia Contributors. 2017. "Libor Scandal." *Wikipedia.* https://en.wikipedia.org/w/index.php?title=Libor_scandal&oldid=816473550.

Williams, Eric. 1944. *Capitalism and Slavery.* Chapel Hill: University of North Carolina Press.

Williams, Robert E., Jr. 2011. "From Malabo to Malibu: Addressing Corruption and Human Rights Abuse in an African Petrostate." *Human Rights Quarterly* 33, no. 3: 620–48. https://doi.org/10.1353/hrq.2011.0047.

Winters, Jeffrey A. 1996. *Power in Motion: Capital Mobility and the Indonesian State.* Ithaca, NY: Cornell University Press.

Woolfson, Charles, John Foster, and Matthias Beck. 1996. *Paying for the Piper: Capital and Labour in Britain's Offshore Oil Industry.* London: Mansell.

World Bank. 2008. *Implementing the Extractive Industries Transparency Initiative: Applying Early Lessons from the Field.* Washington, DC: World Bank.

Wright, Eric Olin. 2010. *Envisioning Real Utopias.* New York: Verso.

Wylie, Sara Ann. 2011. "Corporate Bodies and Chemical Bonds: An STS Analysis of Natural Gas Development in the United States." Thesis, Massachusetts Institute of Technology. http://dspace.mit.edu/handle/1721.1/69453.

Wynter, Sylvia. 1982. *Beyond Liberal and Marxist Leninist Feminisms: Towards an Autonomous Frame of Reference.* Ann Arbor, MI: Institute for Research on Women and Gender.

Wynter, Sylvia. 2003. "Unsettling the Coloniality of Being/Power/Truth/Freedom: Towards the Human, after Man, Its Overrepresentation—An Argument." *CR: The New Centennial Review* 3, no. 3: 257–337.

Yanagisako, Sylvia Junko. 2002. *Producing Culture and Capital: Family Firms in Italy.* Princeton, NJ: Princeton University Press.

Yergin, Daniel. 1993. *The Prize: The Epic Quest for Oil, Money, and Power.* New York: Touchstone.

Young, Alden. 2014. "Measuring the Sudanese Economy: A Focus on National Growth Rates and Regional Inequality, 1959–1964." *Canadian Journal of Development Studies / Revue Canadienne d'études du développement* 35, no. 1: 44–60. https://doi.org/10.1080/02255189.2014.881733.

Young, Alden. 2017. *Transforming Sudan: Decolonization, Economic Development, and State Formation.* Cambridge: Cambridge University Press.

Zalik, Anna. 2004. "The Peace of the Graveyard: The Voluntary Principles on Security and Human Rights in the Niger Delta." In *Global Regulation: Managing Crises after the Imperial Turn,* edited by K. Van Der Pijl, L. Assassi, and D. Wigan. London: Palgrave Macmillan.

Zalik, Anna. 2006. *Re-regulating the Mexican Gulf.* Berkeley: Center for Latin American Studies, University of California, Berkeley.

Zalik, Anna. 2009. "Zones of Exclusion: Offshore Extraction, the Contestation of Space and Physical Displacement in the Nigerian Delta and the Mexican Gulf." *Antipode* 41, no. 3: 557–82. https://doi.org/10.1111/j.1467-8330.2009.00687.x.

Zaloom, Caitlin. 2004. "The Productive Life of Risk." *Cultural Anthropology* 19, no. 3: 365–91.

INDEX

Abaga, Fernando, 15

absolute rule, 7, 9, 174, 177, 197–99, 200, 229. *See also* authoritarianism; dictatorship; repressive regimes

accidents, 63, 67, 72, 184, 193, 254

accounting, 68, 70, 86, 154, 192, 200, 213, 215, 235, 237, 241–44, 251, 253–55, 266, 273; accounting-as-politics, 271; double-entry bookkeeping, 253, 271, 292n3; outsourcing, 179; standardized, 135

accumulation, 23, 26, 50, 150, 171, 213, 228, 230, 232, 243, 280; capitalist, 180; flexible, 177; primitive, 289n4; private, 212, 230–32, 234, 240; rapid, 96

Africa, 2, 16, 24, 26, 45, 49, 54, 94, 107, 132–33, 175, 176, 180, 226; African diaspora, 175, 289n5; Africans, 9, 76, 157–58, 163, 175–76, 195, 275; African states, 24, 32, 29, 102, 160, 211, 230, 275, 282; economies of, 158; law in, 164; postcolonial, 75; women in, 121. *See also* South Africans

African Americans, 92, 174, 187, 189, 196. *See also* Jim Crow; race

African Development Bank, 291n11

African Studies Quarterly, 23

African Union Summit, 208

Africa Unit, 45

aggregator, 252, 263–64, 268, 271–72

agnotology, 6, 238

agriculture, 95–96, 109, 202, 206, 220–21, 242, 290n7, 293n5; cacao farming, 95, 96, 104–5, 133, 177, 190; industrialized, 81, 109, 223; subsistence farming, 223

Algeria, 12, 45, 55

America. *See* United States

Angola, 37, 57, 61, 99, 102, 116, 120, 122, 159, 161, 169, 172, 179, 200–201; diamond mining in, 58; and Equatorial Guinea, 99; insurance companies in, 169

anthropology, 3, 5, 25, 29–31, 134, 141, 164, 211, 253, 282; anthropologists, 22, 25, 29, 84, 93, 133–34, 142, 160, 200, 210, 238 apartheid, 260. *See also* race; South Africa

Appadurai, Arjun, 216

ARAMCO, 27, 94, 134. *See also* Saudi Arabia

archipelago, corporate. *See* corporate archipelago

architecture, 83, 86; academic, 31; modular, 26; moral, 5–6, 278, 280

Argentina, 88, 122

Arthur Andersen scandal, 69

Asia, 176, 245

assemblages, 21, 37, 76, 83, 287n12

Atacama, 99–100

authoritarianism, 63, 245–46, 248. *See also* absolute rule; dictatorship; repressive regimes

banana republic, 165

Bank of Central African States (BEAC), 207

Barclays Bank, 288n4

Barry, Andrew, 27, 30, 32, 35, 82, 112–14, 163, 179, 253–54, 256, 265–66

Bata, Equatorial Guinea, 1, 4, 12, 52, 85–86, 120, 206–7, 216, 219, 222, 227, 240, 259–60, 268, 292n3; airports, 61

Bata Ports, 240

batteries, lithium, 161–64

beaches, 86, 95, 105, 113, 222

Bechtel (oil service firm), 45, 181

Beckert, Sven, 289n4

Biafra, 12

blackness, 83, 175–76, 290n10

blackouts, 204, 210

black radical tradition, 174, 281, 289n4, 292n2; Cedric Robinson and, 285n2

Blair, Tony, 250

boats, 13–14, 109, 216, 222

body shops, 35, 45, 100, 125, 172–73, 177, 183–85, 189–91, 191–93, 196–98

Bonnie Island, 122

BP (British Petroleum), 31–32, 37, 173, 177, 288n7

Brazil, 38, 88, 105, 128, 210; Petrobras, 240

Bremer, Paul, 116, 290n8

Bryan, Dick, 71

budgets, 32, 36, 81, 163, 206, 215, 235, 237, 239–43, 250, 273

Buena Esperanza, 208–9

bureaucracy, 36, 147, 235, 258

Burgess, Anthony, 96, 97

cacao, 95, 96, 104–5, 133, 177, 190. *See also* cocoa

Callon, Michel, 218, 286n3, 287n13

Cambodia, 177

Cameroon, 9, 108, 110, 207

Campos Serrano, Alicia, 63, 100, 183, 292n13

Campo Yaounde, 74

Canada, 24, 38, 61, 116, 168, 195

capital, 14, 32, 34, 81, 95–96, 100–101, 104, 148, 170, 192, 194, 214, 224, 245; fleeing, 49, 158; foreign, 147; international, 7, 249; social, 129

capitalism, 2–6, 21–23, 25–26, 28–29, 31–36, 39–40, 55, 125–26, 139, 140–41, 171, 176–77, 249, 278–83; anthropology of, 22; contemporary, 2, 171; corporate, 5; epistemologies of, 36, 248; global, 3, 23, 41, 82, 123, 171, 174, 178, 203, 213, 279; late, 103, 171, 176–78, 180, 183, 203; liberal, 125, 248; neoliberal, 234, 287n9; petro-capitalism, 5–6, 18, 32–34, 174, 199, 213; racial, 128, 136, 286n2, 292n2; transnational, 29, 210, 292n2

Caracolas, 105, 106, 273

cars, 19, 67, 115, 129, 231

cash, 134, 168, 196, 207, 232. *See also* money

Cayman Islands, 41, 43, 45, 49, 76, 149

Central Africa, 24, 69, 158, 170, 177. *See also* CFA

Central African Economic and Monetary Community (CEMAC), 15, 207, 215

Central African Republic, 207

Central America, 186

Central Europe, 260

CFA (Central African franc), 15, 192–95, 199, 240, 242, 264, 274, 293n9; convertibility of, 207

Chakrabarty, Dipesh, 24

charity projects, 121, 126, 128–30, 290n10. *See also* philanthropy

Chatterjee, Partha, 27

258, 262–64, 266–75, 277, 280; purposes of, 250–61, 266

electricity, 17, 42, 75–76, 80–81, 93, 106–8, 111–12, 132, 138, 156, 164, 192, 196, 204–6, 211, 235; sporadic, 74, 107, 135, 222

embezzlement, 15

employees, 82, 85, 87–88, 93–98, 100, 103, 114, 116, 167, 172–73, 175, 178, 184–85, 193; expatriate, 111, 119, 135; foreign, 81; migrant, 106, 169, 289n2; resident, 86, 109; scheduling, 186; subcontracted, 88. *See also* subcontracts and subcontracting

enclaves, corporate, 6, 17–18, 21, 26–27, 35, 78–88, 93–95, 97, 99–103, 105–21, 126–36, 191, 195, 280; enclaving process, 30, 97, 102, 106; residential/industrial, 27, 211, 85. *See also* ring-fencing

Endurance Corporation, 42–43, 47, 79, 81, 85–89, 91, 95, 97–98, 103–4, 116, 127, 129, 171, 173, 289n2

Endurance Equatorial Guinea Production Limited (EEGPL), 42. *See also* Endurance Corporation

Endurance International Petroleum, 43. *See also* Endurance Corporation

Endurance Oil, 43. *See also* Endurance Corporation

Endurance Petroleum, 43. *See also* Endurance Corporation

England, 19, 122, 132–33

English language, 96, 153, 186–87

Enron, 64, 68–70, 117

entrepreneurs, 7–8, 104, 248, 261

environment, business, 20, 135, 157–58, 202

environment, natural, 7, 51, 67, 76, 83, 94, 103, 132, 139, 154, 171; degradation of, 128, 202; environmental impact, 40, 57; environmentalists, 56; environmental law, 35, 48, 162, 167, 283; environmental standards, 25; urban, 72

epiphenomena, 281

equality, 140, 142, 144, 151, 155; formal, 137; liberal, 213, 281; potential, 246

Equatorial Guinea, 1–12, 14–25, 29–43, 58–63, 83–89, 92–103, 114–23, 126–30, 140–41, 144–63, 169, 171–78, 194–208,

210–32, 236–37, 241–43, 245–46, 248–53, 263, 266, 272–80; and China, 102; economy of, 243; Equatoguinean workers, 40, 42, 60–65, 71–75, 102, 169, 174–75, 201, 206; history of, 100, 263; law in, 42, 50, 162, 267; life in, 83, 102, 117, 279; oil and gas industry in, 3, 16, 35, 98, 100, 103, 118, 186, 212; people of, 13, 19, 51, 76, 114, 118–19, 172, 193–94, 199, 223, 249, 257; terms "Guinean" and "Equatoguinean," 286n5; terrorists in, 203; women of, 120; youth of, 226

Equatorial Guinea, government of, 7, 17–18, 51, 56, 115–16, 140, 150, 174–75, 190, 210, 227, 232, 249–50, 258, 260, 269, 273, 275–76; Hydrocarbons Law of, 291n3; Ministry of Finance and Budgets, 36, 81, 119, 206, 215, 239, 243, 250, 271, 273, 241–42; Ministry of Health and Social Welfare, 169; Ministry of Labor, 190; Ministry of Mines, Industry, and Energy (MMIE), 110–11, 113, 118, 145, 151, 168, 170, 271; Ministry of Planning, 218–19

equipment, 64–67, 72, 84, 101, 109; hospital, 18; personal protective equipment (PPE), 37, 66

Ernst & Young (law firm), 162, 252, 264

ethnography, 3, 22, 33, 176–77, 277, 279; ethnographic accounts, 2, 39, 77, 136, 144, 161, 212, 254; ethnographic data, 85, 243; ethnographic objects, 3, 5, 31, 34, 138, 228, 279, 282; ethnographic thresholds, 174, 203, 282

Europe, 81, 96–97, 108, 275, 285n2; Central and Eastern Europe, 260; European empire, 5, 96, 176, 197

European Union, 265

Excel, 201, 240

exceptionalism, 85

Exclusive Economic Zone, 51–53, 76, 291n3

expatriates, 7, 23, 85–86, 106, 122–23, 130, 132, 138, 193, 196, 198, 201–2, 288n1; expatriate managers, 33, 86, 165, 286n4; expatriate women, 136; Expat Wives' Prayer, 123–25

expendability, 167, 185

contracts, 17, 144, 271; industry, 16–19,
21–23, 25, 30, 39, 41, 47, 49, 70, 137, 167,
175, 183–84, 186. *See also* natural gas; oil
GDP, 17, 156, 205, 213, 228, 246, 293n2
gender, 2, 21, 26, 32, 36, 84, 89, 125–26, 135,
144, 188, 280–81, 286n2
General Electric, 208
genocide, 15
geology, 30, 55, 181; deepwater, 156. *See
also* oil
GEPetrol, 111, 144, 169, 271
GEProyectos, 242
Ghana, 37, 45, 172, 189, 251, 274, 288n4
globalization, 39, 262. *See also* capitalism:
global; licitness
gold, 94, 205, 216, 220
golf, 81, 208; golf carts, 86, 87, 93
Gómez, Juan Vicente, 147
Good Hope, 208
governance, 17–18, 32, 102, 160, 216, 251,
264; corporate, 69, 249; sustainable, 278
grants, 49, 152
graphs, 111, 217–18
Guinea Bissau, 204
Guinea. *See* Equatorial Guinea
Guinea-Conakry, 287n5
Gulf War, 116

Halliburton (oil service firm), 45, 177,
288n7
H&M, 177
Harris, Cheryl, 28, 135, 280
hauntings, 23, 33, 58, 60
healthcare, 74, 75, 81, 138, 211, 247; insurance, 182, 184. *See also* medical care
helicopters, 1–4, 34–35, 37, 39–40, 60–61,
65, 67, 71–72, 74, 85; military, 102
Hetherington, Kregg, 248, 254, 256, 262,
265, 266, 294n1
hierarchies, 27, 42, 92, 121, 126, 129, 158, 163,
174–75, 186–88, 203, 246, 293n7; corporate, 61, 131; discriminatory labor, 181;
historical, 102; occupational, 98; racialized, 62–63, 92, 102; residential, 91
Hispanoil, 156. *See also* Spain
histories: dystopian, 243; legal, 166, 292n13;

liberal, 143; long, 19, 21, 93–94, 98, 100,
105, 123; moral, 253
Ho, Karen, 71, 167, 189, 285n1
homosociality, 129
hospitals, 76–77, 99, 208, 276
hotels, 17, 207, 274
Houseboy (Oyono), 9
house cleaners, 96, 103, 130, 192
housing, 17, 84, 86–89, 91, 93–94, 97, 104,
106, 120, 181; barracks, 89–91; company-
provided, 99, 121; for foreign workers,
82, 106, 120; mansions, 81, 97–98, 207,
223; vacation homes, 86, 174. *See also*
enclaves, corporate; ring-fencing
Houston, 41–43, 60, 80–81, 111–12, 107–8,
114–15, 119, 122, 131, 135, 149, 183, 258,
269
husbands, 67, 75, 103–4, 121–23, 125–26,
130–34, 191, 196, 274, 289n1
hydrocarbons, 3, 6, 22, 30, 38, 58, 60, 72,
76, 145, 244, 271, 275; deposits of, 16, 20,
34, 39, 45, 102, 208, 290n3; hydrocarbon
industry, 18, 28, 50, 135, 152, 166, 168, 173;
legislation regarding, 32, 146, 290n3;
subsea, 20, 60. *See also* oil; petroleum
Hydrocarbons Law (Equatorial Guinea),
290n3

ideologies, 5, 26, 160, 253, 273, 282
ignorance, 8, 25; historical, 93; personal, 132
illness, 103, 107–8; mental, 14. *See also*
healthcare; medical care
IMF (International Monetary Fund), 18,
210, 215–16, 221, 228, 251, 263–65, 273
immigration, 98, 113, 229, 245
imperialism. *See* colonialism
imports, 110, 161, 168, 179, 206, 221; duties
on, 50, 169–70; duty-free, 101
INCESO (social security administration), 197
income, 71, 192, 225, 229; regular, 147, 216;
taxable, 288n6. *See also* taxation
independence, 10–12, 165, 187; demanded,
11; juridical, 45; nominal, 246; political,
246
India, 38, 91–92, 178, 180, 185, 187, 189, 206,
280; Indian Camp, 90–91

Indonesia, 42, 59, 99, 101, 103, 106, 114, 122, 139, 141, 147, 170, 186
industrial revolution, 96
inequality, 129, 140, 142–43, 151–53, 156–57, 168, 171, 174–76, 181, 186, 188–89, 199, 281; gendered, 125; geographic, 62; racialized, 2, 28, 39, 135, 174–75, 189, 203; in salary, 194
inflation, 30, 215
infrastructures, 3, 15–18, 20, 26–28, 32–33, 35, 39–40, 54–60, 63, 94–95, 103, 111–12, 119, 130, 132, 135, 208, 232–33, 242; colossal, 76, 87; export, 220, 224; mobile, 28, 31, 45, 47, 51, 60, 114; offshore, 20, 51, 55. *See also* oil operations, offshore
INPYDE (National Institute for Business Promotion and Development), 241
insurance, 47; health insurance, 61, 71, 75, 169–70, 181–84
International Chamber of Commerce (ICC), 153; International Commercial Terms document, 152
internet, 75–76, 114, 123, 132, 287n11
intimacy, 5–7, 41, 78, 120; intimates, 55, 107, 113; troubled, 83
investment, 10, 17, 32, 45, 70–71, 99, 141, 156–57, 165, 205, 239; foreign, 101–2, 148, 157–60, 245; investors, 9, 43, 68, 140, 148, 158, 165, 183, 226, 251–52; private, 237; public, 232
Iraq, 116, 290n8
Ireland, 42, 106, 122; and Equatorial Guinea, 182; Irish Sea, 37
isolation, 13, 18, 84, 119, 131, 136, 195; cultural, 134; physical, 104. *See also* spaces and spacialization
Israel, 106
Italy, 116, 286n2; mafia wives in, 274
ivory, 96
Ivory Coast, 204

jail, 12–13
Japan, 42–43, 106, 122, 132
Jim Crow, 27, 175–76, 187, 196, 281. *See also* race
jobs, 62, 64–66, 98, 100, 114, 122, 128,

172–73, 175, 190–92, 194, 229, 231–32, 234, 264–65
judicial system, 162, 165, 199, 236, 245–46, 280
jungle, 57, 104–5

Kimball, Solon T., 134, 136
kinship, 27, 35, 82, 84, 121, 164, 229; fictive, 88; gendered, 89; kin-based hiring practices, 231
kleptocracy, 25, 102, 160, 228, 254
knowledge production, 5, 215, 217–18, 238
Kuwait, 186

labor, 17, 20–22, 46–48, 61–63, 68, 70–71, 93–94, 98–101, 149, 180–85, 188–90, 198–99, 225, 289n2; abuses of, 202; costs of, 193; exploitation of, 63, 175, 180, 200; imported, 96, 183; labor brokering, 35, 45, 172; labor history, 63, 177; local, 59, 75; racialized, 21, 76, 211; semi-skilled, 192; slave, 14; specialized, 62; subcontracted, 91, 181, 187; suppression of, 20; sweatshop, 177; unpaid, 13, 223; unskilled, 61, 62, 91. *See also* subcontracts and subcontracting; unions; workers
laborers. *See* workers
landscapes, 4–5, 22, 95, 105; competing regulatory, 163; literal, 232; physical, 17; political, 17; post-oil, 232; resource-rich, 94
language, 129, 169; racist, 11; scientific, 246
Larkin, Brian, 55, 63, 67
Latin America, 59
law, 13, 25, 49–51, 140–43, 151, 160–62, 165–68, 185, 193, 224, 234, 238–39, 267–68, 280–83, 291n3; antilogging, 223; differential, 112; on inequitable pay, 198; lawyers, 190, 205, 237, 292n13; legal theory, 176; national, 69, 101
layoffs, 167, 180
Lebanon, businesspeople from, 111, 185, 204–6
legal theory, 176. *See also* law
liability, 26, 31, 45, 47, 76–77; attenuated, 40, 47, 51, 61, 149, 177, 179; denial of,

shale, 62; transport of, 49, 217. *See also* petroleum

oil industry, 31–34, 70, 72, 93, 112–13, 116–20, 138, 176–79, 183–84, 187–88, 202, 210–12, 232, 244; companies, 5, 19, 69, 92, 100–103, 108–9, 146–50, 160–61, 168–73, 191–92, 198–99, 221–22, 252–53, 259, 271, 274; contracts, 144, 168, 238; infrastructure for, 56, 181; legislation regarding, 146; oil revenue, 18, 30–31, 213, 237, 247, 261, 263, 275; transnational, 7, 9, 23, 31, 36, 40, 42, 99, 144, 169, 172, 175, 180, 243, 247. *See also* production sharing contracts (PSCs)

oil operations, offshore, 2–4, 20–22, 35, 39–41, 43, 48–51, 54–59, 60–65, 73–78, 83, 109–10, 155, 208, 252–53, 277–78, 280; accounts, 25, 102, 138, 145, 150, 158; and corporate sovereignty, 51; drilling rigs, 2, 20, 35, 37–39, 54, 59–68, 71–75, 120, 135, 172, 174, 179, 189, 288n2, 288n11; financial aspects, 39–41, 49–50, 55, 58, 64; FPSOs (Floating Production, Storage, and Offloading vessels), 2, 32, 39, 41, 54, 153; infrastructure for, 39, 47–51, 54–55, 57, 59, 63; platforms, 20, 40, 47–50, 54, 57, 61, 63–65, 73–74, 110, 115, 221–22, 263, 291n9; rental of rigs, 173, 210; safety on rigs, 68; tankers, 2, 35, 39, 49, 54. *See also* infrastructures

Organization for Economic Cooperation and Development (OECD), 228

organogram, 42–43, 144

Orwell, George, 96, 97

Oslo, Norway, 269–70

Oyono, Ferdinand, 9

Pakistan, 91–92, 100, 185–86

Palan, Ronen, 141

palm oil, 95–96

Panama, 41, 49

paranoia, 12, 18, 63, 132, 266

parliament, 16, 145–46, 148, 242, 255

Partido Democrático, 292n5

Pateman, Carole, 35, 137, 144, 171, 176, 194, 203, 281

payments, 150–51, 171–72, 184, 191, 196, 215, 223, 251–52, 263–65, 276; bonus, 271–72; paychecks, 47, 63, 138, 192–93, 195, 197, 241; transfer pricing, 288n6

performativity, 26–27, 36, 40, 115, 135, 246, 286n3

personhood, 94; corporate, 45, 291n7

Peru, 251, 274, 276

Petrobras, 240. *See also* Brazil

petroleum, 16, 39, 42–43, 50, 57, 77, 95, 119, 145–47, 155, 166, 192, 239, 262, 293n5; petro-boom, 205, 207, 210; petro-capitalism, 6, 33, 213; petro-powerhouses, 143, 278; production, 42, 65, 202, 219, 220; revenue from, 18, 128, 232, 239. *See also* oil; oil industry; oil operations, offshore

philanthropy, 27–28, 82, 128, 135–36, 230, 281. *See also* charity projects

Philippines, 38, 45, 60, 62, 89, 100, 186–87, 281; Filipino workers, 61–63, 88, 91–92, 172, 174, 178, 185–89, 194, 196, 203, 281

phones, 74–75, 111, 114, 196, 205–6, 259, 269, 275; iPhone, 7; phone cards, 75

Pierre, Jemima, 175–76, 196–97, 289n3, 290n10, 293n4

Pinochet, Augusto, 150

pipelines, 31, 56–57, 59–60

placelessness, 3, 49, 58, 77, 277

plantations, 11, 95–96; cacao, 95, 96, 104–5, 133, 177, 190; coffee, 14

platforms. *See* oil operations, offshore

Platts Marketwire, 152–53, 171, 189

PNC Financial Services, 150, 291n8

pollution, 25, 57, 113, 154–55; air, 23; offshore, 47, 222–23; visible, 56

Porteous, J. Douglas, 98–99, 100, 104

ports, 50, 110–14, 153, 161, 217, 240, 246

postcolonial era, 2, 10, 24, 29, 36, 128, 142, 164, 166, 196; history, 7, 214, 287n7; inequality, 2, 143, 176, 180, 189, 221, 246, 278, 281

postdevelopmentalism, 101

Povinelli, Elizabeth, 120

precarity, 75, 134, 136

prices, 50, 70, 81, 118–19, 122, 188–89, 196, 206–7, 219; domestic, 215; global, 18; stock, 68, 145, 163. *See also* transfer pricing

PricewaterhouseCoopers, 162, 252, 264

prison, 12, 14, 116, 131

privacy, 21, 132

privilege, 34, 83–84, 102, 104, 121, 129, 143, 185, 190, 213, 231, 283; white, 28, 135, 280, 290n10

production sharing contracts (PSCs), 16, 35, 138–42, 144–46, 149, 151–54, 156–57, 160–61, 168, 170–71, 173–74, 189, 198, 203, 227–28, 272, 281, 290n3

profit, 7, 9, 29, 31, 50–51, 58–59, 76–77, 141–42, 148–49, 158, 163, 168, 186–87, 190, 226–27; maximization of, 47, 112, 156; profitability, 71, 178; profit margins, 17, 58, 167

property rights, 140, 143, 146, 234, 236, 244–45, 246, 290n7

public relations, 68, 156

Punta Europa, 103–5, 290n7

race, 25–27, 35–36, 40, 82–84, 97–98, 125–27, 174–76, 187, 280–81, 285n2; apartheid, 260; Jim Crow, 27, 175–76, 187, 196, 281; racialization, 83, 121, 290n12, 292n2; racial power, 137; racism, 2, 62–63, 128, 189, 196, 222, 281–82

radio, 37–38, 62, 202, 216, 255, 258–59, 270

rank, 8, 102, 158, 174, 203

rationality, 28, 262

recreation, 86, 91–92, 99, 122, 217

recruiting, 61, 183, 190, 192

red tape, 115, 117

reforms, 218, 236–38, 280, 283; neoliberal, 253

Regal Corporation, 43, 69, 272; Regal Energy, 47, 69, 106, 116, 192, 249, 271; Regal Energy Equatorial Guinea, 45

regulations, 39, 41, 48, 50–51, 56–57, 99, 102, 112, 117, 121, 154, 162, 167, 169, 236; external, 50; inability to enforce, 199; worldwide, 154

Reno, William, 59, 94, 102, 112

rent, 104, 106, 192, 196, 207. *See also* housing

reparations, 58

replicability, 3, 26, 77, 154, 171, 179

repressive regimes, 17, 19, 140, 148, 171, 247. *See also* authoritarianism

Repsol (Spanish oil company), 156

Republic of Congo, 207

resource curse, 23–25, 28, 30–32, 211–12, 218–21, 223, 244, 246, 251–52, 263, 275, 282, 293n5; literature on, 30–31

resources, natural, 11, 30, 51, 81, 104, 108, 112, 145, 158, 221, 245, 254; accessing, 195; extraction of, 136; mineral, 9, 148, 150–51, 156, 240; ownership of, 141. *See also* resource curse

restaurants, 19, 81, 86, 120, 204–5, 207, 237, 261

retrenchment, 23, 171, 244

rice, 15, 96, 122

Riggs Bank, 149–51, 175, 212, 231–32, 237–38, 291n8; Riggs Report, 150–51, 234; scandal of, 233

rigs, drilling. *See* oil operations, offshore

ring-fencing, 84, 101, 103, 114–15, 135, 289n2; domestic, 84. *See also* enclaves, corporate

Río Muni, 9, 85, 230, 287

risks, 55, 70–73, 76–77, 118, 252; avoidance of, 65; distribution of, 68, 70–71; financial, 71, 77, 141; management of, 47, 76; mitigation of, 45, 47, 177

rituals, 64, 68, 74, 115, 200, 235; native, 80; safety, 63, 70

rivers, 192, 219

roads, 86, 208; builders of, 173, 178; private, 94

Robinson, Cedric J. *See* black radical tradition

Roitman, Janet, 75

Romania, 38

roustabouts, 61, 64

rubber trade, 287n6

rule of law, 138, 140, 143, 158, 160–61, 165; rule-of-law bureaucracy, 228

running water. *See* water

Russia, 38, 42, 99, 109, 114, 122, 132–33, 139, 182, 203, 210

safaris, 220
safety: equipment for, 37, 72–73; precautions regarding, 63–68; training on, 1, 37, 65, 67. *See also* risk
salaries, 68, 77, 81–82, 92, 169, 184, 191–96, 198–99, 201, 241; parity in, 193; second, 276; for sub-subcontracting, 196
salt, 50, 216, 220
s&p (Standard & Poor's), 62–63, 189
Sarbanes-Oxley Act, 69, 117, 163, 238, 280
Saudi Arabia, 27, 59, 122, 134, 181, 186, 198, 240. *See also* ARAMCO
scalability, 33, 82
SchaeferCorp, 173, 192
schedules, 47, 60, 62–64, 75, 111, 186; rotation, 62, 75, 82, 175, 184–85
Schlumberger (oil service firm), 45, 181
Schmitt, Carl, 48
schools, 10, 12, 15, 94, 98–99, 106, 122, 127, 138, 155, 192, 195, 230; fees, 77, 174, 192
Scotland, 45, 60, 172, 183
seas, 38, 47–49, 50–51, 174, 178, 186–87, 280; sea ports, 245; sea products, 220; seasickness, 60. *See also* oceans
SeaTrekker, 37–38, 54, 172–73, 197
Second National Economic Conference, 212, 214, 250, 259
secrecy, 6, 9, 14, 19
sectors, 30, 236, 239, 250, 278; entrepreneurial, 220; extractive, 278; non-oil manufacturing, 215, 220; social, 217
security, 48, 73, 86–89, 94–95, 116, 120, 122, 130, 291n9; guards, 18–19, 82, 86, 173, 192, 289n2; personnel, 80; social, 77, 184, 192, 197–98
segregation, 27–28, 35, 55, 63, 82–84, 92, 98, 121, 135, 281; domestic, 125; racial, 27, 32, 78, 82; residential, 83, 125; workplace, 27. See also enclaves, corporate; ring-fencing
self-regulation, 70, 100, 154, 165
Sendero Luminoso, 275
Senegal, 204, 205, 218, 219

Senghor, Léopold, 246
Serbia, 38, 61
services, 18, 109, 182, 185, 190, 193, 210, 219, 227, 231, 235–36, 275, 278; ferry, 222; firms providing, 45, 161, 181; phone, 81, 196; social, 138, 214, 235. *See also* phones
sewage, 17, 81, 107–8
sex and sexuality, 84, 120–21, 125; racialized, 123; sex workers, 120–21
shame, 128, 136, 146
shareholders, 43, 68–69, 135, 145, 155; shareholder value, 25, 45, 63, 70, 167, 176–78, 180, 203
Shell, 6, 20, 59, 94, 173
shipping, 41, 48, 96, 152–53, 173, 179, 186–87, 210
sidewalks, 191–92, 204–5, 210
Sierra Leone, 96, 184
Simpson, Audra, 4–5, 31, 282
simultaneity, 4, 136, 188; productive, 40
Singapore, 245–46
singularities, 171, 229; consequential, 143, 278; durable, 145
Sipopo, 208
skills, 104, 192, 201; marketable, 201; specialized, 181
skyscrapers, 17, 204, 207, 208
slavery, 22, 49, 96, 158, 176; slave trade, 9, 95–96
Smith, Adam, 143
Smith Corporation, 85–86, 107, 112, 114, 162, 173, 201; enclave of, 94
social security administration, 169, 197, 240
social welfare, 75, 169
soldiers, 116, 174, 222–23. *See also* military bases
Sonagas (Equatoguinean gas company), 43, 144, 271
Sonangol (Angolan oil company), 200. *See also* Angola
songs, 109, 225–26
Sonrisa (pseudonym), 269–71, 273–74, 277
South Africa, 38, 92, 116, 210, 260; South Africans, 57–58, 88, 62, 92, 260
South America, 69, 95, 181, 185–86, 188

sovereignty, 17, 27, 40, 49–51, 135, 141–44, 160, 170–71, 234, 278; corporate, 35, 51, 113, 202; legal, 50; national, 51, 101–2; postcolonial, 140; strategic, 47

spaces and spatialization, 48, 60, 82–85, 130, 164, 244, 254; conceptual, 160; domestic, 36, 98; enclaved, 119, 139; gendered, 121; imagined, 212, 244; industrial, 94, 98; interstitial, 18, 277; juridical, 50; legal, 35, 40; offshore, 49; securitized, 289n2; segregated, 218, 248; sovereign, 48; unstable, 227

Spain, 9–14, 19, 81, 96, 115, 165–66, 183, 206, 210, 219, 225, 232, 290n7; and Equatorial Guinea, 221; Franco's regime in, 10–14, 215; versus Macías regime in Equatorial Guinea, 11–12; Spanish Civil War, 10–12; Spanish Cooperation, 216; Spanish Guinea, 9–10. *See also* Hispanoil

Spanish Cultural Center, 225

specters, 33, 47, 63, 77, 247

spills, 48, 58

sports, 99, 111. *See also* golf; recreation; tennis

stability, 126, 147, 160, 167, 184, 189, 207, 285; fiscal, 140, 161, 168; institutional, 12; political, 252

Standard & Poor's (s&p), 62–63, 189

standardization, 3–4, 22, 25–26, 35, 77, 82, 112, 115, 152–55, 171, 178–82, 279, 281; modular, 26

Standard Oil, 41

Stanford University, 118

state, 32, 140, 144–47, 168–70, 211–12, 235; corrupt, 236; imagined, 275; independent, 187; modern, 144, 194; power of the, 213, 225; property of the, 208; service-providing, 275; sovereignty of the, 50

statistics, 7, 210, 213–19

steel, 37, 55

Stiglitz, Joseph E., x, 167

stitching, 79, 121, 127

Stoler, Ann Laura, 82–83, 97, 166

subcontracts and subcontracting, 26–27, 100, 125, 171–91, 193, 195, 197–99, 201, 203, 211, 213, 244, 266, 280–81, 289n2;

agencies for, 197, 264; defining and contextualizing, 176–77; discriminatory practices in, 182, 203; history of, 180–83; subcontracting arrangements, 31, 35–39, 48, 173–74, 176–77, 179, 191–95, 201, 227, 280; subcontracting companies, 38, 62, 68, 174–75, 192; subcontractors, 51, 103, 173, 178–80, 184–86, 192, 197, 199, 222, 224, 264. *See also* contracts

subordination, 137, 176, 187, 281; of nonwhites, 123, 176, 287n16

subsidiaries, corporate, 41–43, 45, 51, 144, 148, 161, 169, 173, 286n4, 288n6. *See also* corporate archipelago

subsoil deposits, 20, 39, 141, 146, 291n3

Sudan, 102, 159, 228

supermarkets, 94, 99, 206–7

supply chain processes, 110, 114, 116, 125, 173, 174, 177, 182; labor mobilization, 125, 174. *See also* commodities: commodity chain

Supreme Military Council, 14, 15. *See also* Obiang Nguema Mbasogo

Susan (corporate social responsibility representative for the Major Corporation), 262–63, 266, 268, 271

systematicity, 28, 279–80

tankers, oil, 2, 35, 39, 49, 54. *See also* oil operations, offshore

Tarzan, 164–65

taxes, 25, 39–49, 55, 76, 80, 84, 95, 101, 110–12, 129, 167–69, 173, 183, 191–93, 198, 271, 288n6; deductions, 153; laws, 17, 158, 167; loopholes, 45, 77; tax havens, 40–41, 47–50, 102, 157; tax planning, 280; tax rates, 9, 148, 171, 272; and transfer pricing, 40, 41, 45, 50, 288n6

teachers, 10–11, 94, 99, 134

technocracy, 12, 168, 250

teenagers, 127, 225

television, 67–68, 204, 208; cable, 93; government-controlled, 255

temporalities, 98–100, 153, 177, 180, 184, 203, 272; interfering, 180; personal mareado, 61; unpredictable, 47 tennis, 79, 81, 105, 121, 131

territory, 49–51, 101, 235; national, 51, 101, 239; new, 254; virgin, 226–27

Texaco, 27

Texas, 41, 80, 93, 97, 103, 118, 121–22, 134, 196, 200

Thatcher, Mark, 115

theft, 207, 292n1

Third Country Nationals (TCNs), 180, 189, 281

Third World, 286n2

tidal marshes, 50

time, historical, 183

Tocqueville, Alexis de, 260, 280

tourism, 220–23, 227

training, 20, 59, 62, 77, 127, 195, 198–201, 222, 240, 265, 278; graduate, 224; professional, 285

transfer pricing, 40, 41, 45, 50, 288n6. *See also* taxes

transience, 47, 123

TransOcean, 288n7

transparency, 2, 6, 26, 35, 102, 160, 251–54, 256, 262–63, 266–68, 276–78, 280–81, 282; critiques of, 253; epistemologies of, 36, 247–49; programs regarding, 19, 273; rankings of, 279

Transparency International, 160–61, 228, 253, 291n11

triste memoria, 12, 14, 19, 214

Tsing, Anna, 22, 29, 83, 105, 116, 125, 147, 148, 157, 174, 177, 223, 285

TurboGas, 173

Turkmenistan, 91, 100, 160, 228

typhoid, 74, 80, 135

underdevelopment, 23, 49

unions, 100, 200, 264; deterring membership in, 187. *See also* labor

United Kingdom, 92

United Nations, 53, 211

United Nations Convention on the Law of the Sea, 50

United Nations Development Programme, 216

United States, 31, 38, 57, 81, 92, 142, 148, 150, 232, 234, 280–81, 290n8, 293n7; Americans, 7, 76, 105, 108, 117, 122, 128–29, 131–32, 134, 186, 188–89, 195, 203, 281; banking, 243; Gulf Coast, 45, 48, 91; imperialism of, 174; pension funds, 70; politics in, 158, 229; US companies, 20, 35, 49, 55, 63, 111–17, 144, 146, 148, 175, 194, 258, 271, 286n4; US oil and gas industry, 2, 6–7, 15–19, 20, 26–27, 40–41, 72, 92, 99, 111, 125, 135, 148, 186, 224, 228, 236, 244, 247, 255, 279–80; wives from, 94; workers from, 134, 174, 194

USAID, 210, 216

US Congress, 69

US Navy, 187, 189, 196

USOne, 43

US Senate, 231–32, 234

USSR, 13, 181

Utah, 41, 122

utopias, 216, 224, 227; schizophrenic, 220

Venezuela, 30, 38, 62, 91–93, 100, 147, 181, 186, 281

Vietnam, 92, 106, 109

violence, 5–6, 10, 18, 243, 246, 254; histories of, 6, 19, 21, 270; political, 221–22; routinized, 22; slow, 288n3; state-sponsored, 289n4; structural, 229

visas, 13, 18, 61, 115, 184, 274

visibility, 47, 55–56, 215, 253

Vitalis, Robert, 27, 35, 59, 62, 82, 92, 134, 175, 181–82, 188, 198

voting, 10, 16, 148, 283

vulnerability, 75, 125

wages, 12, 19, 27, 62, 175, 178, 186–89, 198, 224; overtime, 138

Wall Street, 71, 167, 177–81

Washington, 150–51, 216, 291n8

waste disposal, 24

water, 55, 57, 73, 76, 81, 94, 103, 105, 107–8, 132, 138, 155, 189, 196, 204, 222; nonpotable, 135; potable, 60, 93, 107, 211; running, 74, 80, 93, 97, 111, 206, 235, 247

wealth, 29–31, 58, 128, 138, 205, 228–32, 234

welding, 65, 72, 127

West Africa, 112–13, 157, 162–63

Western Europe, 92, 174, 196, 210, 286n4

Westphalian model, 142, 212

whiteness, 26, 28–29, 33, 35, 88, 123, 280–81, 289n3, 290n10, 293n4; gendered, 27; heteronormative, 82; homogenous, 129; married, 135; white masculinity, 83; white people, 12, 128, 132, 136, 176, 196, 222, 286n4; white savior, 127–28; white womanhood, 82, 123, 129, 136

white supremacy, 2, 27–28, 92, 123, 125, 135–36, 144, 280, 283, 287n16; global, 176, 196. *See also* privilege; race

wildcatters, 181

Winters, Jeffrey A., 101

wives, 75, 79–81, 84, 88, 91–92, 94–96, 99, 109, 186, 288n1; migrant, 103, 117–34, 274

womanhood, 88; women's movements, 129. *See also* feminist theory; wives

work: charitable, 127–28; corporate social responsibility, 226; domestic, 187; femininity, 83; volunteer, 128, 130, 134

workers, 33, 35, 37–38, 40, 45, 58–65, 68, 72–74, 91–92, 98–100, 171–72, 174–75, 178–88, 190, 197–200, 202–3; airport, 1; construction, 17, 59; crane operators, 61–62, 64; domestic, 191; local, 96, 99, 172–75, 190; migrant, 85, 96, 99–100, 119–20, 138; rig workers, 39, 64, 75, 194; sex workers, 120–21; subcontracted, 73, 172, 193, 290n1; unionized, 187, 281. *See also* employees; labor; subcontracts and subcontracting

workforce, 91, 178–79; educated, 187; mobile transnational, 186

World Bank, 156–57, 160–61, 210, 216, 221, 228, 250–52, 256–58, 263–65, 274–76, 293n1; and IMF, 265

Wright, Susan, 253, 273

Zafiro oil field, 16

Zaloom, Caitlin, 70–71, 77

Zambia, 99; copper mining in, 99–100

Zimbabwe, 94, 99, 116

Zoback, Mark, 48

zones, 48, 71, 88, 101–3, 105; EEZS (exclusive economic zones), 51; export processing, 48–49, 101; technological, 28, 35, 82, 113; zonal capitalism, 84, 93–94, 117

www.ingramcontent.com/pod-product-compliance
Lightning Source LLC
Chambersburg PA
CBHW030821290525
27270CB00016B/136